AF531431

BIOLOGICAL WARFARE

BIOLOGICAL WARFARE

Editor

DIGUMARTI BHASKARA RAO
DIGUMARTI PUSHPA LATHA
DIGUMARTHI HARSHITHA

'Academy of Communication Culture
Education Science and Service'
1-22-10 Srinivasa Nagar
Guntur 522006
Andhra Pradesh
INDIA

DISCOVERY PUBLISHING HOUSE
NEW DELHI

First Published-2001
Reprint, 2011
ISBN 81-7141-597-0

Published by :

DISCOVERY PUBLISHING HOUSE
4831/24, Ansari Road, Prahlad Street,
Daryaganj, New Delhi-110002 (India)
☎ : 3279245 m Fax: 91-11-3253475
E-mail : dphtemp@indiatimes.com

Printed at : **Mehra Offset Press, Delhi**

PREFACE

The population of the world of today is faced by a challenge that could threaten even its survival in the near future because of biological weapons and warfare. One has to consider that biological weapons could unfortunately be relatively easily available to poor economies and even to terrorists. They are not difficult to produce, relatively easy to hide, and, in the hands of unscrupulous desperate terrorists, could cause incredible damage to large populations anywhere in the world.

'Biological warfare' is the employment of biological agents to produce casualities in man or animals and damage to plants and materials. A 'biological weapon' is an item of material which projects, disperses, or disseminates a biological agent; including arthropod vectors. Biological weapons that are already developed and tested provide a relatively simple and inexpensive means for the attack of people, animals and crops over large areas. The use of biological weapons as a means of mass destruction may lead to man-made epidemics that will introduce bioengineered agents into the human populations and domesticated and wild animals and plants; and in such events, the new agents will have a devastating effect on living organisms and world economy.

Confronted with this menace, the Biological Weapons Convention (BWC) has singled out biological weapons for categorical prohibition, it may be possible to achieve a world-wide system of disincentiveness, deterrants, sanctions, and norms sufficient to forestall the intensive exploitation of biotechnology for hostile purposes.

To protect humans, animals and plants from microbial diseases, a revolutionary approach to develop effective vaccines against epidemic causing agents and certainly against biological weapon agents is needed. Along with conventional vaccines, the synthetic vaccines will allow rapid immunisation procedures in the wake of natural epidemic or the use of biological weapons.

This book is dealing with all aspects that are concerned to biological warfare, biological weapons, bioethics, etc., etc. This book will be of great use to politicians administrators, scientists, educationists, social service activists and young citizens of the global village.

The contributions included in this book are taken from volume 6 of the UNESCO International School of Science for Peace series entitled "First Forum of the International Scientific panel on possible consequences of the Misuse of Biological Sciences" edited by Y. Becker, A Falaschi, V. Kouzminov, M. Martellini and R. Santesso. I am thankful to R. Santesso, UNESCO Venice Office for extending the necessary cooperation on behalf of the UNESCO-ROSTE and Landau Network.

Mrs. D. Pushpa Latha

CONTENTS

CONTRIBUTORS

ADAMS, Stephen
Director for Drug Information and Clinical Services,
Dept. of Pharmacy (MC 2-230),
St. Luke's Episcopal Hospital
and Texas Heart Institute,
6720 Bertner Avenue,
Houston, Texas 77030, USA

ATLAS, Ronald M.
Biology Dept., University
of Louisville,
Life Science Building 139,
Louisville, KY 40292, USA

BECKER, Yechiel
UNESCO-Hebrew University of
Jerusalem - International School
for Molecular Biology and
Microbiology, Dept. of Molecular
Virology, Faculty of Medicine,
HUJ, P.O. Box 12272
Jerusalem 91120, Israel

BERNS, Kenneth I
President, American Society
for Microbiology,
New York, USA

CANEPA, Margherita
Centre A. Volta,
Villa Olmo - Via Cantoni 1,
221 00 Como, Italy

COLIZZI, Vittorio
Professor of Immunology,
Dept. of Biology,
University of Rome "Tor Vergata",
Via della Ricerca Scientifica,
00133 Rome, Italy

D' AMELIO, Raffaele — Directorate Generale for Military Health, Via S. Stefano Rotondo 4, 00184 Rome, Italy

FALASCHI, Arturo — Director General, International Centre for Genetic Engineering and Biotechnology (ICGEB), Padriciano 99, 34012 Trieste, Italy

GEISSLER, Erbard — Max Delbruck Centre for Molecular Medicine, Robert Rossle Strasse 10, 13122 Berlin, Germany

GERIN, Guido — President, International Institute for Human Rights Studies, Via Cantù 10, 134100 Trieste, Italy

GURUGE, Ananda W.P. — Senior Special Advisor to the Director General UNESCO, 8351 Snowbird Drive, Huntington Beach, CA 92646, USA

HARIGEL, Gert — Cern- PPE Division, 69, rue de Lausanne, 1211 Geneva, 23, Switzerland

HUNGER, Iris — Max Delbruck Centre for Molecular Medicine, Robert Rossle Strasse, 10, 13122 Berlin, Germany

LEITENBERG, Milton — Research Scholar, Centre for International and Security Studies, University of Maryland, School of Public Affairs, College Park, Maryland 20742-1811, USA

LEVY, Jonathan	Northwestern University Medical School - 2650 Lakeview 4205 Chicago, IL 60614, USA
MARTELLINI, Maurizio	Secretary General, Landau Network-Centre A. Volta, Via Cantoni 1 - Villa Olmo, 22100 Como, Italy
NETESOV, Sergey	Senior General Director, Institute of Molecular Biology, "VECTOR" Scientific and Production Association, Koltsovo 633159, Novosibirsk Region, Russian Federation
MESELSON, M	United States of America
PASTERNAK, Charles	Oxford International Biomedical Centre, 3 Wingmore Place, London WIH 9DB, U.K.
PETROV, R	Vice-President, Russian Academy of Sciences, Moscow, Russian Federation.
PRAWITZ, Jan	The Swedish Institute of International Affairs, P.O. Box 1253, Il 182 Stockholm, Sweden
SHAPIR, Yiftah	Jaffee Centre for Strategic Studies, Tel Aviv University, Tel Aviv 69978, Israel
VOLKOV, Vladimir	State Research Centre for Applied Microbiology, Obolensk 142279, Moscow Region, Russian Federation
VOLPE, Pietro	University of Rome "Tor Vergata" Chair of Biophysics, Dept of Biology, Via della Ricerca Scientifica 1, 00133 Rome, Italy

ZILINSKAS, Raymond

Centre for Public Issues
and Biotechnology
University of Maryland
Biotechnology Institute
Suite 500, 4321 Hartwick Road
College Park, Maryland, 20740, USA

1

Progress in Biology—Prosperity or Tragedy

—R. Petrov, Russian Federation

Many scientists are not keen on forecasting feasible progress in science. When they do forecast, they nearly always make their predictions optimistic. The optimism in their scientific predictions can be explained by at least three considerations:

First, optimism stresses the power of science; there are no barriers for it; cancer will be conquered; the life span appears to grow by tens of years, etc.,

Second, optimism is considered to be pleasant for society at large, for broad masses of population, for the 'common people'; an optimistic scientist appears to be preferable to a scientist expressing pessimism;

Third, an optimistic prediction is not punishable, since it cannot fail; there is always a way out—the time has not come yet. The problem will be solved later. So, the optimistic forecasters are wiser than their pessimistic counterparts.

In other words science, as a part of culture, has provided modernisation, continued to improve civilisation's quality of life in the past and promises still greater progress for humankind in the future. Of course water and energy problems, health protection and malnutrition will be solved. Biology will play an important role in the solution of problems facing humankind, including energy. It is precisely biology, I stress, and its branches such as genetics, biochemistry, microbiology, cytology with the new synthetic disciplines as molecular biology, gene engineering, biotechnology and immunology that will ensure the solution of major problems of raising the biological quality of life—health, nutrition and longevity. A number of severe diseases will be conquered. New animal or plant species will be created and flowers never seen before will bloom all over the planet. But we need to be careful with an absolute optimism. We must have a little bit of pessimism, *creative* pessimism.

An optimistic future of biology contains two restrictions. One is associated with the political health of society; an ever present determiner of the utilisation of scientific achievements. Science strives to reveal phenomena whereas society utilises the discovered phenomena for good or for evil. Of course scientists, politicians and society have to be cultural and ethical enough to, utilise biology as *any* scientific achievements for *good* not for *evil,* because it would not be an excusable mistake. There exists a second danger in relation to biological sciences, perhaps, more so than to other disciplines. It is inactivity. These two restricting factors tend to poison an optimistic future portrayed for both biology and mankind at large. As an example of an initiative designed to promote a positive political climate and combat inaction, I would like to bring to your attention the growing role of UNESCO in promoting the most up-to-date trends for progress in biology toward humankind's prosperity and increases in the quality of life.

Through the initiative of Professor Federico Mayor, Director-General of UNESCO, work on setting up a new standing UNESCO Committee on Molecular and Cell Biology was begun several years ago. I am the Vice-Chairman of the Committee, Professor Azzi (Switzerland) is the Chairman. The aim of Committee, is to focus biology potentials on the principal and most promising trends of modern biology and to establish a network of leading biological institutions throughout the world. Genetic engineering and immunology have been included in the programme of the Standing Committee on Molecular and Cell Biology. The very fact of uniting scientists working on molecular and cell biology problems under the authority and prestige of UNESCO is widely regard as a significant step toward understanding the social significant of 'hot sport' of contemporary biology.

Humankind must look ahead with its 'biology eyes' open in order to realise the great potential of biology. The title of this paper is 'The Progress in Biology—Prosperity or Tragedy'. With this title, i stress that the potential of biology has two sides.

It is my belief that the potential of biological science helps ensure the prosperity of humankind, despite the growth of human population on the earth and the proliferation of new strains of viruses and bacteriums. We must develop biology and biotechnology if we wish to ensure life and health for everybody. However, we must be mindful of the more alarming aspects of the biological progress. Some of these products must be banned throughout the work and some must be actively controlled. If not, the resulting inactivity would result in a universal vulnerability; danger for everyone everywhere.

I am an immunologist. Immunology is a developing branch of biomedicine, well known today, after AIDS (Acquired Immuno Deficiency Syndrome) has appeared. Everybody has understood the importance of our body's immuno system to defend our health.

Immunology has become extremely beneficial for mankind. The discovery of blood groups and the development of immunisation methods against dangerous infectious diseases is significants examples. Blood transfusions after immunological blood group determination have provided millions of humans with life-saving treatment for a variety of life threatening situations.

Raising specific immunity after vaccination has resulted in conquering many infections, for example, polyomyelitis. Small pox has been conquered as well completely around the world. Many infectious diseases are now under control. However, a variety of problems in conquering infections have not yet been solved. High quality vaccines against influenzae and some parasitic and enteritis diseases have not yet been developed. An anti AIDS vaccine has not yet been introduced. Modern immunology is very close to solving the problems of creating these vaccines. New principles are being applied in biotechnology for vaccine production and the early results look promising.

Molecular and cellular immunology as well as immunogenetics can provide effective methods for controlling the body's immune response. This trend is most significant for the production of a new generation of vaccines. Immunologists are constantly searching for ways of overcoming immune deficiencies. There is evidence that primary, secondary and acquired immune deficiencies are occurring more and more often. The inadequacy of the immune system is a principal factor in the majority of chronic diseases which attack living things. The widespread distribution of those diseases is apparently connected with environmental pollution, urbanisation problems and ecologic shifts. The immune system of our body is considered to be a special system designed to provide a host with homeostasis defense mechanisms as well as forming within the host its intrinsic ecology or so called *body ecology.* Many voices are raised concerning problems of environmental protection, but much less is heard regarding body ecology protection.

The body's immune system, is a key mechanism for maintaining the body's ecology. This system is capable of protecting a person from viral and bacterial invasions, foreign protein attacks, allergic diseases and malignancies. In my view, humankind has two alternatives—either promote research activity in basic immunology and advance as rapidly as possible on a route to overcome immune deficiencies and allergies, or be in active and helpless in the face of extensive proliferation of these disorders throughout the population of the world. Failing to take action should cause the scientists of the world to sound the alarm. If we do not do our best, as researchers, in immunology and allergology, if we simply ignore problems of body ecology, we surely will not stumble upon an appropriate solution for problems of humankind's health protection in the near future. The 'danger areas' located tend to expand quite independently of evil intentions. And all of us have to be vigilant to prevent the *'evil'*.

Finally, there is the problem of cancer immunotherapy and how to overcome a tissue incompatibility barrier; a phenomenon widely known among transplantologists. A key for solving these problems is in the investigation of minimal differences between individual persons on the one hand, and between intact and malignant target cells on the other hand specifically, leaving the healthy tissues intact. We have to answer the question, which protein-based structures (antigen complexes) determine these minimal differences? Then, we need to develop a new generation of specific drugs which would be capable of killing the only malignant target cells. The utilisation of monoclonal technology in this context facilitates solving the problem, and I believe makes the possibility practicable.

The symmetrical evil of this noble problem is the possibility of creating ethnic weapons. In reality, such a possibility is grounded in the recognition of the fine immunological differences between ethnic populations which leads to the creation of chemicals which have selective specific toxicity.

Let's look at some good and evil applications from genetic engineering to illustrate the symmetry of good and evil which are contained in these 'hot-spots' of modern biology. (Table 1)

TABLE 1: 'SYMMETRY' OF GOOD AND EVIL IN IMMUNOLOGY

Problem	Good	Evil
Immunological Supervision (body Ecology)	Immunity Control, The treatment of immuno-deficiencies, The Creation of vaccines of the new generation, Body Ecology	Wide spread of secondary and Acquired Immunodeficiencies, (AIDS) Autoimmune diseases
Allergy	Triumph of Allergy	Total Allergisation
Directed transport of Drugs (Immuno-Toxins)	Immunotherapy of Cancer, The overcoming of tissues incompatibility of transplantation	Ethnic Weapon

A present, the primary stage in the development of genetic engineering has been completed. Methods have been developed for introducing into micro-organisms genes taken from the unrelated species. Among the natural hosts of those genes are viruses and bacteria and donors of the genes are microbes, animals or human beings. The genes introduced were proved to be active in the new host.

Micro-organisms transformed with unrelated genes gain an ability to produce the specific gene-encoded protein in unlimited quantity. This methodology has become an integral part of the microbiological industry technology and is now exploited by manufacturers to produce various high quality proteins or enzymes; hormones (e.g. human insulin); vaccines against hepatitis for example, and certain antibiotics. Further, today's genetic engineering methodology is capable of solving problems in the production of biogas or photosynthetic utilisation of solar energy.

But, there is the other symmetrical side of these achievements which is fraught with danger, I mean the 'evil' that can result from the dehumanised exploitation of gene engineering technology as a technology to create microbiological weapons. Exploiting this technology can produce the evil reality of giving natural micro-organisms the novel qualities to overcome the human body's immunity barrier or produce drug resistance in human beings.

The contemporary stop in the development of gene engineering is the production of 'transgenic' plants and animals. Transgenic means the introduction of some unrelated genes or adding extra genes into genomes of either animal or plant genetic cells. Such a procedure would be expected to produce a birth of an unusual offspring with new qualities. For instance, potatoes, which carry leguminous plant protein encoded genes, are already available. Tubers of such potatoes were found to be highly eniched with the protein. This example illustrates one of the ways of solving the worldwide problems of protein deficiently in both stock-breeding and within the human population. Transgenic plants which acquire the resistance to viral diseases and the producing of anti-insect drugs are now being created.

Some categories of the transgenic animals appear to be able to produce milk or wool, both with special qualities. Manipulations with a somatothropic-encoded gene, that is the introduction of the gene into gametic cells, will permit breeding of giant fishes or extremely big animals. A possibility of raising the body's resistance to infections, diseases, etc., should not be excluded as a real possibility in the near future.

Certainly, manipulations with human gametic cells genes are regarded as beyond the social morality and ethics for the present. However, treatment and correction of some hereditary defects in human somatic cells seem to be becoming a reality. Gene therapy can be applied to the treatment of some genetically predisposed hemopoietic disorders. The diseased hemopoietic cells can be removed from a patient's bone marrow and then be further in *vitro* specifically treated. The imperfect genes found in those cells can be replaced with their healthy counterparts. And then, it is not difficult to bring the restored hemopoietic cells back into the patient's body.

The adoption into practice of the achievements in transgenic plant and animal technology has began. It is my firm conviction that mankind should be careful with these achievements.

I think it is the time to sound the alarm 'Save Our Souls'. The dangers as indicated in the slide are real, expansion of the transgenic plants over the natural environment, or even, the degradation of natural biocenosis with unpredicted consequences. The possibility of monstrous living things in nature, particularly, aggressive animals or insects, cannot be excluded.

Table 2 presents a few examples for immunology of good and evil applications of advances in biological science.

TABLE 2: 'SYMMETRY' OF GOOD AND EVIL IN GENETIC ENGINEERING

Problem	Good	Evil
New Genes Introduced into Micro-organisms	Microbiological industry and Biotechnology; Proteins, Hormones, Ferments, Biogas, Photo-synthesis, Concentration Ore, etc.	Microbiological Weapons, New Variants of Micro-organisms with heightened infection efficacy. Capacity to overcome immunity Resistance to antibacterial drugs, etc.
New Genes introduced into Plants	The creation of Transgenic plants with heightened content of proteins in cells resisting viruses and insecticides and producing natural insecticides, etc.	Ecological Expansion of artificially created plants with the disturbance of natural biocenoses.
New or Additional Genes introduced into Animals	The Creation of Transgenic Animals, Hereditary disease treatment in people production of proteins in Animals, an increase in productivity and resistance in agricultural animals.	Spreading of monster in nature (abnormal aggressive animals and insects) with unpredictable consequences

How to control it? Who has to be a controlling body? A scientist himself? A pharmaceutical or biotechnological industrialist Special National Committee, like FDA in USA? Or may be humankind needs to have a special international body for regulation and controlling gene-engineering activity?

In several countries there are special governmental documents for this reason. Several months ago Russian Parliament issued a law entitled "State politics in gene-engineering activity". Russian Academy of Sciences, Russian Bioscience Commission and myself personally were pushers to create the law. What is a basis for the control of gene-engineering activity? Let me show a scheme to levels and mechanisms of the control.

Of course, we need to have a state social-scientific committee powerful enough to permit or to restrict any steps of gene. But I believe humankind needs to have an international, may be United Nation's body in OEA.

2

Biological Questions in the Third Millennium

—P. Volpe, Italy

Three old but fundamental problems of Biophysics and Molecular Biology, not solved up to now, may represent natural bridges to the basic research of the third Millennium. The first problem concerns the "language" of the *cell differentiation* and *organoformation*. The second concerns the "loss" of this language which could cause a kind of *cell de-differentiation* leading, in turn, to *cancerogenesis*. The third concerns the unique properties of the highly differentiating nerve cells allowing not only short and long-term "storage" but also rapid "reading" of *brain memory codification*. Some global light on these intriguing problems will probably be shed by the *International Programme on the Human Genome*, the success of which will depend upon the emphasis given to basic biology, when compared with that given to applied biotechnology. Any disproportion in the support to these two parts, leading—according to one of the new tendencies—to an exaggerated priority of the research for a fast application, would in the long turn reduce theoretical creativity. Such a disproportion should be considered. Thus, as a heavy form of *misuse* of science.

To begin with, we shall analyse Galileo's point of view, reflected in the *Dialogue Concerning the Two Maximal Systems of the World*: I am walking along a beach—in essence he said—and the sand of this beach is so humble that it is capable of being impressed by my footprint; however, from this humble sand, I will extract glass; and with the glass I will build the telescope through which, finally, I will contemplate the stars to feel myself happier on Earth, because I am searching for the light of Science and its benefits.

Unfortunately, Galileo could not suspect at that time that the logic of his ethical world (implicit in the so-called *Experimental method*) would be inverted in the 20th century, since (as everyone knows) the energy of the atom (investigated by the "Boys of Via Panisperna" in Rome) was not first used for peaceful purposes, as anticipated, but misused, as not anticipated, for the annihilation of whole cities—Hiroshima and Nagasaki.

A number of arguments can be brought up to justify such a historical issue, of course, but in any case several facts are undeniable. Notwithstanding the Inquisition, Galileo found a fruitful intellectual background in Europe to enrich his discoveries with humanistic notions, for on the horizon, the emerging new classes were looking for a coming "Renaissance" or at least some sort of "Innovation", let us say, to provide a *final exit* from the Middle Age, at the cultural level. By a paradoxical comparison, after the Industrial Revolution of the last century, and the controversial events of the last years of this present century, the humanistic significance of Science could be dramatically obscured, because under the strong control of an Economy understood in terms of the simple search for profit there appears to be a trend to orient any research programme simply towards application, *de facto* reducing theoretical creativity. The International Scientific community should therefore struggle for a correct equilibrium between the basic and the applied aspects of Science, particularly in Biology, since basic Biology is not only a symbol of progress *per se* (suffice to mention the achievement by Charles Darwin of the Theory of Evolution) but also at the same time a guarantee for modern ethical thinking. So any "escape" of biotechnological programming through a careless policy of management should be discouraged by Governments, giving back to scientists in all countries both the strength of freedom of research and that of responsibility.

Three bridges to the basic biology of the 21st century

This presentation will draw attention to three questions of Biology which—in my view—represent natural bridges to the basic research of the next century. (i) The first question regards the "language" of *cell differentiation* and *organoformation.* (ii) The second regards the "loss" of this language, leading *vice versa to cell de differentiation,* namely to cancer. (iii) The third regards the properties of the highly differentiated neuronal cells allowing the storage and the reading of *brain memory.* These intriguing questions, taken together, expect a global answer from the *International Programme on the Human Genome* which will continue for at least another 25 years. Within its framework, I will summarise the research performed in our laboratory to reveal the *internal structure of the eukaryotic gene* and to decipher a number of "worlds", written along it, as signals of a "supercode"—overlapped on the main genetic one—responsible for *modulation of gene expression and, thus, for regulation of cell differentiation.*

The internal design of the eukaryotic gene

At the beginning of the 70s, there was a problematic discussion not only about the size and the origin of the eukaryotic pre-m RNA and mRNA [1-4], but also about the internal structure of the eukaryotic transcriptional unit [5,6]. The background experiment performed in our laboratory to clarify both these points originated from previous studies regarding the timing of DNA methylation [7,8] and the non-random genetic scattering of 5-methylcytosine (5mC) along a DNA strand [9]. This experiment showed a preferential methylation of promoter [9] (as reviewed in [10] and of all those sequences that do not code for mRNAs [9] (as reviewed in [11]. In brief, in HeLa cells, the genomic single strand (ss) DNA fragments hybridised with preprocessed high molecular weight mRNAs (purified from nuclei) contain a large number of 5mCs, while the genomic ss DNA fragments hybridised with processed low molecular weight mRNAs (purified from polysomes) contain few, if any, 5mCs[9]. Since polysomal mRNAs are much shorter than nuclear pre mRNAs [6], such results suggested for the first time that, while the eukaryotic transcriptional unit has to be thought of as an intermittence of coding and uncoding sequences [9] (five years before the description of "splicing" [12]), methylation does not significantly involve the coding regions, namely the translatable exon sequences, but preferentially involves the uncoding regions, namely the intervening intron and signal sequences complementary to those parts of mRNAs which are removed during processing [9].

The family of eukaryotic genes regulated by methylation

The scattering of methylation on the promoter and on the

introns, in contrast with hypomethylation of the exons [9], suggested that 5mC might function as a negative signal for transcribing DNA-dependent RNA-polymerase [9]. This suggestion was supported by the observation that, in synchronised HeLa cells [13], there was an inverse correlation between the genomic DNA methylation, occurring during the S-phase [7,8], and the bulk of transcription [14] and translation [15], occurring during the phases G_1 and G_2(as reviewed in [16]). Such an inverse correlation (suggested for the first time by us at the International Congress of Biochemistry held in Stokholm, in 1973 [17], and rediscussed, in 1974, in *FEBS-Letters* [9] and, in 1976, in *Horizons of Biochemistry and Biophysics* [18]) was confirmed by Pierre Chambon in France, in 1979 (for the isolated *ovoalbumin gene* [19]), and by Walter Doerfler in Germany, in 1980 (for the *early and late genes of the adenovirus A12* [20]). At the present time, the genes known to be regulated by methylation are more than one hundred: essentially they are *housekeeping* (HK) and, particularly, *tissuespecific* (TS) genes (the hypermethylation of their promoter always causes a switch-off of transcription) [21]. This led us to ask whether the genes, which are regulated by methylation, share a "language" in common [21,22].

The language of the genes regulated by methylation

To speak briefly about the work made to decipher this eventual language [21-23]: in agreement with previous information concerning the Sea urchin DNA [24], the digestion of the genomic DNA extracted from syncronised HeLa Cells with the pancreatic DNAse released nine dinucleotides: five of them were unmethylated, because they did not contain cytosine (which is the sole methylatable nitrogen base in the eukaryotic DNA [25]; the other four were methylated to a different extent as a function of the cell cycle [26]. A very low methylation appeared only on the CpG dinucleotide, during the M-phase; in G_1 and G_2 although extremely modest, some methylation was completed on all four dinucleotides (CpT, CpG, CpC and CpA); an intensive methylation involved particularly the Cpg dinucleotide, during S [26]. The chromatographic separation could not distinguish the "direction" of these dinucleotides in the double helix, which is characterised by antiparallelism. Actually, we detected four pairs of dinucleotides, each read in the two possible 3' and 5' directions: In total, eight "words" 23,26]. Two of them were palindromic and symmetrically di-methylated;

six of the were non-palindromic and asymmetrically di-methylated [23,26]. The palindromic di-methylated words were produced by a *maintenance DNA methylase,* active during the S-phase in particular along the CG-rich sequences (to transmit the methylated pattern, semiconservatively[7,8,27], from one cell generation to the next); the non-palindromic mono-methylated words were produced, during the whole cell life cycle, by a *de novo DNA methylase* active in particular along the AT-rich sequences [26, 28-30].

The molecular modeling of the methylated 'worlds'

The *molecular modeling* analysis, performed in our laboratory recently, showed the fine tridimensional structure of these words [23]. The symmetrically di-methylated 5′-CpG-3′/3′- GpC-5′ dinucleotide pair is strongly maintained by six hydrogen bonds, with well compacted methyl groups in the double helix. The asymmetrically mono-methylated 5′-CpC-3′/3′- GpG-5′ dinucleotide pair is also maintained by six hydrogen bonds; but its two parts have a different molecular weight: two lighter pyrimidines against two heavier purines. The asymmetrically mono-methylated 5′-CpT-3′/3′-GpA-5′dinucleotide pair is not only less consistent, because maintained by five hydrogen bonds; it is, in addition, structurally heterogeneous, because constituted of four different nitrogen bases.

The same properties characterised the asymmetrically mono-methylated 5′-CpA-3′/3′-GpT-5′ dinucleotide pair: again five hydrogen bonds maintain the whole structure; and again a full heterogeneity of nitrogen bases characterises it.

The direct sequencing of the methylated dinucleotides along the cloned PCR products of a DNA pre-treated with a busulphite salt

The methylated dinucleotides can be sequenced directly [23], exploiting the following new methodology [31]: at the beginning, the reaction of a ss DNA chain with a bisulphite salt (in the presence of hydroquinone)—through a stepwise sulphonation, hydrolytic deamination and de-sulphonation—converts the DNA cytosines (Cs) into DNA uracils (Us); then, through a PCR amplification, the DNA Us become DNA thymines (Ts). The DNA 5mCs do not react with the bisulphite salt; thus, they can be detected as DNA Cs. So, the sequencing of the cloned PCR products strikingly confirms the existence in eukaryotic DNA of four methylated dinucleotides [23]:

since the former DNA Cs, after conversion, migrate together with the DNA Ts, the DNA 5mCs migrate as the DNA Cs (this DNA 5mC is frequently detected in the CpG dinucleotide); in the restriction site for HpaII (C5mCGG->TCGG), one can read, in one direction, again a 5mCpG dinucleotide and, in the other direction, a 5mCpC dinucleotide; in addition, one can read, in the EcoRII site (C5mCTGG->TCTGG), a 5mCpT dinucleotide; outside any restriction site, one can detect a 5mCpA dinucleotide.

Location of the methylated dinucleotides along the eukaryotic gene

Let us now take a look at the location of these methylated dinucleotides along the eukaryotic gene. In 1993, we published the restriction analysis of ten human genes regulated by methylation: among them, there were three HK and seven TS genes [21,22]. All they really revealed a "methylated language" in common, which was well represented by the example of the gene for calcitonin: the promoter was characterised by a high concentration of the 5mCpG dinucleotide; the introns were characterised *vice versa* by a significant concentration of the Gp5mC dinucleotide; a small concentration of 5mCpC, 5mCpT and 5mCpA dinucleotides characterised the exons [21,22].

The mystery of brain memory codification

But what can be envisaged about the *brain memory codification?* It would be naive to believe that it could also be stored in a particular sequence of monomers in a given macromolecule, as in the case of the genetic memory or as in that of the supercode, herein considered, for modulation of gene expression. One should accept the idea that the mechanism of storage of memory in the nerve cells is based absolutely on other, not yet known, mechanisms. Since nobody knows *where* and *how* the brain memory is written, any reasonable model should be taken into consideration, in the hope that the next century will provide some explanations. At the present time, just to exploit some available classic electrophysiological information, one could suppose that, first, a wave of depolarisation of a given shape and intensity could be created on the external part of the neuronal membrance by a nerve signal of a given entity; then, a corresponding wave of positivity, on the internal part of the membrane could orient a macromolecular "ball" to "print" the "translated information" on a certain support present inside the nerve cell. But how could such a support store the information for a long time, since it is known that all parts of the cell are subjected to a *turnover* in any case ? The stored memory would be lost very soon. So, the problem is certainly dramatically open.

The correct use and the misuse of the scientific discoveries

Returning back to what one should expect from the International Programme on the Human Genome, I would like to conclude by shedding some more light on two aspects of the work performed in our laboratory: the first aspect concerns basic Biology, namely the problem of cell differentiation and de-differentiation; the second aspect concerns applied biotechnology, namely the problem of genetic disorders.

(a) *Cell differentiation and de-differentiation.*Several years ago, at the International Institute of Genetics and Biophysics, in Naples, we showed that, during the early stages of the sea urchin development, an increase of DNA methylation takes place up to a saturation [32]. In view of the inverse correlation between DNA methylation and gene expression [17-20], discussed before [16], this behaviour should be interpreted in terms of a progressive exclusion from transcription of the genes which are not necessary for maintaining the differentiated state [23,32]. Anyway, in the *in vitro* experiments, this methylation pattern can be cancelled by low does of radiation [27,33-35]. In fact, let us suppose that, in a hypermethylated promoter [9,23], a symmetrically di-methylated palindromic 5'-CpG-3'/3'-GpC-5' dinucleotide pair [23] is flanking a radioinduced TT-dimer [34,35]. After the digestion of the damaged sequence, in the reconstructed sequence the *excision-repair* mechanism replaces the former 5mC with a simple C (because in the soluble pool of the nuclear triphosphonucleosides the methylated dCTP does not exist). The result is, therefore, a local *de-methylation* [35]. Because of the inverse correlation between DNA methylation and transcription, this non-enzymatic de-methylation in principle could reactivate some uncontrolled gene expression [36] which might even signify a cancer transformation [34,35]. In such a case, a misuse of biology would be the strategic induction of cancer through relatively low doses of radiation. This danger is real, since the natural background of radioactivity, which never should be higher than 5×10^{-8} Gy per hr, is now largely surpassed in many regions of the world.

(b) *Biotechnology of genetic disorders.* An example of a correct use of biology is implicit in the study of genetic disorders. We are considering the regulation of the FMR1 gene which, if mutated, can cause mental retardation[23]. This gene is characterised by a repetition of CGG trinucleotides in the promoter region: if the number of these trinucleotides oscillates from 1 to 47, the gene remains umethylated, the pre-mRNA is transcribed, and the protein (*an informofer*) is expressed (the person is healthy); if the repetition of CGG trinucleotides arrives at 200, 1000 or even 2000, there appears a methylation on the same repeated sequence and on the promoter which blocks the gene expression (the person becomes retarded) [37]. Biochemical Engineering, able to de-methylate *ad hoc* with low does of radiation the repeated CGG trinucleotides, could lead to a recovery.

A concluding remark about the optimism of basic biology

In conclusion, I Think that—whatever the case is—the correct use of biological sciences should not be, simply; the limitation of a danger (we have taken into consideration the example of low does of radiation). The correct use of Biological Sciences should be first of all the search for an advance in knowledge itself (as part of the intrinsic progress of the same Biology). This can be naturally warranted by the basic approach only. Exobiology should be—in the coming next century—an example of the highest expression of a non-profit, optimistic, science, being oriented towards the achievement of knowledge about the origin (or origins) of life (or lives) in the Universe.

References

1. Georgyev G.P. and Mantieva V.I.: The isolation of DNA-like RNA and ribosomal RNA from the nucleolo-chromosomal apparatus of mammalian cells, Biochim. Biophys, Acta 61, 153-161 (1962).
2. Volpe P. and Giuditta A.: Kinetics of RNA labelling in fractions enriched with neuroglia and neurons. *Nature* 216, 154-155 (1967).
3. Darnell J.E.: Ribonucleic acids from animal cells. *Bacteriol. Rev..* 32, 262-290 (1968).
4. Eremenko T., Benedetto A. and Volpe P: Poliovirus replication during the HeLa cell life cycle. *Nature* 237, 114-116 (1972).

5. Georgyev G.P.: On the structural organisation of operon and the regulation of RNA synthesis in animal cells. *J. Theor, Biol.* 25, 227-231 (1969).

6. Darnell J.E., Jeleneck W.R. and Molloy G.R.: Biogenesis of mRNA: Genetic regulation in mammalian cells. *Science* 181, 1215-1221 (1974).

7. Volpe P. and Eremenko T: Nuclear and cytoplasmic DNA synthesis during the mitotic cycle of HeLa cells. *Eur. J. Biochem,* 32, 227-232 (1973).

8. Geraci D., Eremenko T., Cocchiara R., Granieri A., Scarano E. and Volpe P: Correlation between synthesis and methylation of DNA in HeLa Cells. *Biochim. Biophys. Res. Commun.* 57, 353-361 (1974).

9. Volpe P. and Eremenko T.: Preferential methylation of regulatory gene sequences in HeLa cells. *FEBS-Lett.* 44, 121-126 (1974).

10. Maclean M. and Hilder V.A., Mechanism of chromatin activation and repression. *Internat. Rev. Cytol.* 48, 54-97 (1977).

11. Tentravahi U., Guntaka R.V., Erlanger B.F. and Miller O.J.: Amplified ribosomal RNA in a rat hepatoma cell line are enriched in 5-methylcytosinces. *Proc. Natl. Acad. Sci.* USA 78, 489-493 (1981).

12. Chambon P.: Split genes. *Scient. Amer.* 244, 60-71 (1981).

13. Volpe P. and Eremenko T.: A method for measuring the length of each phase of the cell cycle in spinner cultures. Exptl. Cell Res. 60, 456-458 (1970).

14. Volpe P., Menna T. and Eremenko T.: RNA transcription and processing as a function of the cell cycle. *Bull. Mol. Biol. Med.* 1, 18-28 (1976).

15. Eremenko T. and Volpe P.: Polysome translational state during the cell cycle. *Eur. J. Biochem.* 52, 203-210 (1975).

16. Liau M.C., Chang C.F., Saunders G.F. and Tsai Y.H. : S-adenosylo mocysteine hydrolases as the primary target enzymes in androgen regulation of methylated complexes. *Arch. Biochem. Biophys.* 208, 261-272 (1981).

17. Geraci D., Eremenko T., Granieri A., Scarano E. and Volpe P.: The *in vivo* and *in vitro* methylation of DNA in synchronised HeLa cells. *9th Internat. Congr. Biochem.,* Stockholm, Abstr. 3m64, 190 (1973).

18. Volpe P.: The gene expression during the cell life cycle. *Horizons Biochem. Biophys.* 2, 285-340 (1976).

19. Mandel J.L. and Chambon P.: DNA methylation organ specific variations in the methylation pattern within and around ovalbumin and other chicken genes. *Nucleic Acids Res.* 7, 2081-2092 (1979).

20. Sutter D. and Doerfler w.: Methylation of integrated adenovirus type A12 DNA sequences in transformed cells is inversely correlated with viral gene expression. *Proc.Natl, Acad. Sci. USA* 77, 253-256 (1980).

21. Volpe P., Esposito C., Iacovacci P., Butler R.H. and Eremenko T.: Language of genes with inverse correlation between methylation and transcription. *Macromol. Funct. Cell* 7, 59-71 (1993).

22. Volpe p., Iacovacci P., Butler R.H. and Eremenko T.: 5- methylcytosine in genes with methylation-dependent regulation. *FEBS-Lett.* 329, 233-237 (1993).

23. Volpe P.: A code for genes regulated by methylation. Ital. Biochem. Soc. *Transactions* 9, 93-96 (1997).

24. Grippo P., Iaccarino M., Parisi E. and Scarano E: Methylation of DNA in developing Sea urchin embryos. J. Mol. Biol. 36, 195-208 (1968).

25. Vanyushin B.F., Tkacheva S.G. and Belozersky A.N.: Rare bases in animal DNA. *Nature* 225, 948-949 (1970).

26. Volpe P. and Eremenko T.: A language of DNA modification during replication. *14th Internat. Congr. Genet.*, Moscow, Abstr. 1, 194 (1978).

27. Eremenko T., Palitti F., Morelli F., Whitehead E.P. and Volpe P: Hypomethylation of repair patches in HeLa cells. Mol. Biol. Rep. 10, 177-182 (1985).

28. Eremenko T., Granieri A. and Volpe P.: Organisation, replication and modification of the human genome: I. Differential methylation of two classes of HeLa nuclear DNA separated on Ag^+/Cs_2SO_4 gradients. *Mol. Biol. Rep.* 4, 163-170 (1978).

29. Eremenko T., Granieri A. and Volpe P;: Organisation, replication and modification of the human genome: II. Temporal order of synthesis and methylation of two classes of HeLa nDNA separated in $Ag^+Cs_2SO_4$ gradients. *Mol. Rep.*4, 237-240 (1978).

30. Eremenko T.: Timofeeva M.Y. and Volpe P;: Organisation, replication and modification of the human genome: III. Temporal order of replication and methylation of palindronmic, repeated and unique HeLa nDNA sequences. *Mol. Biol. Rep.* 6, 131-136 (1980).

31. Clarck S.J., Harrison J., Paul J.L. and Frommer M.: High sensitivity mapping of methylated cytosines. *Nucl. Acids Res.* 22, 2990-2997 (1994).

32. Volpe P. and Eremenko T.: Cell and virus macromolecular modification during the cell life. *Macromol. Funct. Cell* 2, 179-195 (1982).

33. Delfini C., Alfani E., De Venezia V., Oberholtzer G., Tomasello C., Eremenko T. and Volpe P.: Cell-cycle dependence and properties of the HeLa cell DNA polymerase system, *Proc. Natl, Acad. Sci.* USA 82, 2220-2224 (1985).

34. Volpe P. and Eremenko T: Repair-modification and evolution of the eukaryotic genome organisation. *Cell Biophys.* 15, 41-60 (1989).

35. Volpe P. and Eremenko T.: The gene repair-modification system in eukaryotes. In *Proceedings of the 16th FEBS Congress"* (Ovcinnikov Y.A., ed.), Science Press, Utrecht, Vol.3, pp. 123-129 (1985).

36. Volpe P. and Eremenko T.: Resumption of transcription following inhibition of DNA methylation. *Macromol. Funct. Cell* 3, 59-69 (1984).

37. Timshenko L.T. and Kaskey T.: Trinucleotide repeat disorders in humans: discussions of mechanisms and medical issues. FASEB J. 10, 1589-1597 (1996).

3

Biology for Peace

—A. Falaschi, Italy

Introduction

The population of the world of today is faced by a challenge that could threaten even its survival in the near future. The gap between the minority of the privileged population and the great majority that lives in total destitution is increasing and could bring to unforeseen levels of violence, made even more destructive than in the past by the technological means available today. In this context, biologists have a specific responsibility, in two ways: on the one hand, with their science they can contribute to solve some of the greatest problems of the developing part of the world and thus reduce or abolish the reasons for tension; on the other hand, they could offer means of global destruction of human people and of the environment so powerful as they never were in history.

In this presentation, I shall make a brief analysis of the way in which biology can have a positive effect on the future of mankind and describe in particular the organisation I am responsible for which intends to play a hopefully significant role in helping the humanity of the near future to avoid those dangers.

The Problems

Hunger comes foremost to the mind as possibly the greatest adversity pervading large fractions of the population, particularly in

Africa, South East Asia and certain areas of Latin America. Very often crops suffer from a great variety of stresses, either due to climatic conditions or to qualities of the earth in which they grow as well as the stress due to pests such as insects, fungi and viruses. Also, the scarcity of fertilisers causes very poor yields in the crops, whilst the presence of a biological pest, be it a population of insects or the arrival of a novel virus, may wipe out the crops of entire regions.

Secondly, an element of misery for the developing world, is the spread of **diseases**, some of which are typical of tropical areas and of difficult living conditions. In this context one thinks immediately of malaria, possibly the greatest health problem of mankind: approximately half a billion people suffer from this disease in the tropical belt of the planet, and the number of deaths it causes in one year is estimated to be to the order of three million, half of which are children. Other infectious diseases are particularly rampant in the developing world, such as those due to schistosomas, trypanosomes and other protozoa, and to viruses, like the hepatites (A, B, C, D, and E) and AIDS. However, one must not consider only infectious diseases as problems of the developing world: disorders which are considered typical of industrialised countries, present, in fact, huge problems to the Third World: for instance cervical cancer (the main cause of which is also mostly viral) represents the principal cause of death by cancer in women in Africa and Southern Asia.

In the context of health, and referring to the diseases which are more specific to the tropical world, one observes likewise a great dearth of research addressed to them, in view of the limitation of the health market of those countries, in financial terms: thus, diseases like malaria and hepatitis E (not to mention schistosomiasis and trypanosomiasis) can be considered orphan diseases.

Also, the adversities touched upon briefly above, cannot be taken out of the context of the socio-economical conditions of that part of the world: **poverty and unemployment** are elements which offer the foundation to those problems and prime, so to speak, a vicious circle that, through the difficulties in nutrition and health, aggravates itself. The causes of poverty and unemployment are many, and among them one can certainly consider the scarcity of natural and energy resources, capital and the presence of entrepreneurial spirit.

Also, **pollution** of the environment disproportionately plagues many areas of the developing world, worsening therefore the nutrition and health problems that are discussed above.

On top of all that, one has to consider that the **biological weapons** could unfortunately be relatively easily available also to "poor" economies. They are not difficult to produce, relatively easy to hide, and, in the hands of unscrupulous desperate terrorists could cause incredible damage to large populations.

The Remedies—1.

Biology for development

Biology can offer today very powerful means, even if not the only means, for solving many of the problems that plague such a large portion of mankind. Through science novel methods can be utilised by the citizens of the countries which suffer most to alleviate and possibly abolish the worst hardships which are due to the present, difficult conditions. Thus, in the field of **nutrition**, the ability to manipulate, almost at will, the genome of higher plants and animals offers a great variety of possible applications to several aspects important for human nutrition. In agriculture, we can improve the nutritional values of the plants of common use in developing countries: for instance, several plant species are relatively poor in proteins as far as quantity and composition are concerned; one can plan to introduce genes in these plants which allow the production in great quantities of proteins as nutritional as human milk. Similarly, one can think of increasing the production of plant species of greater use (such as rice, wheat and maize) by making them genetically resistant to different stresses, such as climate conditions (extremes of temperature, drought, high salt concentration) or harmful biological agents (insects, fungi, viruses or weeds).

Animal breeding can also draw great advantages from genetic engineering, both for the protection of animal health and for the possible introduction of foreign genes in higher animals, thus improving their nutritional capacity as well as their resistance to infectious agents or to particular environmental conditions. Furthermore, both plants and animals can be "engineered" in order to allow them to produce molecules of particular value, such as drugs of complex protein structure: in this way, one could extract expensive drugs from plants cultures at low price, or from the milk of engineered animals, to assure a continuous, high level and cheap production of useful molecules.

As far as the **health** problem is concerned, genetic engineering approaches offer in the first place, not only a highly increased possibility of studying the infectious agents and their interactions with the human organism, but also, thanks to this knowledge, they allow for the preparation of novel diagnostics, vaccines and drugs for such diseases. Therefore, for these countries, the application of genetic engineering technologies to the diagnosis and therapy of the diseases prevalent therein, as well s to the study of their molecular basis, is no less important.

Additionally, drugs of great importance for many widespread diseases can be obtained in a safe, efficient and economic way by the genetic engineering approach, such as insulin, human growth hormone, erythropoietin, etc. These biotechnology based products require limited investments and have a high added value. From this consideration, it follows that biology can give a very effective hand to solve the **socio-economic problem** of this less developed part of the world by offering new ways to industrial development and therefore to the creation of wealth and employment. In particular, the new biotechnologies based on genetic engineering can offer both new industrial products and new methods to obtain traditional products, in conditions particularly fitting to developing countries. In fact, biotechnology-based products and processes are typically low-demanding in terms of capital investment, energy, and strategic raw materials, whereas they depend essentially on qualified personnel and on raw materials of biological origin. In this context we may quote, in the first place, the industrial production of new diagnostics, new vaccines and new drugs. In a similar way the industrial processing of food or feed and the most effective utilisation of agricultural waste-products can be markedly improved by the new biotechnologies.

The **chemical industry** may also be partly renewed by the utilisation of bioreactors, that is of chemical reactors based on biological organisms or on biologically derived molecules. This can give rise to a totally new chemical industry which is much less demanding in terms of capital investment and energy requirements than the traditional one, while being at the same time environmentally friendly. Also, the **mining industry** can be positively affected by the new biotechnologies, by improving the properties of micro-organisms currently utilised for the concentration of important metals, typically copper (mineral leaching).

Furthermore, the **Protection of the environment** can be effectively aided by novel scientific approaches: the use of specific micro-organisms may, for instance, remove heavy metals from the polluted areas in a way similar to those utilised for mineral leaching. More importantly, several micro-organisms or micro-organism-derived molecules (biosurfactants) may prove essential for the reduction or removal of oil spills and for the cleaning of oil residues in tankers. Finally, one must consider the use of biopesticides to substitute the chemical pesticides which are widely used in many tropical countries and which may cause several problems for the environment. Another approach is to introduce the genes for biopesticide production directly in the plants, making them genetically resistant to the pest, and thus reducing, or abolishing, the need for chemical pesticides.

Ultimately, the scarcity of **energy** also plaguing many parts of the Third World can be addressed by the novel biotechnologies: the utilisation of agricultural waste by genetic engineering could offer a way of exploiting important renewable energy sources in an efficient way. Likewise, the improvement in fermentation procedures can offer ways to more efficiently utilise energy sources (such as ethanol, methanol or methane) derived from renewable material or from organic waste. The utilisation of micro-organisms capable of partially metabolising the hydrocarbons can allow for the exploitation of the remaining oil from the well (the so-called tertiary recovery of oil). This can potentially multiply the extent of available energy resources derived from fossil hydrocarbons. Also, the bioleaching approach can also be applied to the cleaning of coal sources rich in heavy metals: this would make an important energy source available by rendering it less aggressive to the environment.

Finally, as far as **biological warfare** is concerned, the modern biotechnologies offer tools of unprecedented specificity and sensitivity to monitor the respect of the Convention on biological Disarmament: PCR-based methods, with carefully selected primers, following adequate concentration of the atmosphere and dust of a given environment, could allow the detection of traces at level of the single molecular fragment of the passage of a dangerous organism.

The Remedies—2.

The International Centre for Genetic Engineering and Biotechnology

The International Centre for Genetic Engineering and Biotechnology, ICGEB, was set up in 1982 with the aim of exploiting the new sectors of genetic engineering and biotechnology, in order to help raise the living standards of peoples in developing nations and alleviate suffering through Scientific research in the-sciences. Operating through twin centres in Trieste, Italy and New Delhi, India, ICGEB today counts 41 Member States, many of which host Affiliated Centers. This organisation has its roots in a decision taken in 1981 at the United nations agency UNIDO, the organisation concerned with industrial development. Recognising the vast potential of advances being made in genetic engineering and biotechnology, not least for application in developing countries, scientists at UNIDO proposed the establishment of a centre of excellence for fostering biotechnology, considered a key to alleviating disease, hunger, and to economic progress of the developing world. This concept was approved the following year, and the statutes of the Centre were signed by 26 countries at a meeting in Madrid.

The immediate outcome was ICGEB, whose establishment and subsequent development have been guided, under the aegis of UNIDO, by a panel of scientific advisers and by a preparatory committee of representatives of the Member States. In 1994, the Statutes entered into force, and signatory countries which fully complied with the requirements of international law to be party to an international agreement became Member States. The following year an Agreement between ICGEB and UNIDO for the transfer of assets signalled the beginning of full autonomy for ICGEB as an **International Autonomous organisation** composed now of 41 Member States. At present there are 56 Signatory Countries which have subscribed to the ICGEB Statutes—the difference represented by nations yet to ratify the Statutes. The decision for ICGEB to become autonomous has meant that the Centre, whilst closely collaborating with UNIDO, is today a novel organisation under the general UN umbrella, but administratively independent.

During the preparations for this phase, activities had already begun, starting in 1987, while full operation in terms of scientific research, production of results, training programmes and advisory services, was reached in 1992. These activities are designed to achieve the ICGEB goals of increasing awareness in biotechnology, and offering programmes that strengthen a nation's R & D capability, specifically by:

- providing developing countries with a necessary "critical mass" environment to pursue advanced research in biotechnology;
- training schemes and collaborative research with Affiliated Centres to ensure that significant numbers of scientists from Member States are trained in state-of-the-art technologies, in areas of direct relevance to the specific problems of their countries;
- acting as the coordinating hub of a network of affiliated Centres that serve as local nodes for the distribution of information and resources located at ICGEB.

Located in Trieste, Italy and New Delhi, India, the Centre forms an interactive network with Affiliated Centres in Member States. In its two Components, ICGEB now has many different research groups working in spheres related to health, nutrition and the environment. Research projects on topics such as hepatitis, gene therapy, cancer, malaria, transgenic rice and cotton are targeting the special needs of the less-developed countries. Work published this year alone, and including techniques to combat malaria, treatment of anaemia, HIV-1 dynamics in infants, and improvement of cotton and rice crops, demonstrates just how critical the research is globally, and in particular to developing countries. It is important, too, that through the Affiliated Centres, ICGEB is working directly in the field, bringing help precisely where it is needed, and boosting the research capability of developing countries.

At present, ICGEB in its two Components together hosts over one hundred fifty scientists, including senior and junior scientists, and trainee fellows from the Member States. In addition, there are around sixty technicians and thirty-five administrative staff.

The scientific activities of ICGEB are reviewed and monitored by a Council of Scientific Advisers of eleven members (including two Nobel laureates) who also report their advice to the board of Governors, the governing body of ICGEB formed by one representative per member country.

ICGEB has a comprehensive biotechnology programme in Trieste and New Delhi: intensive scientific research is performed in a total of eleven research groups. In addition to research, there are several training and other schemes in action, such as the awarding of fellowships for post-doctoral researchers and for Ph.D programmes organised in both Components as well as the organisation and sponsoring of meetings and courses.

The main areas of the research programme are:

- human health, with emphasis on infectious disease control, vaccine production, and genetic diseases;
- plant biology, with emphasis on crop improvement.

The New Delhi Component concentrates its activity on plant genetic engineering, in malaria and on hepatites. At the Trieste Component, research is focused on molecular and cell biology, virology, microbiology, protein structure and function, molecular pathology, molecular immunology. Scientists working in Trieste collaborate with biologists at SISSA (International school for Advanced Studies), an international university in Trieste and the International Centre for Theoretical Physics, and also have access to a powerful source of X-rays generated by the synchrotron Elettra, located close to ICGEB.

Of course, the operation of ICGEB goes far beyond the Components of ICGEB, in Trieste and New Delhi. In fact, these two laboratories form an active network with the Affiliated Centres in Member States, established research institutes which have attained, or have the potential for, research at high standard. The Centres of this international network host many of ICGEB's training activities and channel the resources and services of ICGEB to local institutions. In addition, a Collaborative Research Programme encourages joint effort between ICGEB and the Affiliated Centres, with the development of new research programmes of specific relevance to participating countries. Another initiative that ensures the transfer of know-how to where it is needed and useful in developing countries involves the drawing up of several agreements between ICGEB and the industrial sectors in member States.

In all its activities, ICGEB promotes co-operation with international agencies, to bring together and accelerate common programmes. In addition to strong ties with UNIDO, ICGEB has close contacts with WHO, FAO, UNEP, UNESCO, IAEA, OECD, the Commission of the

European Communities, and the European Molecular Biology Organisation and Laboratory.

The more basic aspects of the scientific programme of ICGEB offer technologies and knowledge essential for the overall scientific programme. Also, it might be stressed that the most important aspect of the ICGEB scientific programme is the **capability building:** that is, the training of high quality scientists, and the help in their return to their country and not only performing appropriately high quality research there but also providing the local industries with the state-of-the-art know-how essential for either development of original products or management of acquired licences. This capability building is, arguably, the most cost-effective way of improving the productive capacity of developing countries. In fact, ICGEB denotes a large part of its energy to long term training (that is pre-and post-doctoral fellowships for Member Country scientists who are involved in the ICGEB scientific programme for a minimum period of one year) as well as the short term training, that is the participation of Member Country scientists in short courses (theoretical and practical) workshops etc. Consequently, and subsequently, the Collaborative Research Programme will help those scientists to establish themselves again in their country and perform high quality work there.

It is obvious that, in terms of its scientific programme, the ICGEB is not unique: several laboratories all over the world, in fact, perform similar or related projects to those approached by ICGEB; what makes ICGEB unique is the fact that this Centre is the property of its Member Countries, which are mainly developing countries: The Member Countries, in fact, run it and can influence its activity: the scientists who come from the Member Countries to work at the ICGEB are not simply guests, but are participating in a project which is also their own. It is also expected that any useful results coming from this scientific programme, as well as the knowledge acquired by the scientists who come to work in ICGEB are all addressed to the advantage of the developing Member Countries: any new technology developed in ICGEB will be made available, even if protected by an ICGEB owned patent, to the Member Countries and to their scientists, with the principle of assuring them the maximum scientific and productive advantage from the use of advanced science. Considering that the ICGEB is the only operating laboratory in the field of genetic engineering and biotechnology with in the realm and with the ideals of the United Nations, the Centre has to be considered a unique scientific resource for the developing world.

The Possible Role of the ICGEB in Assuring Biological Disarmament

In an effort to diminish or abolish the hazards of the possible use of biological weapons, an International Convention has been established that intends to ban completely all research and production activity aimed at the obtainment of biological weapons. Three main reasons can be advanced for considering a positive interaction between the Convention and ICGEB and the instrumental role the latter could play:

1) ICGEB's unique nature and its spectrum of activities;
2) ICGEB's increasing role in international fora, of which the BWC is an important aspect;
3) ICGEB' pro-active character as a centre of excellence for research and training in modern biology;

1) ICGEB's unique nature and its spectrum of activities

The fact that ICGEB started its operations when the spirit of reform was already blowing throughout the United Nations system has given it a unique advantage over older, larger and more bureaucratic international institutions in so far as its management has been careful to run its operations in an extremely efficient way while minimising the administrative load and reducing the cumbersome bureaucracy. Moreover, the Centre has maintained the peculiarities of a modern research institution. So, while ICGEB is governed by UN rules and regulations and keeps its formal link with the System, it is able to maintain a high degree of operational flexibility on a day-to-day basis.

In this context, it is useful to recall the objectives spelt out in article 2 of the ICGEB Statutes and see how, on the one hand, they have been translated into action through the main activities carried out in the last 10 years by the Centre and to consider, on the other hand, the relevance of these objectives and activities for the primary role ICGEB can play in the transfer of biotechnology for peaceful purposes and the role it could play in the implementation of article X of the Biological Weapons Convention.

"a) To promote international co-operation on developing and applying peaceful uses of genetic engineering and biotechnology, in particular for developing countries;

b) to assist developing countries in strengthening their scientific and technological capabilities in the field of genetic engineering and biotechnology;

c) to stimulate and assist activities at regional and national levels in the field of genetic engineering and biotechnology;

d) to develop and promote application of genetic engineering and biotechnology for solving problems of development, particularly in developing countries;

d) to serve as a forum of exchange of information, experience and know-how among scientists and technologists of Member States;

f) to utilise the scientific and technological capabilities of developing and developed countries in the field of genetic engineering and biotechnology; and

g) to act as a focal point of a network of affiliated (national, subregional and regional) research and development centres."

Moreover, ICGEB has already been quoted by the Convention's Third Review conference in the context of the international assistance to facilitate biotechnology research and transfer to developing countries. In addition, the substantial role ICGEB could play in support of a verification mechanism of the Convention has been stressed in a study prepared by a group of experts on all aspects of verification for the 50th session of the General Assembly of the United Nations.

2) The concept of clearing house and the increasing role of ICGEB in international fora

In the light of the provisions contained in Article X of the BWC and the subsequent transfer of techologies, especially to the developing countries, the need to create a sort of clearing house mechanism entrusted to an international Authority has become more evident in the past few years.

On the other hand, in the present global context, the international vocation of ICGEB as a "two-way" gateway through which the flow of information and know how could be accessible to all member

Countries, and eventually to other non-member developing countries, is becoming more and more crucial.

Thanks to its networking arrangement with its affiliated national laboratories, through its short and long term training activities and with its already well established information systems (see below), could have a major role to play within the overall strategy of a clearing house approach. The centre would be able to effectively contribute to information sharing mechanism through which both developed and developing countries would exchange information on a fair and equitable basis, a mechanism through which ICGEB could also assist its member Countries on matters relating to biosafety and intellectual property rights.

If properly utilised, the importance such an instrument would have for the international community, cannot be underestimated.

3) The pro-active character of ICGEB as a centre of Excellence for research and training in modern biology addressed to the needs of the developing world.

The fact that ICGEB remains first and foremost an international organisation dedicated to research and training in modern biology gives it the particular characteristics of modern scientific institution when the activities already well underway in some fields of specific interest for the implementation of Article X of the Convention would only need to be extended or increased for them to impact significantly on the major issues relevant to Article X of the Biological Weapons Convention, by way of illustration:

Training Activities: an enlargement of the annual course organised by ICGEB on biosafety could be instrumental for providing technical assistance aimed at the gradual upgrading of national biological safety practices in the States party to the Convention while constituting a proper framework for the enhanced involvement of donor countries in the field of biosafety.

Collaborative Research Programme: an increase of this programme, in connection with more precise information on the outbreaks of emerging diseases, production of new vaccines and diagnostic reagents, would allow for an enhanced circulation of information, thereby improving the conditions of life in certain countries and, more generally; raising the level of confidence among the Parties to the Convention.

Scientific Information Services: apart from offering advice for scientific programmes, ICGEB also provides it member Countries with access to the most improtant biological data bases through the UCGEB net, a bioinformatics network and through BINAS (Biosafety Information Network Advisory System) developed in conjunction with UNIDO. ICGEB net and BINAS, apart from being important research tools could also be adapted to the needs of the Parties to the Convention and become major repositories of data on good manufacturing practices, safe laboratory procedures, biological containment. Besides, they would give to the Parties and to the concerned authority the access to information on releases of genetically modified organisms and on other experiments which could be related to biological weapons or which could entail environmental risks beyond national borders.

To conclude, these are only a few examples to show how, by enlarging the scope of its already well established activities, ICGEB could make its facilities and expertise available through its programmes whilst exploiting its unique nature of a universally recognised independent scientific institution within the UN system.

Building upon its expertise and capability, ICGEB would be in a position to offer a high quality input to the clearing house mechanism and its participation would also minimise the costs involved since it would be possible to keep the general expenditure at a minimum level.

4

Biological Warfare Convention (BWC)

M. Canepa & M. Martellini, Italy

The Biological Warfare Convention (BWC) was started on 10 April 1972 and came into force after its ratification by the United States, the Soviet Union and Great Britain on 26 March 1975. The BWC bans the development, production and storage of warfare biological agents (bacteria, viruses and toxins) and provides for the destruction of existing ones. As a matter of fact, this convention is a further step after the important Geneva Protocol, signed on 17 June 1925, which banned the use (but not the holding) of gases and of any other poisoning or toxic substance at war. The BWC integrates and complements the Geneva Protocol, since all the signatory states to the convention are banned from developing and producing warfare biological agents and toxins. Article 10, in particular, provides that the signatory states shall carry out the conversion of laboratories and use acquired know-how for peaceful purposes, and call for international collaboration.

Unlike the Non-Proliferation Treaty (NPT) of 1968), allowing five states (namely the USA, former Soviet Union, France, Great Britain and China) to hold nuclear weapons, the BWC does not favour holders of biological weapons. When the convention came into force, there were two official holders of biological weapons: the United States and the Soviet Union. However, it should be recalled that in 1969, under Nixon, before the signing of the Convention, the United States unilaterally gave up the holding and development of these weapons, launched a programme for their destruction and started to convert defence laboratories into medical research centres.

The American Government estimates that only 4 nations had an offensive bacteriological programme in 1972, rising to 10 in 1990 and apparently 12 today. Some of these nations have signed the BWC and two of them are known to be able to weaponise biological agents and toxins. According to a document issued in 1993 by the US Armed services House of Representatives Committee "the nations that have officially declared that they have a biological warfare programme are: the USA, the former Soviet Union and Iraq; other countries listed in the US Government archives are China, Iran, Libya, North Korea, Syria and Taiwan. Countries which may potentially develop a biological warfare programme, but which are not listed in western nations' documents, are Cuba, Egypt and Israel". This data alone show how difficult it is to state in certain terms that a country holds a biological warfare programme or has the ability to develop one. The reason for this is twofold. Pathogenic biological agents have a dual nature: many of them, as the anthrax bacilllus and botulina bacteria, are naturally present and can cause spontaneous epidemics. Moreover, some viruses are normally used in medical research to produce vaccines.

Another reason why it is so difficult to identify potential holders of biological warfare programmes with certainty, is that a great amount of military biological research can be legitimated as defence research, which is not banned by the Convention. Finally, another especially delicate aspect is linked to the monitoring and control procedures: the quantities of biological agents involved are extremely limited, and the intrusive nature of inspections may violate laboratories' rights to privacy. Nevertheless, a sweeping monitoring programme, based on samples taken from the population, may harm individual ethics and generate substantially wrong results.

The chapter of monitoring methodologies is especially difficult to work out and is a controversial one; so far, the BWC does not have its set of control and monitoring procedures, since when it came into force, back in 1975, the Soviet Union refused to open the doors of its laboratories to international inspections. Since 1994 various ad hoc groups have been formed to introduce the necessary control and monitoring procedures. The recent example of UNSCOM inspections in Iraq shows that even direct controls can be unable to detect signs of biological warfare research and development. Furthermore, the BWC does not provide for a specific list of offensive or poisoning biological agents, since, to avoid the Convention, nations may develop new genetically engineered agents, which would threaten the control over the proliferation of biological weapons.

Given that potentially weaponisable biological agents are "dual threat agents" (DTA), the only way to limit their misuse is the implementation of Article 10 of the BWC. As already mentioned, this Article calls for the maximum cooperation and transparency between signatory states in biological and technological research. At present there are three proposals for the partial implementation of Article 10:

1) One is known as PROMED, which stands for "Programme for the Monitoring of Emerging Diseases". It was proposed by Mark Wheelis and Stephen Morse in 1992 and consists in an action of international monitoring for an early identification of unusual epidemics in order to avoid their dissemination. It relies on the creation of data bases on endemic diseases, prompt information about anomalous epidemics and the issue of recommendations in trade and tourism;

2) PROCEID is a "Programme of Containment of Emerging Infectious diseases", this, too, developed by Wheelis and Morse, based on prophylaxis, diagnostics and therapy;

3) the "Vaccine for Peace" Programme by Erhard Geissler of 1992 stresses the need to develop research in the production of vaccines against "dual threat agents".

Recently, a fourth solution was proposed by the molecular biologist Yechiel Becker to prevent the proliferation of biological warfare—the production of genetically engineered synthetic vaccines" bases on recent knowledge in the field of DNA synthesis. The experimental phase has already started and the first results are encouraging, although so far they have been limited to the struggle against HIV-1.

International security

From the point of view of international security, it should be mentioned that two countries (the former Soviet Union and Iraq) have developed considerable means in biological warfare R&D.

The Former Soviet Union

In 1970, a special "Committee on Research and Technological Development for Biological Warfare" was set up under the control of the Ministry of Medical and Microbiological Industry and the Defence Ministry. The former managed the Committee through an industrial complex, BIOPREPARAT in charge of the production of military biological agents, but without neglecting the production of medicines and biological components for civilian purposes. BIOPREPARAT had plants all over the territory of the USSR and controlled from 20 to 30 civilian and military research institutes. It had six production plants and it employed about 25,000 scientists and engineers (human resources involved were seven time higher than in the United States).

In October 1987, the Soviet Union officially revealed the existence of only five laboratories equipped for the implementation of a biological warfare programme, depending on the Defence Ministry, namely in Leningrad (now St. Petersburg) Kirov, Sverdlovsk (now Ekaterinburg, where a leak of anthrax spores in 1979 caused almost a hundred casualties), Zagorsk and Aralsk. In 1987, the Soviet Union was yet to declare the existence of the laboratories belonging to BIOPREPARAT, among which the best known where the laboratory of Koltsovo, near Novosibirsk, in central Siberia (named State Research Centre on Molecular Biology) and the laboratory of Obolensk, in the Moscow area (named State Research Centre on Applied Microbiology).

It has been estimated that about 90 per cent of the former Soviet Union scientists who worked in biological warfare R & D left Russia for Europe, the United States and Israel. The number of Russian scientists who may have moved to other Middle East countries is unknown. At the same time, though, since 1979, many laboratories belonging to the BIOPREPARAT complex, such as those in Obolensk and Koltsovo, have started a wide conversion programme geared to public health and environmental protection. In conclusion, Russia is carrying out a serious conversion programme and many of its laboratories are actively collaborating with American laboratories. It would certainly be advisable that a larger part of the American "Gov to Gov" and "Lab to Lab" financing be earmarked for the biological and not only for the nuclear sector.

Iraq

In 1973, Iraq signed the Biological Warfare Convention, but many doubts remain over its observance of the Convention. Iraq's propensity to use biological weapons became clear during its eight-year war against Iran. In 1984, Iraq attacked the Majnoon Islands with chemical weapons based on Mustard gas, Tabun gas (a nerve gas) and mycotoxines. Many people died and were injured and many survivors had respiratory troubles for a long time after the attack.

At that time there was no international condemnation of the attack, also because the Convention on Chemical Weapons (CWC), banning their development, use and possession, and also asking for their destruction, was opened for signing only in 1992 and came into force as recently as in April 1997.

Between 1985 and April 1991 (end of the Gulf War) Iraq had already developed anthrax bacilli, botulina bacteria and mycotoxines for biological warfare. During the same period, Iraqi scientists had selected five strains of bacteria, one mycotic agent, five viruses and four weaponised toxins, at a cost of some 100 million USA dollars according to UNSCOM sources. Biological agents were selected in the laboratories of Al Hakam, Al Manal and Salman Pak, on the region of Baghdad (by June 1996 these centres were destroyed by UNSCOM troops) and were carried to the Muthanna state plant, where they were placed into 200 biological bombs weighing 400 pounds (100 charged with botulina bacteria, 50 with anthrax and 7 with toxins) 25 of which were to be used as warheads for 600 Km-range Scud missiles. All these weapons, detected by UNSCOM troops at the end of the Gulf war" were already in place and ready for attack.

In the 6 years following the Gulf War, the Iraqi Government has produced 8 documents providing, from their point of view, final and complete information on Iraqi biological warfare potential. UNSCOM, however, has always rejected the reliability of these documents, which it considers inaccurate and insufficient. And it is namely as a result of Iraq's alleged non-fulfilment, that the UN Resolution 687 on economic embargo has been maintained. Evidence of Iraq's non-compliance with the BWC came in August of 1995 (after Iraq ratified the Convention in 1991, at the end of the Gulf War) when, following General Kamel's defection, Iraq was forced to officially admit that it held biological weapons and that it continued to condust the relevant research.

This autumn's events are well known, but, in short, we could recall that the crisis started when UNSCOM inspectors asked to visit 20 sites, The so-called Presidential Palaces, in the basements of which, according to intelligence sources, offensive biological warfare experiments are under way. Recently the Iraq Government has confirmed to UNSCOM the US suspicious about the past production of at least 100 tons of VX nerve gas but it was also declared that this gas has been destroyed.

Yet, biological warfare seems to hold a special appeal for Middle East countries: Egypt and Syria have signed but have not ratified the BWC, Iraq and Iran have ratified the BWC but they are suspected of infringing it, Israel has never signed, nor ratified the BWC. This situation seems to indicate that biological weapons are the atomic bomb of the poor countries, but a more accurate analysis would be needed to understand the reasons for preferring this offensive potential, which is, in fact, unreliable and unpredictable from the military point of view and highly immoral, since it does not distinguish between civilians and military targets.

5

Chemical and Biological Weapons: If it's so Useful...

Y. Shapir, Israel

Introduction

Chemical and biological weapons are considered a major threat throughout the world. American stated strategy cites chemical and biological weapons a threat to both to the American continent and to US forces overseas.[1] The chemical and biological threat was emphasised during the past decade, for two main reasons. (a) The end of the Cold War created the impression that the danger of nuclear war declined. (b) The extensive use of chemical weapons during the first Gulf war and the revelations of the extent of the Iraqi chemical and biological programme immediately after the second Gulf war.

Biological weapons are not a new threat. In the book of Exodus (7:19 and onward) we find the use of the ten plagues—most of them would be considered, in modern terms as "biological weapons": frogs, lice, wild beasts, plague, locust and boils.[2]

Definition

What are biological weapons? What is the line that separates them from chemical weapons? The BWC, for example, does not define the term "biological weapons" or "biological warfare agents". It simply forbids the development, production etc. of "microbial or other biological agents, or toxins whatever their origin or method of production...". There is no further definition of the term "toxin" or what the "other biological agents" are.

An old US military manual defines biological warfare (BW) as " the military use of living organisms or their toxic products to cause death, disability or damage to man, his domestic animals or crops"[3]. The same manual notes that the term "biological warfare" is preferred to "bacteriological warfare". That is, because according to its authors, there are 5 types of biological warfare agents:

- Micro-organisms (bacteria, viruses, fungi etc.);
- Toxins (microbial or other);
- Vectors of disease (insects, acarids);
- Pests (of animals and plants);
- Chemical anticrop compounds.

Most modern analyses of biological warfare is still limited, like the biological Weapons Convention (BWC), to the first two categories: micro-organisms and toxins. A modern US military manual, for example, brings the following definitions, which are as stated in the NATO Military Agency for Standardisation publication on agreed terms, AAP6[4]:

- *Biological Agent (BA).* The NATO definition of a biological agent is: a microorganism (or a toxin derived from it) which causes disease in man, plants or animals or which causes the deterioration of material;
- *Biological Warfare (BW).* Biological warfare is the employment of biological agents to produce casualties in man or animals and damage to plants or material. The NATO definition then continues, to include, "or defence against such employment";
- *Biological Weapon.* A biological weapon is an item of material which projects, disperses, or disseminates a biological agent; including arthropod vectors;
- *Toxin.* A posonous substance produced or derived from living plants, animals, or microorganisms; some toxins may also be produced or altered by chemical means. Compared with microorganisms, toxins have a relatively simple biochemical composition and are not able to reproduce themselves. In many aspects, they are comparable to chemical agents.

It is evident from these definitions that on one hand they are concerned almost solely with micro-organisms and toxins, but on the other hand we can see from the definition of biological weapons, that anthropods, for example, are still considered as weapons. Some abiguity still remains about the line that seperates chemical weapons and toxins. It should be noted, that the Chemical Weapons Convention (CWC) defines "chemical weapons" as "toxic chemicals" and their precursors...", and since toxins are "toxic chemicals" they are covered by the CWC as well as the BWC. At least one toxin is explicitly mentioned in the CWC-Ricin, which is listed in Schedule I in the annex on chemicals of the CWC.

A Historical Overview

Almost every historical overview of biological weapons begins with the story of the battle of Kaffa—a town in Crimea where, in 1347, the besieging Tatars catapulted the cadavers of the victims of plague, to disperse the plague among the besieged.[5] This case is illuminating, because it also showed the dangers in the use of biological weapons: This event marked the beginning of the outbreak of the "black death" that wiped out one third of the population of Europe.

Kaffa was neither the first nor the last attempt to use poisons or spread disease as means of warfare. There are recorded incidents through the ages, from ancient Greco-Roman world to the 19th century. But chemical and biological weapons came of age during the great war: different chemical warfare (CW) agents have been used on both sides of the front. As for biological weapons—the story is less clear. There is evidence that the Germans contemplated the use of bacteriological agents, though it is not clear whether these plans were actually carried out[6].

Chemical weapons have been used between the two world wars on several occasions, and all the fighting states during the second world war had large stockpiles of chemical and biological agents. It is interesting to notice, though, that even during the most difficult moments in the war—these weapons have not been used on the battlefields. (The Germans used gas to murder Jews in the concentration camp, and the Japanese conducted experiments on prisoners in Manchuria, by the notorious unit 731).

Since the end of WW-II there were very few cases of the use of Chemical and Biological Weapons (CBW) agents in warfare: There were allegations that toxins have been used in Vietnam and Laos (the "yellow rain"), Egypt used CW agents in Yemen (1962-1966)[7] and the Iraq used them during the first Gulf war (1980-1988)[8].

History of chemical and biological warfare would not be complete without mentioning the events of terrorism with CBW. The most well known of these is the employment of sarin nerve agent by the Japanese sect Aum Shinrikyu in Tokyo, in 1995. In the realm of biological weapons there is a good example in the case of the outbreak of Salmonellosis in Oregon, that was caused by contaminating salad bars by a religious sect[9].

The history of the use of biological (and chemical) weapons is intertwined with the history of the efforts to prevent this use, or at least to limit this use, by legal means. Even in ancient times, poisons were considered as "unfair" or "unmanly" and there were laws and customs against their use. However, the practice of negotiating international, bi-or multilateral agreements to limit certain types of weapons is fairly recent.

One can point to the Declaration of St. Petersburg of 1868 as a possible starting point of modern arms control. In it the parties agreed to renounce among other weapons, projectiles that the "charged with fulminating or inflammable substances". This effort was continued in two conferences that convened in the Hague, in 1899 and in 1907. These conferences drafted conventions that restricted, among other things, "asphyxating gases"[10].

These restriction did not prevent the use of chemical weapons during the first world war. Few years after the war, another agreement was signed—the 1925 Geneva Protocol (which is still in force today). This protocol prohibits the use in war of "asphyxating, poisonous or other gases...", and also prohibited the use of "bacteriological methods of warfare...". This protocol did not prohibit the production or stockpiling of these weapons, and included no provisions for verification.

Arms control negotiations after World War II produced the 1972 Biological Weapons Convention (BWC), which prohibited the use, production and stockpiling of bacteriological and toxin weapons. It was the first international convention which actually attempted to eliminate a whole class of weapons. The BWC lacks effective provisions for verification—a deficiency which 3 review conferences tried to rectify. The last and most important treaty was the "Chemical Weapons Convention" (CWC) which was signed in 1993 and entered into force in April 1997. This convention was signed during the peak of the world wide euphoria that followed the end of the Cold War and the revelations about the Iraqi CW programme. It is a bold experiment in arms control—with the most intrusive verification regime ever devised into an international treaty. This treaty entered into force in a time when the atmosphere in world capitals is less euphoric, more sober, and only time will tell us whether the world will get used to intrusive verification or not.

A Poorman's Atomic Bomb?

Why do states need chemical and biological weapons at all?

Just another weapon?

From the outset, these kind of weapons were perceived as just another type of weapon, to be used in the battlefield like any other weapon, at the military commanders' discretion and judgement. This is exactly how chemical weapons have been employed during the first world war. More recently, Iraq used chemical weapons in the battlefields against the Iranians and against the Kurds. It is possible that the different plans to use biological weapons during both world wars were made under similar way of thinking. Even in the west there are still few proponents of the tactical use for CB weapons.[11]

CBW deterrence

Nuclear weapons, and the experience of the massive bombings of World War II have introduced the idea of mass destruction into the strategic thinking. And with it came the idea of massive retaliation and of deterrence. The great powers, and few smaller powers acquired nuclear power, and the threat of mass destruction is said to have prevented wars.

The idea of deterring war, or severe damage in case of war, did not escape the minds of other leaders of states. Some of them, being unable to acquire nuclear weapons, turned to chemical and biological weapons as a substitute deterrent.

It should be stated quite clearly: biological weapons, deadly as they can be, are a far cry from nuclear weapons in terms of their destructive power. There is no defense available against nuclear weapons, whereas there are several ways to counter the effects of biological weapons. It should be remembered that while biological weapons might kill many lives, a nuclear threat is a threat of total annihilation.

Nevertheless, biological and chemical weapons can be used for mass killing of civilian population, therefore they might be kept as a deterrent. In simple words, this means that state A can warn state B that if B attacks A, A would retaliate, and cause a damage which will be unacceptable to B.

Without going too deeply into deterrence theory, it should be noted that:

- Deterrence can be achieved by threatening to damage non-military targets,` mostly civilian population. Biological and chemical weapons are quite suitable for that purpose.
- Threats of massive destruction can deter, but rarely can they compel the other side to do something it does not want to do.
- Deterrence works as long as the weapons are not employed. Employing the weapons means that deterrence has failed.

Tit for tat and the samson option

Between the two extreme reasons for owning biological and chemical weapons (that is: tactical weapons for the battlefield, strategic deterrence), there are two more reasons. First, as a means of punishment, in the case that deterrence had failed. This is important if the other side owns biological or chemical weapons, and the potential victim wants to be able to retaliate in-kind. Second—as a weapons of last resort: if A faces an imminent defeat by B, he might use biological and chemical weapons in an attempt to prevent this kind of defeat or destroy the enemy while commiting a national suicide ("the Samson option").

Prestige and technological level

There are other motives behind the desire of countries to acquire weapons of mass destruction, that should be mentioned here, although they are hardly relevant for biological weapons: first, there is the national prestige that comes with special kind of weapons.

Second, there is the desire to push forward a country's technical and industrial level. Both motives are very relevant for the proliferation of nuclear weapons and ballistic misiles. The second one was relevant for chemical weapons as well during the 1920's-1930's but less so today. Biological weapons, however are too abhorring to be considered prestigious. Most BW programmes are clandestine, anyway.

Terrorists and CBW

In the previous paragraphs we discussed the motives of states for acquiring weapons of mass destruction. But CBW might be attractive to non state entities. The Aum Shinrikyu attack in Tokyo in 1995 was a good example for a subversive organisation, that used chemical weapons to spread terror.

Many terrorist organisations are interested in causing death and terror in order to get the public attention to their cause. Many analysts think that for such subversive groups, CBW can be a convenient option, equivalent or even better than bombs. These analysts point to the ease of getting the weapons, their destructive power, and a possible shift in their motives.[12]

If it's so good...

In our historical overview we mentioned quite a few cases where CBW have been used. We did not mention all of them of course, but a full list of all cases is not inconceivable. Compared to the number of cases where conventional weapons have been used—chemical and biological weapons have been almost insignificant in warfare.

The conventional wisdom is, that states are constantly in quest for power, and tend to use if when they have it. This view of the human nature is prevalent in the study of history and the relations between nations, since Thucydides. According to this realist view, if CBW are such good weapons, we should have seen a great deal of these weapons in warfare.

Therefore, we face a difficult question: why have CBW not been used?

There is no one answer to that. We can only try and put forward several, probable reasons for that. In order to be able to do that, we have to try to enter the minds of military planners and figure out what their perceptions are like. Similarly, we have to try to figure out how do terrorists think and assess their options.

Unpredictability

Both chemical and biological weapons are difficult to use because it is difficult to assess their effect in advance, since it is subject to many factors, like the direction of the winds, temperature, humidity and other environmental factors, which might change unexpectedly.

Biological weapons are even more susceptible to these conditions. Living organisms are sometimes sensitive to solar radiation, and factors like humidity and temperature can change their viability dramatically. Biological weapons are also very different from one another, and data collected about the operational behaviour of one type of organism, is not necessarily relevant to another.

Biological weapons have another unpredictable factor—the possibility that the effects of a contagious agent would propagate, and cause an uncontrollable outbreak of a disease (the plague of the 14the century is always in people's minds!).

Battlefield use

Biological weapons are probably deemed by military planers as totally unsuitable for battlefield use. Probably the most important aspect is, that the effect of most BW agents is delayed, by hours or days. Thus, they can be ruled out as a weapon for the opening of an offensive: to make them effective they would have to be employed long before the beginning of an offensive maneuver, an action that would jeopardise the surprise effect. They can also be ruled out as a weapon to be called upon in the heat of the battle, as the need arises—a situation that probably would call for a prompt reaction.

Chemical weapons are, of course different, and they have been employed in battles—from WW I to the first Gulf war. But they too put constraints on the side that employs them, as well as the victim. This is usually the case during an offensive: if the attacking side plants to enter the affected area, he will have to do so with forces limited by the protective gear.

Safety

Both chemical and biological agents are extremely dangerous materials, and very nasty to handle. This is, of course, a major problem during the development, testing and production stages. But even when we deal with weaponised, ready to use agents, there could be many safety problems for the operators. Even when these problems are manageable, a commander might deem them to be an impediment to a decision to use them.

Fratricide

A problem which is related both to the problem of safety and of battlefield usability of CBW is the possibility that friendly forces will be affected. In the case of volatile agents this could happen if the wind changes direction. If the agents are persistent (like VX), friendly forces might enter the affected area while it is still contaminated. With biological agents these problems are valid, and in addition there is the problem that an outbreak of a disease might cross borders, and affect friendly forces or even friendly civilian population.

Mutual deterrence

Mutual deterrence is probably the best explanation for the lack of use of CBW during WW II. All the billigerents had stockpiles of chemical and biological weapons, and every one was ready to retaliate in-kind if and when the other side would use them. Yet no one wanted to fight in the mess of chemically or biologically affected areas, and everyone knew that once one side 'breaks the line' everyone else would, too. Therefore—no one did.

Moral considerations

The realist outlook on issues of war and peace tends to discredit moral considerations. Yet these considerations do exist. Chemical and biological weapons are considered as less moral than bullets, shells and bombs. Probably they are also considered as "less manly", "treacherous". All these might not be the main issue when the use of CBW is considered but they can swing the balance in favour of not using them.

Effectivity

A military leader has to consider the effectivity of the weapons he employs, and calculate the costs and benefits. All the points raised here show that the cost might be high. Does the benefit outweigh the cost? Those who believe that biological weapons are almost comparable to nuclear weapons would probably think it does. But chemical weapons under combat conditions are probably not much more effective than conventional weapons. They certainly do not have the efficacy of PGMs. Biological weapons are even more problematic, because the costs are much higher, and the benefits are practically unknown.

Terrorists and CBW

CBW were presented as an ideal choice for terrorists, yet there have been very few cases of actual use. It seems to be that the technical problems of making, testing and employing CBW are much higher than most analysts would admit. Besides, there are other important considerations: most terrorists groups prefer a strict control over the level of violence in their actions, because they have a political agenda which they do not want to jeopardise by too much violence. They usually have constituency of supporters, which they do not want to alienate, and many groups have moral considerations as well.

There is also the question of taking responsibility, and getting the media coverage for them and their cause. It is easy to get it with a big bang, and much more difficult with an epidemic that no one can be sure whether it was an act of terror or a natural phenomenon.

It might be added that the only type of terrorist groups that might use CBW are religious, apocalyptic sects, who are usually cut off from the main stream of society, and believe that the world is coming to an end anyway.

Conclusion

This paper does not intend to claim that CBW are not dangerous weapons, nor that they are never going to be used. It does not claim that there is no need to try and prevent their use.

It does attempt to look at the cases of acquisition and use of these weapons and try to understand the motives and impediments to such actions. A technical outlook on the subject might show us that scientific developments have made these weapons more potent than they were before. But a weapon system is but a tool. One should consider not only the tools but the people who make use of them, and they have not necessarily changed with the weapons. The reasons for acquiring such weapons and for using them remain unchanged. The question of proliferation is a political question, and therefore political measures are required to address the problem. This is the significance of arms control efforts. They cannot guarantee the total elimination of these weapons, nor promise us that they will never be used. But they can lower the probability for such a mishap.

Annex: Current Situation

Biological and chemical weapons have been used in past wars, as we have just demonstrated. In recent years many states renounced the use of these weapons, and joined the appropriate international arms control regimes. Many other states never even considered them—whether from lack of capability, lack of will or any other reasons. Few states, however still consider these options as viable. In some cases there is strong evidence that a state has chemical or biological programmes. In other cases there have been unconfirmed suspicion and allegations—usually in conjunction with allegations of nuclear or ballistic missile programmes.

Former Soviet Union

The demise of the Soviet Union has left large quantities of chemical and biological weapons in the hand of the new states (as well as strategic nuclear weapons—which are much more important but out of the scope of this paper). Russia has the largest and most advanced chemical weapon programme in the world. It has admitted having a stockpile of some 40,000 tons of chemical weapon agents. Russia is obligated to destroy this arsenal, by a bilateral agreement with the US. It is a signatory of the CWC which it has not yet ratified. Planned efforts to destroy these weapons have not yet commenced, and unlikely to do so in the near future. There are concerns that Russia still maintains a biological warfare capability, even though the USSR is a party to the BWC.

China

China has a substantial chemical warfare programme, and probably maintains a biological programme as well. (It had an active programme prior to its accession to the BWC). China is also as proliferation concerns, since its export control are very lax, and it was willing to transfer technologies for weapons of mass destruction, that no other supplier was willing to sell.

North Korea

North Korea, notorious for the past decade for its ballistic missile programmes and its assistance to other Ballistic Missile programmes (in Egypt, Syria and Iran), had a chemical weapon programme since the 1960's. Today it is capable to produce blister agents (like mustard), nerve and blood agents. Its biological programme also started at the early 1960's, and it is believed to be able to produce limited amounts of the "standard" biological and toxin weapons.

South Asia—India and Pakistan

The rivalry between India and Pakistan encouraged both states to develop nuclear capability. India has never admitted having offensive chemical and biological capability. It does have a substantial chemicals industry and produces many of the dual-use chemicals that can serve as precursors to chemical warfare agents. Pakistan has a good chemical infrastructure too. Both countries are parties to the CWC. Both countries have the infrastructure for biological research. There were no allegations of an offensive biological programme on India, but Pakistan is suspected of conducting a programme with potential offensive applications. Both are parties to the BWC.

The Middle East

Most of the cases of chemical and biological proliferation in recent years occurred in the Middle East. The region has also witnessed the only cases of actual use of chemical weapons since the second World War.

Egypt

Egypt was the first country to use chemical weapons in the region—during its participation in the war in Yemen during the 1960's. During 5 years it used mainly mustard and phosgene.

In the 1980's Egypt changed course, and became an advocate of arms control and disarmament. Nevertheless, there were serious allegations that its capability to produce advanced chemical agents was enhanced since the early 1990's and that it gained self sufficiency in producing the precursors for an offensive CW programme. Egypt refused to join the CWC.

Syria

Syria has a chemical weapon programme with a very long history. Its stockpile was originally based on Soviet made bombs, acquired from Egypt in the 1960's. It has now at least two plants for the production of chemical weapons. It deploys nerve agents—sarin and recently VX as well, and it has developed the capability to deliver them in bombs and in the warheads of ballistic missiles. It is assessed that Syria is developing biological weapons capability as well.

Libya

Libya is yet another country in the region that actually used chemical weapons (in Chad, in 1987). It strives from the mid 1980's to build its own chemical weapons production capability—first in the Rabta plant and later, in Tarhuna. It is assessed that it does not have the infrastructure for a biological weapon programme.

Iran

Iran has a biological weapon programme since the 1980's, and it is assessed that Iran probably has managed to produce biological warfare agents and even weaponised a small quantity of them. Iran also produces a large variety of chemical warfare agents, including mustard and nerve agents. American intelligence sources estimated that it has produced several hundred tons of them. It should be mentioned that Iran is a party to the CWC, and has been very active in the Organisation for the prohibition of Chemical Weapons (OPCW) in the Hague.

Iraq

Iraq used chemical weapons extensively during its 8 year war with Iran. During several battles it employed blister and nerve agents (mustard and sarin) against Iranian soldiers. Worse than that—it used chemical weapons against civilian population in Kurdistan—and the case of the village of Halabja became well known. Iraq has a large chemical industry, and it has the capability to produce many of the precursors for the chemical warfare agents it produced in the past. When the inspectors of the United Nations Special Committee (UNSCOM) entered Iraq in 1991, following UN Security Council resolution 687, they discovered very large quantities of weapons, as well as large storage of chemical warfare agents in bulk. These included mustard, sarin and VX.

Evidence for the vast Iraqi biological warfare programme was found only in 1995, although suspicions existed long before that. The Iraqis admitted the production of some 90,000 liters of Botulinum toxin, 8,300 liters of anthrax, 1,850 liters of aflatoxin, and smaller quantities of other agents and toxins.

Although Iraq is still under severe monitoring regime, it is suspected that not all the past activities have been revealed, and that some capability is still present.

References

1. See for example: *Proliferation: Threat and Responses* Office of the Secretary of Defense, April 1996.
2. One does not have to accept the biblical story at face value to realise that whoever committed this story to paper (or vellum) considered a mass use of lice, plague etc. as a weapon.
3. "*Military Biology and Biological Warfare Agents*", TM-3-216/AFM 355-6, Washington DC January 1956.
4. *Hand book on the medical aspects of NBC Defensive Operations* FM 8-9 Washington DC: Department of the Army February 1996.
5. See for example: Harris, Robert and Jeremy Paxman, "A *Higher Form of Killing*" New York: Hill and Wang, 1982.
6. Hugh Jones, Martin. Wickham-Sted and German Biological Warfare Research. *Intelligence and National Security Vol. 7* No 4 (1992) pp. 379-402.
7. Shoham Dany. *Chemical Weapons in Egypt and Syria: Evolution, Capabilities, Control* (Hebrew) BESA center for strategic studies, Security and Policy Studies No. 21, June 1995. On the issue of Egyptian use of CW he relies mostly on SIPRI - the problems of Chemical and Biological Warfare, 1971.
8. Report of the Secretary General on the Implementation of the Special commission for the ongoing monitoring and verification of Iraq's compliance with Relevant parts of section C of Security Council Resolution 687,S/ 1995/ 864, United Nations, March 1995.
9. Torok, Thomas J et al., "*A large out break of Salmonellosis Caused by Intentional Contamination of Restaurant Salad Bars*" JAMA 278:5 6 aug 1997 pp. 389-395.
10. For an overview of arms control treaties and agreements see Goldblat Jozef "*Arms Control: A guide to Negotiations and Agreements*" Oslo, Norway: PRIO 1995.
11. For one example of many see McCall, Sherman "A Higher Form of Killing" *US Naval Institute Proceedings* February 1995.
12. Davis, Zachary "*Weapons of Mass Destruction: New terrorist threat*" *The Monitor* Center for international trade and security, University of Georgia, Vol 3 No 2 spring 1997.

6

International Criminalisation of Biological Weapons

——————————M. Meselson, United States of America

Introduction

Every major new technology—metallurgy, explosives, aviation, electronics, nuclear energy—has come to be exploited not only for beneficial purposes but also for hostile ones. Will this be the case for biotechnology, surely destined to be a dominant technology of the 21st century?

Biological weapons that are already developed and tested provide a relatively simple and inexpensive means for the attack of people, animals and crops over large areas. A world in which the constraints against such weapons have broken down would be a world in which a multitude of states and non-state entities could acquire otherwise unattainable capabilities for overt and covert mass killing and destruction.

Over the longer term, a world in which biotechnology could be freely exploited for hostile purposes would be a world in which the very nature of conflict had radically changed. As our understanding of life-processes becomes increasingly profound and as biotechnology continues to advance, it will become possible not simply to destroy life, but to manipulate it, including the processes of cognition, development, inheritance and reproduction. Through such manipulation could lie terrible new opportunities for violence, coercion, repression or subjugation. Movement towards such a world would distort the accelerating revolution in biotechnology in ways that could vitiate its vast beneficial application, with inimical consequences for the future course of civilisation.

Confronted with this menace, the international community faces a momentous challenge. Now that the Biological Weapons Convention (BWC) and the Chemical Weapons Convention (CWC) have singled out biological and chemical weapons for categorical prohibition, it may be possible to achieve a worldwide system of disincentives, deterrents, sanctions, and norms sufficient to forestall the intensive exploitation of biotechnology for hostile purposes. Doing so would constitute a turning-point in the development of civilisation, distinguishing the accelerating revolution in biotechnology from all previous major technologies.

The prohibitions embodied in the Biological Weapons Convention (BWC) and the Chemical Weapons Convention (CWC) are directed to the actions of States, not individuals. Although the CWC and, less explicitly, the BWC contain provisions obliging each State Party to prohibit persons under its jurisdiction from undertaking activities that are prohibited by these treaties, such provisions fail to deal with the situation in which an individual who orders, directs or knowingly commits an act prohibited by the BWC or the CWC enjoys official protection of the state, or is in a state that lacks jurisdiction to prosecute or fails to prosecute for any reason.

These gaps in the existing regime against biological and chemical weapons can, to a significant extent, be addressed by the creation of appropriate international criminal law that holds individual perpetrators responsible for prohibited activities. This is illustrated by two draft conventions currently under discussion under the auspices of the UN General Assembly, both directed against the use (but not the development production, acquisition or transfer) of biological and chemical weapons. The first of these draft conventions would create an International Criminal Court (ICC), with jurisdiction over war crimes, including the crime of using biological or chemical weapons by anyone, including leaders and officials of states. The second, known as the "Bombing Convention", would also prohibit the use of such weapons but would apply exclusively to terrorists. A third draft convention, applicable to anyone, is being developed by the Harvard

Sussex Programme and would make it a crime under international law not only to use but also to develop, produce, acquire, stockpile, retain or transfer biological or chemical weapons.

Convention to Establish an International Criminal Court

In June 1998, a diplomatic conference of plenipotentiaries will meet in Rome to finalise and adopt a convention on the establishment of an International Criminal Court. This follows years of study and drafting of a statute for such a court by the International Law Commission, a body of experts appointed by the General Assembly to codify and develop international law followed by further study and drafting by an ad hoc committee and then a Preparatory Committee established by the General Assembly in December 1995.

The International Criminal Court would differ from the existing International Court of Justice in The Hague in that the latter advises or adjudicates disputes only between states, while the ICC would have jurisdiction over individual person accused of certain specified crimes, including war crimes. The ICC would also differ from the ad hoc tribunals created by the Security Council to deal with crimes in the former Yugoslavia and in Rwanda in that the ICC would be permanent, would have jurisdiction over certain categories of crime wherever they occur and would draw its authority not from the Security Council but rather from the large number of states that would be parties to the convention that creates the Court.

The present rolling draft text of the statute of the Court includes among the crimes proposed to be under its jurisdiction the crime of using "bacteriological (biological) agents or toxins for hostile purposes or in armed conflict" and of using "chemical weapons" as they are defined in the CWC. Regarding individual responsibility for crimes covered in the draft statute, the rolling text provides that "a person is criminally responsible if that person commits, orders, induces, or knowingly facilitates the commission of such a crime which in fact occurs or is attempted." Moreover, the statute is to be applicable to "all persons without any discrimination whatsoever". Further, "official capacity as Head of State of Government, or as a member of a Government or parliament or as an elected representative, or as a government official shall in no case exempt a person from his criminal responsibility under this Statute, nor shall it [*per se*] constitute a ground for reduction of the sentence."

The existence of an International Criminal Court operating under such a statute could have profound implications for the prohibition of biological and chemical weapons, extending beyond actual use to include attempted use and covering perpetrators and all who order or knowingly aid them in the prohibited activity, regardless of official position. Faced with the possibility of indictment and, if brought before the court, conviction, sentencing and punishment, those contemplating biological or chemical weapons activities, from terrorists to heads of state, might well reconsider.

It is not appropriate here to attempt to address the many political and technical issues connected with the creation and operation of the ICC. But it should be noted that the knowledge and perspectives of those experienced in CBW arms control efforts will be needed in the further development of the ICC statute as it applies to CBW weapons if the applicable provisions are to be effectively worded.

Information and documents pertaining to the International Criminal Court may be found on the World Wide Web at amnesty it and at igc.apc.org/icc.

Convention for the Suppression of Terrorist Bombings

UN General Assembly resolution 51/210 of 17 December 1996 established an Ad Hoc Committee to elaborate an international convention for the suppression of terrorist bombings. On 11 February 1997, France, on behalf of the group of Seven Industrialised States and the Russian Federation, submitted to the General Assembly a draft convention on terrorist bombings which is now before the above-mentioned Ad Hoc Committee.

The draft Bombing convention makes it an offence to deliver, place, or discharge any device that releases "biological agents or toxins" with the intent to cause death or serious bodily injury, or to attempt to do so, or to organise or direct others to do so. The draft convention includes similar provisions dealing with explosive, incendiary, and toxic chemical bombs and devices. The draft bombing convention applies only to acts committed by terrorists. Article 3 specifically states that it does not apply to acts by "military forces of a State in connection with their official duties."

The draft Bombing convention would obligate a State party in which an alleged offender is found, without exception and whether or not the offence was committed on its territory, either to extradite the person or to submit the case to its competent authorities for prosecution in accordance with the laws of that state.

Convention for the Prevention and Punishment of the Crime of Developing, Producing, Acquiring stockpiling, Retaining, Transferring or Using Biological of Chemical Weapons

This proposed convention seeks to create international criminal law that would hold individual perpetrators criminally responsible for activities that are prohibited by the biological and chemical conventions. It defines activities prohibited by the BWC or the CWC as criminal offenses and obliges each of its States Parties: (i) to establish jurisdiction with respect to such crimes extending to all persons in its territory, regardless of the place where the offence was committed and the citizenship of the offender and (ii) to prosecute or extradite any such offender found on its territory. A person who, for example, orders, directs or knowingly participates in the development, production or use of biological weapons in a State "A " which, for any reason, fails to apprehend or prosecute the person would face the risk of apprehension, prosecution and punishment or of extradition should that person subsequently be found in a State "B" that supports the proposed convention.

The same obligations, to establish criminal jurisdiction and to extradite or adjudicate (*aut dedere aut judicare*), are included in international conventions now in force for the suppression and punishment of aircraft highjacking and sabotage (1970; 1971), crimes against internationally protected persons (1973), hostage taking (1979), theft of nuclear materials (1980), torture (1984) and crimes against maritime navigation (1988).

The proposed convention also provides that jurisdiction with respect to the offences it sets forth may be exercised by any international tribunal that may have jurisdiction in the matter. Such a tribunal would presumably have the authority to issue an indictment and an international arrest warrant even when the offender is beyond its reach at the time.

The proposed Convention includes provisions that oblige its States Parties to provide assistance to one another in the adjudication of offences and that guarantee fair treatment and due process of law in all proceedings.

Adoption and widespread adherence to the proposed convention would provide a new dimension of constraint against biological and chemical weapons by applying international criminal law to hold individual offenders responsible and punishable wherever they may be and regardless of whether they act under or outside of state authority. Such individuals would be regarded *as hostis humanis generis*, enemies of all humanity. The norm against chemical and biological weapons would be strengthened, deterrence of potential offenders, both official and unofficial, would be enhanced, and international cooperation in suppressing the prohibited activities would be facilitated.

A model treaty criminalising biological and chemical weapons under international law has been drafted by the Harvard Sussex Programme on CBW Armament and Arms Limitation and has been further developed with the advice of a committee of legal experts. It is attached as an appendix.

APPENDIX

Draft 10 March 1997

Draft Convention on the Prevention and Punishment of the Crime of Developing, Producing, Acquiring, Stockpiling, Retaining, Transferring or Using Biological or Chemical Weapons

Preamble

The states Parties to this Convention,

Recognising that any use of disease or poison for hostile purposes is repugnant to the conscience of humankind;

Believing that the maintenance and reinforcement of the worldwide norm against biological and chemical warfare is essential to international and domestic peace and security;

Considering that biological and chemical weapons pose a threat to the well-being of all humanity and to future generations;

Recognising that the development, production and use of biological and chemical weapons are within the capability of States as a result of the decisions and actions of individual government officials and personnel, as well as within the capability of other entities and of individuals;

Considering that knowledge and achievements in the fields of biology, chemistry and medicine should be used exclusively for the health and well-being of humanity;

Desiring to encourage the peaceful and beneficial advance and application of these sciences by protecting them from adverse consequences that would result from their hostile exploitation;

Recalling that the development, production, stockpiling, acquisition, retention, transfer and use of biological and chemical weapons are prohibited by the Geneva Protocol of 1925, the Biological Weapons Convention of 1972 and the Chemical Weapons Convention of 1993, and other international agreements;

Seeking to ensure and enhance the effectiveness of the Biological weapons convention and the Chemical Weapons Convention, to which the majority of States are parties or signatories;

Recalling that every State Party to the Biological weapons Convention is obliged to take any necessary measures to prohibit and prevent the development, production, stockpiling, acquisition or retention of biological weapons within its territory or under its jurisdiction or control anywhere and that every State Party to the Chemical Weapons Convention has that same obligation as regards chemical weapons as well as an obligation to prohibit natural and legal persons anywhere on its territory or in any other place under its jurisdiction from undertaking any activity prohibited to a State Party under the Convention and that every State Party to the Chemical Weapons Convention has accepted the obligation to extend its penal legislation to any such activity prohibited to a State party under the Convention undertaken anywhere by natural persons, possessing its nationality, in conformity with international Law;

Noting that many states have made it an offence for persons under their jurisdiction to engage in activities prohibited by the Biological Weapons Convention or the Chemical Weapons Convention, and providing penalties for such offences;

Determined, for the sake of human beings everywhere and of future generations, to eliminate the threat of biological and chemical weapons;

Have agreed as follows:

ARTICLE 1

1. Any person commits an offence if that person:

   (a) orders, directs, plans or knowingly participates in the development, production, acquisition, stockpiling, retention, transfer or use of biological or chemical weapons; or

   (b) attempts to commit any offence described in sub-paragraph (a); or

   (c) assists, encourages or induces, in any way, anyone to engage in the development, production, acquisition, stockpiling, retention, transfer or use of biological or chemical weapons; or

   (d) threatens to use biological or chemical weapons to cause death or injury to any person or in order to compel a natural or legal person, international organisation or State to do or refrain from doing any act.

2. It shall not be a defence against prosecution or extradition for the above offenses that a person acted in an official capacity or under the orders or instruction of a state, a superior officer, a public or private authority or any other person or for any other reason.

3. Notwithstanding the above provisions of this Article, nothing in this Convention shall be construed as prohibiting activities that are not prohibited under the Chemical Weapons Convention or the Biological Weapons Convention or that are directed toward the fulfillment of a State's obligations under such convention and that are conducted in accordance with its provisions.

ARTICLE II

1. Biological weapons are defined, in accordance with Article I of the Biological Weapons Convention, as:

   (a) Microbial or other biological agents, or toxins whatever their origin or method of production, of types and in quantities that have no justification for prophylactic, protective or other peaceful purposes;

(b) Weapons, equipment or means of delivery designed to use such agents or toxins for hostile purposes or in armed conflict.

2. Chemical weapons are defined as in Article II of the chemical Weapons Convention.
3. Person means any natural person or, to the extent consistent with municipal law, any legal entity.

ARTICLE III

Each State Party shall make the offenses set forth in Article 1 punishable by appropriate penalties which take into account the grave nature of those offenses.

ARTICLE IV

1. Each State party to this Convention shall take such measures as may be necessary to establish its jurisdiction over the offenses set forth in Article I which are committed:
   (a) in the territory of that State or on board a ship or aircraft registered in that State;
   (b) when the alleged offender is a national of that State; or
   (c) when, if that State considers it appropriate, the alleged offender is a stateless person whose habitual residence is in its territory.
   (d) when the offence is committed to harm that State or its citizens or to compel that state to do or abstain from doing any act.
2. Each State party shall likewise take such measures as may be necessary to establish its jurisdiction over the offenses set forth in Article I in cases where the alleged offender is present in its territory and it does not extradite such person pursuant to Article VI and VII.
3. Each State Party shall inform the Depositary of the legislative and administrative measures taken to implement this Convention.
4. Each State Party shall designate a contact point within its government to which other States Parties may communicate

in matters relevant to this Convention. Each State party shall make such designation known to the Depository.

5. This Convention does not exclude any criminal jurisdiction exercised in accordance with national law.

6. Jurisdiction with respect to the offenses set forth in Article I may also be exercised by any international tribunal that may have jurisdiction in the matter.

ARTICLE V

1. Upon being satisfied that the circumstances so warrant, any State party to this Convention in the territory of which the alleged offender is present shall take such person into custody or take other measures to ensure the presence of such person for the purpose of prosecution or extradition.

2. The custody and other measures shall be as provided in the law of that State Party but may only be continued for such time as is necessary to enable any criminal or extradition proceedings to be instituted.

3. Such State shall immediately make a preliminary inquiry into the facts.

4. When a State, pursuant to this article, has taken a person into custody, it shall immediately notify the States which have established jurisdiction in accordance with Article IV and, if it considers it advisable, any other interested States, of the fact that such person is in custody and of the circumstances which warrant the detention of that person. The State which makes the preliminary inquiry contemplated in paragraph 3 of this article shall promptly report its findings to the said States and shall indicate whether it intends to exercise jurisdiction.

ARTICLE VI

1. The offenses set forth in Article I shall be deemed to be included as extraditable offenses in any extradition treaty existing between States Parties. States Parties undertake to include such offenses as extraditable offenses in every extradition treaty to be concluded between them.

2. For purposes of extradition between States Parties, the offenses set forth in Article I shall not be regarded as a political offence or as an offence inspired by political motives.
3. If a State Party which makes extradition conditional on the existence of a treaty receives a request for extradition from another State Party with which it has no extradition treaty, it may, if it decides to extradite, consider this Convention as the legal basis for extradition in respect of the offenses set forth in Article I. Extradition shall be subject to the other conditions provided by the law of the requested State.
4. States Parties which do not make extradition conditional on the existence of a treaty shall recognise the offenses set forth in Article I as extraditable offenses as between themselves subject to the conditions provided by the law of the requested State.
5. The offenses set forth under Article I shall be treated, for the purpose of extradition between States Parties, as if they had been committed not only in the place in which they occurred but also in the territories of the States required to establish their jurisdiction in accordance with paragraph 1 of Article IV.

ARTICLE VII

The State Party in the territory of which the alleged offender is found shall, if it does not extradite such person, be obliged, without exception whatsoever and whether or not the offence was committed in its territory, to submit the case without delay to competent authorities for the purpose of prosecution, through proceedings in accordance with the laws of that State. Those authorities shall take their decision in the same manner as in the case of any other offence of a grave nature under the law of that State.

ARTICLE VIII

1. States Parties shall afford one another the greatest measure of assistance in connection with criminal proceedings brought in respect of the offenses set forth in Article I.
2. The provisions of paragraph 1 of this article shall not affect obligations concerning mutual judicial assistance embodied in any other treaty.

ARTICLE IX

States Parties shall cooperate in the prevention of the offenses set forth in Article I, particularly by:

(a) taking all practicable measures to prevent preparations in their respective territories for the commission of those offenses within or outside their territories;

(b) exchanging information and coordinating the taking of administrative and other measures as appropriate to prevent commission of those offenses.

ARTICLE X

Any person regarding whom proceedings are being carried out in connection with any of the offenses set forth in Article I shall be guaranteed fair treatment at all stages of the proceedings, including:

(a) assistance in communicating immediately with appropriate authorities of the State entitled to exercise rights of protection;

(b) the right to be visited by a representative of the State of which that person is a national.

(c) enjoyment of all the rights and guarantees provided by the law of the State in the territory of which such person is present; and

(d) treatment in accord with internationally guaranteed human rights.

ARTICLE XI

1. Each State party shall, in accordance with its national law, provide to the Depositary as promptly as possible any relevant information in its possession concerning:

   (a) the circumstances of the offence; and

   (b) the measures taken in relation to the offender or the alleged offender, and, in particular, the results of any extradition proceedings or other legal proceedings.

2. The State Party where the alleged offender is prosecuted shall communicate the final outcome of the proceedings to the Depositary, who shall transmit the information to the other. States Parties.

ARTICLE XII

Any dispute between States Parties concerning the interpretation or application of this Convention which is not settled by negotiation shall, at the request of one of them, be submitted to arbitration. If within six months from the date of the request for arbitration the parties are unable to agree on the organisation of the arbitration, any one of those parties may refer the dispute to the International Court of Justice.

ARTICLE XIII

1. Five years after the entry into force of this Convention, or earlier if it is requested by a majority of Parties to the Convention by submitting a proposal to this effect to the Depositary, a Conference of States Parties shall be held at [Geneva, Switzerland], to review the operation of the Convention with a view to assuring the purposes of the preamble and the provisions of the Convention are being realised.
2. At intervals of seven years thereafter, unless otherwise decided upon, further sessions of the Conference shall be convened with the same objective.

ARTICLES XIV et seq. on modalities, to include the following:

Amendment

Duration

Signature and Ratification

Entry Into Force

Depositary

Authentic Texts

7

The Consequences of a Biological War: Can We Protect Humans and Animals by Synthetic Peptides and DNA Vaccines?

—Y. Becker, Israel

The use of biological weapons as means of mass destruction may lead to man-made epidemics that will introduce bio-engineered agents into the human populations and domesticated and wild animals. In such events, the new agents will have devastating effect on living organisms that may also destroy the world ecology. The threat of the use of BWs will damage the efforts of medicine to provide maximum health to all the inhabitants of planet Earth. One way to prevent the development of BWs is the Vaccine for Peace Programme (1).

The transition from production of natural antigens of viruses and bacteria to synthetic peptides and DNA vaccines may solve the major problem of large scale production of infectious agents for vaccines production. With synthetic peptide and DNA vaccines, it will not be possible to conceal BWs activity as vaccine production. It will be necessary to develop an international consortium for the research, development and production of synthetic vaccines for international use.

The Consequences of Microbial (virus, bacteria) Epidemics in Humans and Animals

The evolution of vertebrates required the development of cellular mechanisms for protection against diseases caused by microbial agents. In humans, specialised cells were developed that constitute the three arms of the known immune system. The encounter of human populations with a new and a highly pathogenic agent caused epidemics with only few survivors, as was the case of malaria in West Africa. The local inhabitants are those who survived malaria epidemics due to the selection of immune system genes that were able to use an antigen from the malaria parasite to induce immune protection that enabled them to survive. Europeans in the same geographic region with an immune system genes that cannot induce protective immunity against malaria succumbed to the disease. Although influenza viruses are present in humans and in birds, horses and pigs, all humans can be immunised against the local influenza strains, but influenza epidemics and pandemics were recorded during the 20th century, starting with the Spanish flu epidemic that after World War I killed more than twenty million people in a disease that left the victim only 3-4 days before ending in death. The appearance of mutations in two of the influenza virus genes led to new world wide pandemics. Thus, if we use virus epidemics to predict the consequence of the spread of a biological weapons (BWs) containing pathogenic viruses or bacteria into human populations it is correct to assume that a biological war will develop into a killing disease in the population that will be attacked by BWs and will quickly spread to all continents as an epidemic due to the movement of the infected people. The infection of humans with BWs may lead to the distribution of mutants of the agents used, to which vaccines are not available, that may spread around the world populations without restraint. The highly pathogenic smallpox virus is still kept in several national laboratories while the younger human population of the world is not immunised anymore and are therefore vulnerable to this deadly virus. The consequences of the use of this virus by terrorists is one of the threats that we are facing.

In a discussion of the consequences of the spread of BW agents on urban populations, we must consider the smallpox epidemics that spread in Europe during the 18th and 19th centuries with devastating effects on the urban populations, leaving a path of death and mutilation wherever the disease struck. It should be noted that the smallpox virus is heat resistant and can remain in the immediate environment of humans for years, also in the tropics. Infection by the virus starts by the entry of the virus into the respiratory tracts of people, quickly causing an internal disease, with the virus also reaching the skin to replicate in vesicles that erupt and spread the virus to the environment. In 1976 Edward Jenner published his first human experiment that showed that cowpox from vesicles on the skin of milk maids protected them against smallpox also protected a boy from smallpox. It took almost 180 years to successfully eradicate the smallpox disease in humans by immunising with vaccinia virus all the populations in the endemic regions in India and the Horn of Africa. The immunisation of the world populations with the vaccine strains of the vaccinia virus was, consequently, stopped, as recommended by the World Health Organisation. The situation in the world today is that our younger population do not have immune protection against the reemergence of the pathogenic smallpox virus. Since wild type strains of the pathogenic smallpox were not destroyed as recommended by expert virologists who suggested to destroy all remaining stocks of the pathogenic virus since the nucleotide sequences of the smallpox genomes are available (2), there is no need to keep the highly pathogenic wild type virus for experimentation. The logic of the decision to stop vaccination of the world population against smallpox is questioned since in Zaire, the poxvirus of monkeys, designated monkeypox virus, which was limited to infections of monkeys, had recently started to infect humans and to cause a disease with clinical symptoms identical to those of smallpox in human patients. Luckily, the monkeypox virus is not a yet deadly virus. However, several new mutations in critical genes that contribute to pathogenicity may occur during subsequent passages of the virus through infected people. Such mutations may generate a highly pathogenic monkeypox virus that may develop in to an epidemic. The experience acquired from the AIDS epidemic clearly shows that in HIV-1 mutations are introduced into the viral nucleic acids during the infection of a human host. Therefore, the public should be aware that in BWs, the biological processes that cause viruses to mutate can cause the emergence of new virus diseases due to man-made mutations in viruses and bacteria. Therefore, the protection of the population against pathogenic viruses and bacteria should be a continuous effort.

An example of the environmental consequences of the use of BWs may be learned from Australian attempts in the 1950s to use a South American Rabbit Myxomavirus, and in 1996 to spread the Rabbit Hemorrhagic Disease (RHD) virus to restrict the number of feral rabbits (1). The viruses spread quickly in the rabbit populations and markedly reduced the number of wild rabbits. However, the mass killing of the feral rabbits led to the adaptation of the surviving rabbits to the myxomavirus and the frequent reemergences of the pathogenic myxomavirus. However, almost 50 years later a new pathogenic virus was needed to control the rabbits. Presumably, the myxomavirus was not effective. If we project this knowledge to the consequences of a biological war it is obvious that BWs are inhumane and should be eliminated.

In addition to the consequences of BWs viral and bacterial agents for mass destruction of humans and animals leading to disease and death, the impact of BWs on the environment should be a major concern. The agents of BWs spread to contaminate the water resources and the habitat over large distances and will persist in the environment thus having long term affect that will cause difficulties for the protection of populations even in places distant from the affected region. We have more than enough reason to eliminate the agents of BWs and to prevent the stockpiling of Biological Weapons of mass destruction. The world is fortunate that the United Nations inspectors (UNSCOM) are still searching for the stockpiles of BWs in an UN member state.

Protection of Humans and Animals with New Vaccines

To protect humans and animals from microbial diseases (viruses or bacteria) a revolutionary approach to develop effective vaccines against natural epidemic causing agents and certainly against BW agents is needed. The production of current vaccines requires factories to produce large quantities of virus. It is difficult to differentiate between vaccine production and the preparation BW agents. Therefore new vaccines are needed to achieve the goal of "Vaccine for Peace" suggested by E. Geissler (1) and collaborators to protect populations against emerging diseases and to prevent the production of BWs. The current developments in the field of synthetic peptide vaccines and synthetic DNA vaccines (Table 1) provide chemical, molecular and biotechnological procedures for the production of the viral and

TABLE 1: NEW PERSPECTIVES IN THE DEVELOPMENT OF SYNTHETIC VACCINES AGAINST PATHOGENIC VIRUSES, BACTERIA, PARASITES, TOXINS AND VENOMS TO PROTECT LARGE HUMAN POPULATIONS AGAINST EMERGING MICROBIAL DISEASES AND BIOLOGICAL WEAPONS.

Current and Future Vaccine	Year	Technology	Mode of Delivery	Advantages/disadvantages
1	2	3	4	5
A. Current vaccines				
1. Live virus vaccines				
Vaccinia virus	1797	Embryonated eggs Cell cultures	A scratch on skin	Protects well against smallpox Retains pathogenicity
Yellow fever	1934	Embryonated eggs	Injection	Protects well
Poliovirus(Sabin's)	1960	Tissue culture	Oral	
2 Inactivated virus vaccines				
Rabies	1900			
Polio (Salk)	1956	Tissue and cell culture for virulent virus production	Injection	Special laboratories with safety devices. Rigorous Safety control
Pathogenic viruses		Tissue and cell culture for virulent virus production	Injection	Special laboratories with Safety devices. Rigorous Safety control
Bacteria		Inactivation	Injection	Biological factories for the production of the pathogen
3. Proteins subunits	current research	Production of pathogens to produce the protein		Biological factories for large scale production of proteins

cont...

1	2	3	4	5
		antigens or biotechnologically cloning expression of genes and purification of antigens		
B. Future vaccines				
1. Synthetic peptides research	current	Chemical synthesis of short antigenic peptides of all viruses, bacteria, protozoa, toxins		Chemical factories for large scale production of synthetic peptides
2. Synthetic DNA	current research	Biotechnology facilities for gene cloning	Injection into skin, muscle	Biotechnology factories
3. Peplotion (Synthetic peptides and synthetic DNA in lotion)	current research	Chemical synthesis of peptides and synthetic DNA coding for peptides	A lotion to spread on skin by vaccines	Chemical synthesis factories

bacterial vaccines that will eliminate the need to grow large quantities of viruses that will eliminate the need to grow large quantities of viruses and bacteria. The development of new synthetic vaccines and methods for vaccination of large populations (like the Peplotion Vaccine (2-4) in which the synthetic antigens will be applied to the skin by the vaccines) may allow rapid immunisation procedures in the wake of a natural epidemic or the use of BW by terrorists. The induction of the cellular arm of the immune response by synthetic peptides can be achieved within 2-5 days.

The Advantage of the Development and Use of Peptides and DNA Synthetic Vaccines

The transition from production of natural antigens of viruses and bacteria for the production of vaccines to synthetic peptides and DNA vaccines will solve a major problem of monitoring large scale productions of infectious agents, since BWs activity can not be concealed as vaccine production. It will be necessary to develop an international consortium to conduct the research development and the chemical and viral production of sythetic peptide vaccines as well as biotechnological laboratories for the production of synthetic DNA vaccines. Computer analyses of bacterial proteins and the knowledge of the molecular basis, of the immune processes will support the production of the synthetic vaccines to protect against diseases. The development of an anti HIV-1 vaccine that will induce the cellular cytotoxic arm of the immune response could be the first synthetic Vaccine for Peace.

References

1. E. Geissler, 1994. Arms control, health care and technology transfer under the Vaccines for Peace Programme in control of dual threat agents: The Vaccine for Peace Programme p. 10-37 editors: E. Geissler and J.P. Woodall, Oxford University Press.

2. S.N. Schelkunov Functional organisation of Variola Major and Vaccinia Virus Genomes, Virus Genes 10, 53-71, 1995.

3. Y. Becker, 1994. Dengue fever virus and Japanese encephalitis virus synthetic peptides, with motifs to fit HLA class I haplotypes prevalent in human populations in endemic regions, can be used for application to skin Langerhans cells to prime antiviral CD8+ cytotoxic T cell (CTLs)—a novel approach to the protection of humans. Virus Genes 9, 33-45.

4. Y. Becker, 1994. An analysis of the role of skin Langerhans cells (LC) in the cytoplasmic processing of HIV-1 peptides after "Peplotion" transepidermal transfer and HLA class I presentation to CD-8+ CTLs—an approach to immunisation of humans. Virus Genes 9, 133-147.
5. Y. Becker, 1994. A novel approach to vaccination of humans against AIDS with an "HIV-1 Peplotion vaccine for transepithelial transport of HIV-1 proteins/ peptides to epidermal Langerhans (dendritic cells) to prime antiviral CD8+ cytotoxic T cells for long term protection. UNESCO-Roste Technical Report 17, of UNESCO Net work "Man Against Virus", 125-138.

8

Understanding Peace in the Context of Science: a Historical and Ethical Look at Biological Warfare

—J. Levy, *United States of America*

The history of biological weapons and defense development is filled with important ethical decisions that shed light to a significant modern concern—the role of science in the development of peace. The unifying moral and ethical principles inherent in religion may provide the necessary link to foster proper consciousness of the threat of biological warfare.

Science for Peace. The value of understanding science in relation to concepts of peace is often under-emphasised. The world has begun shifting into a global society. A society that builds development on advances in science and technology. Ironically, science and peace making have rarely been placed into the same arena. It is unusual to see science structured in terms of peace. The institutions and research organisations have for years focused on developing technology and science for the advancement of society.' Often, the efforts have helped peace development . But rarely have efforts directly been made at using science for the unique purpose of developing peace. What is needed is a comprehensive analysis of Science for Peace. The history of biological warfare development is filled with some of the most important arguments and considerations common to both science and peace. It is here that science and peace meet. It is here that Science for Peace can begin to the understood. It is here that an exercise of Science for Peace can be performed. What follows is an examination of biological weapons development in the context of ethical issues that are over and above what is being undertaken technically by scientists and politically by the international community.

However, an ethical understanding is not enough. A modality of action needs to follow. Past efforts have concentrated on international documents and agreements and the analysis of scientific social responsibility to foster awareness and control of biological weapons development. However, the fact that biological weapons continue to be developed illustrates the need for a more integrated approach. The world public opinion needs guidance toward a more permanent solution. Such guidance may only be possible by fostering major changes in universal conscience. Religious or spiritual, ethical, and political systems offer three of the most influential avenues for fostering universal change. If these three systems can be integrated into a single comprehensive approach, the change can begin.

Religion and Morality

An analysis of the religious dimension of morality will help in understanding the underlining ethical issues inherent in a discussion of biological weapons development. After all, central to every document, agreement, and discussion of biological warfare are the issues of ethics and moral values. In an utopian moral and ethical world, not only would the need for biological weapons be avoided, but the motives for development would also disappear. In developing biological weapons, the scientist is placed in a situation where certain morals must be compromised. Conscious development of weapons of mass destruction is a test of true ethics and morals. Understanding moral and ethical standards in relationship to religion may help to develop the integrated approach needed for a universal change.

The question of whether religion has a direct or indirect effect on morality has been debated for years. Nonetheless, the overwhelming opinion of philosophers and religious thinkers is that the two are linked and constantly influence each other. Virtually every society has established some form of myth or religion to standardise and explain morality. The first explanations of morality began with identification with the divine. Supernatural powers were responsible for dictating codes and laws that standardised morality for that particular society. Examples such as the Code of Hammurabi and the Ten Commandments illustrate such a dependence on the divine. However, the development of society and religions resulted in an integration of philosophical morality, independent from divine doctrine. What remained constant was the interrelationship between religion and morality. Each religious doctrine reflects ethical, moral, and social values: the humaneness of Confucianism; personal holiness of Judaism; agape, or love of Christianity; compassion of Buddhism; and brotherhood of Islam.[130] As the Encyclopaedia of philosophy explains, "the human needs that molality serves, nonaggression and cooperation, are everywhere the same; and it is not surprising that intelligent beings, reflecting on their experience, have evolved broadly similar codes for meeting them" (Nowell-Smith)[129] If all religions contain stresses on nonaggression and cooperation, it is necessary then, to find ways to take advantage of these principles.

It is at this junction that biological weapons and universal morality converge. For a biological weapon to remain a threat, aggressive countries must preserve the biological threat. Thus, cooperation is avoided, and secrecy is required. True biological warfare is thus secretive and aggressive by nature, the antithesis of the common morals central to all religions: cooperation and nonaggression.

Keeping in mind the universal principles of morality common to religions, it is now necessary to understand the historical development of biological weapons.

Historical Analysis

Biological weapons have been present for centuries. Biological historians cite cases of biological warfare as early as 300 B.C. (Poupard and Miller)[1] when the Greeks used diseased corpses and beasts auto polute water supplies of enemies. While the methods of disease transmission were far from being understood, early societies soon learned the power of disease. The history of infectious agents is filled with dramatic effects on human society and institutions. Civilisations have risen and fallen; wars have been won and lost; religions have gained and lost strength; all at the microbes hand. However, it is modern society that has been challenged most by biological weapons. Although military strategists in the late 1960s dismissed biological warfare as having little strategic value in battlefield situations and even less in the age of nuclear weapons (poupard and Miller),[2] biological weapons have now been recognised as a tremendous threat. President Richard Nixon, in 1969, labeled biological weapons "repugnant to the conscience of mankind," with the potential of causing "massive, unpredictable, and potential uncontrollable consequences" (Code).[3] It was with that statement that President Nixon renounced biological warfare entirely and pledged to destroy the United States' biological arsenal. But the true potential of biological weapons has only recently been fully accepted. The reality of the biological arsenal of Sadamm Hussain during the 1991 Gulf War, the growing biological programmes of many developing countries, and the attractiveness of biological warfare as an alternative to resource dependent nuclear warfare, have resulted in a consciousness of the intense threat of biological warfare. The potential danger of this type of warfare is made worse by a few convenient features. Biological warfare is invisible, silent, widely available, and easily obtained.

The Geneva Protocol

For many, however, the threat of biological warfare is nothing new. Recognising the dangerous potential of chemical weapons after the onslaughts in World War I, several attempts at controlling its development were made. Following a series of treaties mentioning chemical warfare dating back to the mid 19th century; (See Appendix I), the first major attempt was made in 1925. In response to the general awareness of the potential of biological and chemical weapons, the Geneva Convention produced the 1925 Protocol for the Prohibition of the Use in War of Asphyxiating, Poisonous or Other Gases, and of Bacteriological Methods of Warfare (Poupard and Miller),[4] known as the Geneva Protocol (see Appendix II). Though originally the Protocol was drafted only against chemical warfare, lobbying by the Polish delegation resulted in the inclusion of biological warfare restrictions (Mobley).[5] While the true motivation for the Geneva Protocol was a response to the use of chemical warfare in World War I, the agreement banned the use of both chemical and biological weapons in war. Yet, there was no prohibition of development, manufacturing, or stockpiling of agents as retaliatory weapons. The Geneva Protocol outlawed "first use", but not retaliation. The United States did not immediately ratify the Geneva Protocol due to pressure from the US chemical industry and the US Army Chemical Warfare Service to allow the preservation of their developments. In fact, ratification did not take place until 1975 (Goldsmith).[6]

The tense political atmosphere after post-World War I led to several weakness in the agreement. Charles Flowerree of the committee for National Security cites three major weakness of the Geneva Protocol[7]:

1. It says nothing about research, development, or stockpiling of chemical or bacteriological weapons, creating a situation in which many countries "adhered" while maintaining the right to respond in kind if their forces are attacked with these weapons;
2. The term "bacteriological" may not include all possible types of biological warfare agents (viruses, for example);
3. There are no provisions for investigating or ensuring compliance.

What soon resulted from the lack of provisions was the early development of biological warfare research facilities. In 1929, the USSR opened a facility north of the Caspian Sea, and in 1934, the UK and Japan officially followed. The Japanese biological research programme was the most developed and by far the most controversial.

Japanese Biological Warfare Programme

Even before the official 1934 decision to develop biological weapons, the Japanese had begun their attempt to become the world's preeminent biological warfare power (Harris).[8] The Japanese programme was unique. The Japanese military recruited the top scientists, physicians, dentists, veterinarians, and technicians to participate. Together, these researchers began an ambitious attempt to become the world power of East Asia. What resulted was a true test of ethical consciousness, as men, women, and children were used as experimental subjects. They were subjected to horrifying experiments, and those who survived were sacrificed when they no longer were useful. The case of the Japanese Biological Warfare Programme is a useful lesson of ethical considerations for the scientist as well as the statesmen.

The beginnings of the Japanese Biological Warfare Programme began with Koizumi Chikahiko, the Japanese Army's Surgeon General, who, as early as the mid-1920s, supported studies on biological and chemical warfare. It was Koizumi's belief that biological and chemical warfare would allow Japan to fulfill its ambitious expansionist goals. Koizumi soon discovered Major Ishii Shiro, an army doctor who would soon lead the Japanese Biological Warfare Programme. Ishii realised the potential power of biological warfare immediately. After reading the Geneva Protocol, he saw a great opportunity in developing a weapon that was being prescribed. In 1930, Ishii was appointed Professor of Immunology at the Tokyo Army Medical School, where he first began his research on biological warfare. His research in Tokyo showed potential, but it was Ishii's belief that animal experimentation would not be enough. The only way to accurately determine the true utility of biological weapons was to develop a comprehensive programme of human experimentation. Ishii Shiro never was educated in ethics. He had grown up in a country filled with ultra- nationalists proclaiming nationalism and ethnocentricity (Harris).[9] For Ishii the ethical question of human experimentation was secondary to military success.

Towards the end of 1932, Ishii set up a laboratory in the industrial city of Harbin, Manchuria, the newly acquired Japanese colony. Known as the Togo Unit, the laboratory location proved to be unsuited for the large scale human experimentation that Ishii desired. Secrecy was too difficult. The decision was thus made a year later to move the laboratory to the village of Beiyinhe, 100 km south of Harbin. The largest of the 100 buildings, in Beiyinhe, the Zong Ma Castle, was a combination of small holding prison for future subjects, laboratories, crematorium, and ammunition dump. In 1935, the prisoners rioted. After an ammunition dump exploded mysteriously a year later, the camp was ordered destroyed (Harris).[10]

By the fall of 1936, Ishii's programme had moved to Ping Fan, a suburb of Harbin. The local population was informed that a lumber mill was being built. Ironically, the human subjects at Ping Fan were regarded as "marutas," or "logs." Ishii was given command of a new unit, known from 1937-1941 as the Ishii Unit, and later as Unit 731 (Harris)[11].

Ping Fan was the largest of the research centers, but it was not the only one. A biological research center was established in

Changchun. Known as Unit 100, the Changchun research center was led by Major Wakamatsu Yujiro, a military veterinarian, and operated independently from Ping Fan. Another center was established in Manchuria's most important industrial city, Mukden. The research facility in Mukden was led by Kitano Masaji, a Professor in the Manchurian Army Medical College, who would later replace Ishii as head of Unit 731 in 1942 (Harris).[12]

The human experimentation was essential to each of the Japanese biological weapon research facilities. Humans were tested with virtually every known pathogen, chemical pesticide, and poison. Diseases included: plague, typhus, smallpox, yellow fever, tularemia, hepatitis, gas gangrene, tetanus, cholera, dysentery, glanders, anthrax, scarlet fever, undulant fever, tick encephalitis, "songo" or epidemic hemorrhagic fever, whooping cough, diphtheria, pheumonia, typhoid fever, epidemic cerebrospinal meningitis, venereal diseases, tuberculosis, salmonella, and other diseases that affected local populations in Manchuria and China. Over 10,000 human subjects were used (Harris).[13]

In July 1937, large scale tests began. Soon, reports from China described extensive Japanese use of chemical and biological weapons. From 1939-1942, Japan conducted twelve major field tests. In October of 1940, the town of Chuhsien was sprayed with rice and wheat mixed with plague-infested fleas. The result was 21 deaths. In January, a similar attack was made on Ningpo. The children of Nanking were given chocolates infected with anthrax in July of 1942 (Mobley).[14] By 1943 China convinced President Franklin D. Rossevelt to condemn the Japanese actions. Roosevelt threatened to retaliate against Japan if the "inhumane form of warfare" did not end (Harris).[15]

The response to the Japanese actions made by the United States was surprising. At the end of World War II, trials for war criminals were held. Major Ishii Shiro did not attend. Instead, a secret agreement was made with the entire Japanese Unit 731 (Freeman).[16] Interviews were made with the Japanese scientists to collect biological warfare information. The American representatives sent to Japan ensured the scientists that any information released would not be used as evidence in war crime trials. The interviews continued for three years. The United States was well aware of the human experimentation that was performed. However, neither ethical nor moral issues were ever discussed. Over 5000 Japanese were tried as war criminals. Not one high level Japanese biological warfare expert was charged with a crime. Ironically, the data gathered held little value. The United States Biological Research Programme soon surpassed the Japanese. The information from human experimentation lost its valued utility (Harris).[17]

World War II

Although the Geneva Protocol had banned he bacteriological weapons, the 1940s witnessed the most intense development and stockpiling of biological weapons. Joining the Russia, the U.K., and Japan were the United States and Canada, who, by 1941, had opened their biological weapons development centers. It was a period of instability in the world. Throughout the 1940s, plans for biological warfare and defense were being prepared by most of the world powers.

At the start of World War II, biological warfare programmes of several countries were ready for action. The Allies were highly suspicious of the state of their enemies biological warfare programmes. By 1942, the Allies had reached the same conclusion that the Japanese had made from their experimentations: anthrax was the choice agent for biological warfare. Anthrax spores were rugged, and could endure the high temperatures of explosions, making them an easy attachment for conventional warheads. In 1942 and 1943, under the joint effort of British, Canadian, and United States biological research programmes, the Microbiological Research Establishment at Porton experimentally infected Gruinard Island, off the north-west coast of Scotaland, with anthrax (*bacillus anthracis*).[18] that the island was off-limits to civilians for nearly a half-century. In the early 1980s, it was reported that anthrax could be detected only in the limited area of experimentation. Only recently has Gruinard Island been Ôre-opened.' On May 1, 1990, the island was sold back to the original owner for £ 500 (Aldhous).[19]

Additional preparations were made by the Allied forces. 5,000,000 anthrax-infected cattle cakes were prepared for usage by Britain. The plan was to infect cattle, resulting in a war-time food shortage. By 1943, the Allies were preparing 500 pound anthrax bombs for war.

The Gruinard Island experiments, as well as Allied preparations in general, were a result of fears that Germany had also developed a comprehensive biological warfare programme (Aldhous),[20] and of intelligence reports of German and Japanese intentions (Doyle and Lee).[21] Intelligence reports in 1944 that Germany was prepared to launch powerful V-1 "buzz bombs" with biological agents prompted even more fear. The threat was made by Hitler in his Danzig speech on September 19, 1939. Hitler spoke of a secret weapon that could not be used against him (Mobley).[22] The German high command was getting desperate. They had reached a strategic crisis, and saw the use of biological warfare as a mean of gaining a permanent advantage (Bernstein).[23] As a precautionary measure during the D-Day invasions, 100,000 Allied troops were given "self-Inoculating syringes" against the possibility of a biological attack. Additionally, Allied soldiers invading Italy wore shirts drenched with DDT, aimed at preventing typhus (Press).[24]

There exist one final documented use of biological weapons in World War II. In 1942, Reinhard Heydrich was assassinated by the Czechoslovakian underground. The method of assassination has been documented as being biological warfare. A grenade, manufactured in Porton, England, was infested with "X," the British code name for botulism toxin, and used in the assassination. While Heydrich suffered only minor wounds in the chest and spleen, his death several days later matched the symptoms of botulin poisoning (Mobley).[25]

Regardless of all the preparations that were made, biological warfare never entered World War II in the large scale that was anticipated. The German biological arsenal was harshly exaggerated. For some unknown reason, Hitler had barred all offensive biological warfare research (Bernstein).[26] Allied forces never used the anthrax bombs. The two million anthrax infested cattle cakes never reached the battlefield. In the end, the focus turned to biological warfare's historical alternative- nuclear warfare.

The United States Biological Weapons Programme

The United States Biological Weapons Programme began in 1941 under the auspices of the US Army Chemical Warfare Service. By the middle of 1942, the War Research Service, headed by George Merck (president of Merck & Co., Inc.) took over the active research. From 1942 to 1943, the War Research Service allowed the Chemical Warfare Service to expand their own biological research. The Chemical Warfare Service was given millions of dollars to build research facilities, including the 500 acre Camp Detrick (Frederic, Maryland), the 2000 acre Horn Island (Pascagoula, Mississippi), the 250 square mile site known as Dungway Proving Ground (Utah), and the 6,100 acre plant in Terre Haute, Indiana. By 1943, the United States was playing an active role in preparing for World War II, producing anthrax and botulism bombs for Allied forces (Bernstein).[27]

After World War II, the United States led the way in biological warfare research and development. The Cold War atmosphere left Russia the United States' major competitor. In 1947, President Truman officially withdrew the Geneva Protocol from possible ratification, claiming that current developments had already invalidated the principles of the treaty (Poupard and Miller).[28] The United States soon began what would become a series of experimental tests in the non-laboratory setting. Towards the end of the 1940s, United States worked with Britain and Canada, releasing pathogenic microorganisms from ships in the Caribbean (Poupard and Miller).[29] Open air experiments were conducted in Norfolk, Virginia by the Navy in September 1950. And, in a series of experiments performed by the Army one year later, *Serratia marcescens and Bacillus globigii* were released by ships in the San Francisco Bay area. The intent was to discover truly how vulnerable American cities would be to biological attacks. At the time Serratia marcesens was viewed as having no pathogenic potential, however, later experiments revealed the contrary (stainier).[30] A mini-epidemic broke out at Stanford Hospital, resulting in the death of Edward Nevin.

In 1964 and 1965, the Army conducted experiments to determine the vulnerability to smallbox virus. Experimenters dressed in plain clothes wandered the North Terminal bus station and the National Airport in Washington, D.C. Travellers were sprayed with *Bacillus subtilis* through aerosol sprayers concealed in suitcases. The bacteria was thought to be harmless, however it was later determined that the bacteria can interfere with the immune systems of elderly and those suffering from debilitating diseases.[31]

Another documented testing occurred in the New York City Subway in 1966. The Chemical Crops Special Operations Division filled light bulbs with *Bacillus globigii*. The light bulbs were dropped into ventilator shafts of the subway from a moving train. The bacteria was soon carried to the ends of the subway tunnels. The tests were performed on three major lines in the mid-Manhattan area (Huxsoll, Parrott, and Patrick).[32] *Bacillus globigii* was not found to be pathogenic.

The conclusions made by these experiments indicated the vulnerability of American cities to biological attack. However, the fact that the American public was oblivious to the testings has led to considerable mistrust of military biological warfare research.

Accusations: the case of the Korean War

The United States has not been the only country to perform tests. The history books are filled with accusations by countries on the possible use of biological warfare in both testing and war situations. The theme of accusations dates back to the fourteenth century outbreak of Black Death. Many European Christians passed the blame to Jews, who were accused of poisoning wells with plague (Moon).[33] During World War I, Germany was accused of using cholera in Italy and plague in St. Petersburg. While these allegations were never proven, Germany never denied them. On the other hand, accusations made by England, that Germany had used plague bombs, and others made by France, that disease-infected toys and candy were used in Romania, were completely denied by Germany (Poupard and Miller).[34] In 1933, the French journalist Henry Wickham Steed claimed that the Germans had released clouds of *Serratia marascens* into ventilation shafts of the Paris Metro and near French forts. These allegations were denied by the German government (Robertson).[35]

Allegations surfaced again at the onset of World War II. China made several complaints of Japanese biological warfare attacks. Originally, these accusations were dismissed as propaganda attempts to attract world sympathy. Soon, however, the world realised the extent of the Japanese attacks. Unfortunately, it took until 1943 for the United States take action against the Japanese regime (Harris).[36]

Accusations peaked in the 1950s during the height of the Korean War with what came to be known as "black propaganda" (Rolicka).[37] In 1951 and 1952, the Communist leadership of North Korea and China claimed that the United Nations Command was using biological warfare. What followed were a series of accusations and denials. Beginning on February 22, 1952, propaganda campaigns were attempted by Communist Korea. According to the propaganda, the United States had launched a biological weapons campaign against North Korea, attempting to start widespread epidemics. Various vectors such as flies, fleas, lice, locusts, spiders, mosquitoes, and crickets were used to disseminate diseases such as plague, anthrax, typhus, smallpox, cholera, dysentery and encephalitis (Rolicka).[38] However, the speedy and effective North Korean response helped to spoil the United Nation's plan. China immediately responded with a much-needed public health campaign, using the American threat of biological warfare as a catalyst to encourage hygiene. The charges of germ warfare proved to be a way of getting things done domestically. To support the propaganda, witnesses, laboratory tests, and autopsies were produced to prove the realness of the attacks.

These accusations were soon denied by the United Nations Command and the US Department of State. The first denial was made by the US Secretary of State, Dean Archeson on March 4, 1952. He challenged North Korea to allow the International Committee of the Red Cross to perform an impartial investigation of the situation (Moon).[39] General Matthew B. Ridgway called the allegations a cover-up of the domestic problems of North Korea. He felt that North Korea made the accusations to hide the government failure to deal with the disastrous domestic health situation (Moon).[40]

The situation of accusations never developed beyond the propaganda level. Several attempts were made for impartial investigations, yet none were performed. At the 18th International Red Cross meeting in Toronto (1952), a special commission was suggested to investigate the charges. However, North Korea failed to respond to the request, based on the feeling that Western influences were attempting to spy. An additional attempt was made by the World Health Organisation, which was once again, refused. Instead, the Chinese chose the International Association of Democratic lawyers, who was already documenting the atrocities of the American forces, to investigate the biological warfare evidence (Mobley).[41]

Archive research has revealed a significant amount of information clearly pointing to the false propaganda attempts of North Korea. In 1956 a Hungarian journalist, was in touch with the Hungarian Hospital in Korea, published a paper expressing his doubts regarding his original observations of germ warfare attack. Similar evidence was found about Korean hospitals in general. Polish doctors who had been working at the Polish hospital in Korea, which had been established in 1952, concluded that the Korean hospitals were understaffed, the medical staff was unqualified, and many medics functioned as doctors. Such a situation made it rather easy to falsify information (Rolicka).[42]

The theory suggested by General Ridgway has much support. This suggestion rests on the assumption that epidemics did develop. A 1984 article by Albert Cowdrey supports this argument. Cowdrey feels that the epidemics did occur throughout Korea and China due to "a natural disease environment of singular variety...ills usually associated with cold, moderate, and warm climates flourished—epidemic typhus, malaria, Japanese B encephalitis, cholera, hemorrhagic fever, typhoid, and smallpox, among others."[43] In the beginning of 1951, smallpox and typhus were reported throughout Korea. Mass inoculations followed. However, according to UN Command technical intelligence officers, the diseases seemed to be excessively present in the north. Cowdrey felt that for North Korea, the germ warfare accusations took a "Janus-like form." The propaganda would encourage hatred of the United States, and serve as a motivation to prevent any recurrence of the epidemics that broke out in 1951, through a complex, domestic clean-up (Cowdrey).[44]

Regardless of whether there were epidemics, there is no evidence that the United States used biological warfare in the Korean War. In fact, recently recovered documents reveal that the US was not even prepared for biological warfare. In July 1952, the Chairman of the R & D Board declared that the United States probably could not "create a real capability for offensive antipersonnel BW by 1954" (Moon).[45] By the end of the Korean War, the US did not possess any biological agent "of high infectivity and satisfactory half life." Additionally, no satisfactory delivery device existed. (Moon).[46]

While it can be argued that allegations during the Korean War were the clearest abuse of biological warfare accusations, the allegations did not end there. In the 1970s, accusations were made concerning "yellow rain. In 1979, accusations were voiced that the Soviet had released anthrax in Sverdlovsk. In 1980, the Cuban government argued that an outbreak of swine fever was the result of CIA action. In October of 1985, the Soviet official Valentin Zapevalov claimed to have found the true origin of AIDS. He said that scientists working for the US biological research programme in fort Detrick had created the virus and began the spread by injecting humans and animals (Seale).[47] A Moscow Radio commentary on December 26, 1985 noted, "Dr. John Seale of Britain has concluded that the AIDS virus has been artificially created and its appearance is possibly the result of a human error. This conclusion supports the view that the

AIDS epidemic has been caused by experiments with humans carried out in the USA as part of the development of new biological weapons" (Medvedev).[48]

The past examples have hinted to the potential danger of biological warfare accusations. Biological agents are often undetectable, thus, accusations are often dismissed due to the nature of the use of biological warfare. Verification of biological warfare use is also difficult. Rarely does one come across a 'smoking gun' leading to an obvious conclusion. Attempts to prove accusations, are, therefore, a vigorous challenge. Often enough, the accusations are ravish attempts to condemn an innocent country. However, in certain situations, exemplified by the Chinese case in World War II, the accusations were true. What results from the large number of accusations is what I will call the "Cry-Wolf Theory." The immediate world reaction to biological warfare accusations is doubt. Countries making accusations are viewed as either attempting to gain sympathy from other nations, or using the accusation for some ulterior motive. When a "true" biological warfare allegation is made, it is often dismissed and overlooked; an unfortunate result of the abuse of biological warfare allegations.

Developments Leading to the Biological Weapons Convention

In 1957 the United Kingdom made a revolutionary decision in regards to their biological weapons programme. The decision was made to destroy all offensive biological weapons, allowing only for defensive research to continue. With the United States and several other world powers still actively developing their biological weapons programmes, the international arena was not quite ready to follow in the British footsteps.

However, on November 26, 1969 the history of biological weapons development reached a milestone:

"Mankind already carries in its own hands too many of the seeds of its own destruction. By the examples we set today, we hope to contribute to an atmosphere of peace and understanding between nations and among men" (Press).[49]

It is with this speech that President Nixon officially ordered the destruction of the American biological weapons arsenal. Under UN Supervision, all American offensive biological weapons were destroyed.

While this decision seemed to be a revolutionary accomplishment by President Nixon, many argue that Nixon's declaration was based on a purely political rational, without consciously considering the uncontrolability and unpredictability of the weapons (Freeman).[50] Regardless, with the United States leading the way, several other nations reconsidered their biological weapons policies. The political atmosphere was finally ready for a strong international effort to prevent biological weapon development.

In 1972, following president Nixon's renouncement of biological weapons, the Convention of the Prohibition of the Development, Production and Stockpiling of Bacteriological (Biological) and Toxin Weapons and on Their Destruction was held, otherwise known as the Biological Weapons Convention (BWC) (see Appendix III). Representatives of 80 nations at the BWC agreed to destroy or convert to peaceful use all biological warfare organisms, toxins, equipment, and delivery systems (Poupard and Miller).[51] The agreement encourages the dynamic exchange of scientific and technological information, materials and equipment to aid research in biological agents and toxins for peaceful use and treatment of disease (Huxsoll et al).[52]

Periodic Review Conferences were scheduled at the time to update the agreement, and allow for diplomatic developments to be compensated. Since 1972, four Review Conferences have met. As an October 10 issue of *Science* indicates, the treaty is regarded as "the world's first disarmament treaty, as it is the first treaty to outlaw the production and use of an entire class of weapons of mass destruction."[53]

The major criticism of the BWC is the lack of verification mechanisms to ensure compliance. Charles Flowerree of the Committee for National Security believes that the reason for this was that at the time of the BWC there was a "general belief that biological warfare was of little or no military significance" (Flowerree),[54] The Review Conferences have attempted to solve this problem, however a feasible solution has yet to be found.

The largest reason for the current insecurity in the BWC arises from the resurfacing evidence of continued offensive biological research, a product of the verification problem. Each accusation and every discovery of biological weapons development is a direct challenge to the BWC. Between Nixon's declaration in 1969 and 1986, 40 allegations of biological warfare have emerged (Orient),[55] the most significant being the "Yellow Rain" controversy and the 1979 Sverdlovsk accident. Since then, several others have emerged, including the discovery of the large biological arsenals of Saddam Hussein during the 1991 Gulf War.

Yellow Rain Controversy

Beginning in the late 1970s, reports began to surface of biological and chemical warfare in the areas of Laos and Kampuchea. Allegations were made by the United States that the Soviets were using what was described as "yellow rain" as a bio chemical war agent. Several witnesses testified to the fact that they had seen or been the victim of such warfare. On September 13, 1981, Secretary of State Alexander Haig made the following declaration to the Berlin Press Conference:

"For some time now, the international community has been alarmed by continuing reports that the Soviet Union and its allies have been using lethal chemical weapons in Laos, Kampuchea and Afghanistan. We now have physical evidence from Southeast Asia which has been analysed and found to contain abnormally high levels of three potent mycotoxins—poisonous substances not indigenous to the region and which are highly toxic to man and animals" (Meselson, et al.)[56]

Matthew Meselson led the investigation of the mysterious yellow rain. After travelling to Southeast Asia, he began in intense investigation of the accusations made regarding "yellow rain." He established a comprehensive analysis consisting of interviews, chemical analyses, and descriptions. Most interesting was the chemical analysis.

Investigations and testing of the mysterious substances were performed by the US Army, the University of Minnesota, and Rutgers University. Chester J. Mirocha, of the University of Minnesota reported five of six samples to the positive for the poisons. The other positive tests came from Robert Rosen and Joseph Rosen of Rutgers University concluded that while "it has been implied that the mycotoxins in the State Departments samples analysed by Mirocha could have been naturally occurring... the information obtained in this study, in which four mycotoxins and a synthetic material were found in the same sample, suggests otherwise,"[57] However, the US Army, after testing more than 80 samples, failed to find a single sign of the toxins.

Meselson concluded that the accusations placed on the Soviet Union were false. In fact contrary to the State Departments claims, the mysterious "yellow rain" was not a form of biochemical warfare, but rather the feces of Southeast Asian honeybees. The reports of yellow clouds of biowarfare, Meselson explains, were in fact showers of honeybee feces that occur often in the Tropics of Southeast Asia, "the yellow rain: it is a phenomenon of nature, not of man" (Meselson, et al.).[58] Despite attempts to re-establish the theory that "yellow rain" was a clear act if warfare (Rosen),[59] Meselson's theory remains the accepted theory.

Sverdlovsk Anthrax

Around the same time that accounts were beginning to surface regarding "yellow rain," two events were taking place. On March 1980, the First Review Conference of the BWC was held. It was during the March conference that the United States revealed information alleging that the April 1979 outbreak of anthrax in Sverdlovsk was a result of the release of anthrax spores from a Soviet military complex. The complex, they claimed, was working on the development of biological weapons. The outbreak resulted in ninety-six cases of human anthrax and sixtyfour deaths (Meselson, et al).[60]

The first report of the anthrax outbreak was reported by the East German magazine , *Posev* (Orient).[61] Soviet news agencies and officials admitted to the outbreak, but attributed the spread to contaminated meat sold on the black market. However, US intelligence made claims to the contrary. The US felt that the outbreak was consistent with an accident at the Military Compound 19 in Sverdlovsk, however, the Soviet story never changed. At the time of the publication of the July 21, 1984 issue of *The Lancet*, both Matthew Meselson and Zhores Medvedev (a noted critic of the Soviet military) had accepted the Soviet version,[62] however many questions remained unanswered.

In an attempt to uncover evidence to the cause of the 1979 anthrax outbreak, Meselson renewed previously unsuccessful attempts to have independent scientists investigate the Sverdlovsk outbreak in 1986. The request was granted in 1988. The reports from the Soviet scientists indicated the number of infected individuals and deaths. In 1990, a series of articles published by Russion scientists regarding the 1979 outbreak surfaced. It was revealed that in 1979, research on an anthrax vaccine was being performed. In May 1992, Yeltsin acknowledged that "the KGB admitted that our military developments were the cause." Meselson still felt that further investigation was needed. In June 1992, Meselson was granted an invitation to visit Sverdlovsk and begin an on-site study. The conclusions stated that " outbreak resulted from the windborne spread of an aerosol of anthrax pathogen... the source was at the military microbiology facility... the epidemic is the largest documented outbreak of human inhalation anthrax" (Meselson, et. al.)[63]

Combined with the reports of "yellow rain," the Sverdlovsk anthrax outbreak of 1979 was a devastating blow to the First Review Conference and the entire BWC efforts. After all, the USSR was one of three Depositary Governments established by the BWC to ensure ratification of the Conference. The two events clearly demonstrated the extent of the verification problem inherent in the BWC. Both events were an indication of active development and stockpiling of biological weapons, in violation of both the Geneva Protocol and the BWC.

The 1991 Persian Gulf War

The entire international community went through a tremendous eye-opening experience during the 1991 Persian Gulf War. As Middle East editor of *Newsweek* Christopher Dickey put it, "for several days at the end of 1990 and the beginning of 1991, the world stood on the brink of the first holocaust ever caused by biological weapons."[64] Twenty-two nations joined the United States against Iraq, a member of both the Geneva Protocol and the Biological Weapons Convention. The war marked the first time that complex offensive and defensive biological weapons systems were prepared. By studying the expectations and preparations made for the Gulf War in comparison to the reality of the Iraqi biological arsenal, it is possible to understand the extent of biological warfare possibilities.

By the beginning of the Gulf War, intelligence sources had known about the Iraqi biological warfare programme for two years. The Iraqi programme had been actively researching botulism, anthrax, typhoid, cholera, tularemia, equine encephalitis, as well as several other biological agents. Biological weapons were being developed in research laboratories in the villages of Samarra and Salman Pak. Although a new Genetic Engineering and Biotechnology Research Center was opened in Baghbad, it was understood that the technology was not yet advanced enough to develop effective genetically engineered biological weapons. In August of 1990, it was reported that "there is good reason to believe that Iraqi President Saddam Hussein has boasted to other Mideast leaders that biological warfare is what will ultimately stop the United States should it decide to attack Iraq (Knudson).[65]

With the knowledge of the extent of the Iraqi biological arsenal, several preparations were made. Specially designed mission Oriented Protective Posture (MOPP) gear was designed for both biological and chemical warfare. However, according to Major Gregory Knudson of the Area Microbiology Service at Letterman Army Medical Center, while this protective gear might be sufficient for chemical warfare, it would not be sufficient for the biological warfare that Saddam was prepared to use.[66] Thus, researchers in Fort Detrick, Maryland, home of the United States Biological Defense Research facilities, began developing a variety of vaccines, drugs, and other means of combating the biological threat. Intelligence and researchers felt that the biological agent most likely to be used would be *Bacillus anthracis* and botulinum toxin. Thus, troops were given vaccinations against anthrax and botulinum *toxin*, and several other vaccines were stockpiled for possible use including tularemia, Query fever, Rift Valley fever, and Argentine hemorrhagic fever (Marwick).[67] While most held the possible biological warfare attack as a distinct possibility, others felt that the likelihood of Iraq using these weapons was doubtful. Advocates of the latter view feel that "heavily defended US positions, with troops in MOPP gear using gas-detection devices and decontamination kits, atrophine injectors, and medical support... [as well as] the threat of massive retaliation" is enough to deter any possible Iraqi attack. However even advocates of this view agree that if this deterrence fails, the large number of civilian and military casualties would be massive (Marwick).[68]

In retrospect, estimates of the Iraq's biological arsenal were significantly miscalculated. A GAO report noted that at the onset of the Gulf War, the United States Army's vaccine stockpile for botulism and anthrax was significantly less than what was needed (Horgan).[69] In September 1995, Iraq admitted the extent of their biological weapons programme. At least twenty-five missile warheads containing 5,000 kilograms of biological agents were prepared. Additionally, an extra 15,000 kilograms of biological agents were armed inside bombs to be dropped from airplanes or to be reserved for other weapons.

If these weapons had been used in addition to the Scud missiles which reached Tel Aviv, "the result could have been as horrifying as Hiroshima or Nagasaki" (Dickey).[70]

The most likely reason that Saddam Hussein chose not to tap into his biological arsenal was the forceful warnings that were made by the United States. In January 1991, Iraq was warned that the "most extreme measures" would be chosen it weapons of mass destruction were used. Iraq most likely interpreted this to mean that the United States would retaliate with nuclear measures (Dickey).[71] Additionally, the preparations made by the United States may have helped in deterring an Iraqi attack. Nevertheless, Saddam Hussein's verbal threats of a biological weapon attack were never substantiated. No practical preparations were made, and Iraqi troops were never immunised against any of the agents (Goldsmith).[72] Regardless of the fortunate outcome, the 1991 Persian Gulf War illustrated the realness of the threat of biological weapons.

Informed Consent and the Nuremberg Code

An interesting development occurred during as a side-effect of the 1991 Gulf War. Just before the war began, the US military faced a significant dilemma. In order to successfully prepare for the possibility of chemical and biological warfare in the war, US troops were to be administered several vaccinations. However, the military foresaw a potential problem—the possibility of some troops declining the vaccinations. According to US statutory law, informed consent is needed for all "investigational use" of drugs, with the exception of when consent is "not feasible" or professionally determined to be "contrary to the best interests of such human beings."[73] Thus, the ultimate decision would rest in the soldier. If the soldier decided against the vaccination, then based on his rights of informed consent, the soldier would have the option of non-administration. However, if every soldier was not given the important vaccinations, the entire Desert Storm Operation might have been disrupted. Thus, the military pursued a waiver of requirements for informed consent for the use of these drugs. The Food and Drug Administration (FDA) granted the military request by issuing rule 23 (d), allowing waivers of informed consent on a drug-by-drug basis. The FDA explained that consent was "not feasible in a specific military operation involving combat or the immediate threat of combat."[74]

A large debate broke out in reference to the ethics of this decision. Opponents to FDA 23(d) used the Nuremberg Code as their major argument. The Nuremberg Code was devised to established ethical and humane guidelines for human experimentation. The first, and most well known of the ten-point code emphasises the voluntary and informed consent of the subject being absolutely essential. Since some of the agents being used in the Gulf War were considered experimental agents, by ignoring the voluntary consent of the troops in the Gulf War, the United States was in clear violation of the Nuremberg Code. Opponents of FDA 22 (d) argue that the Nuremberg Code has no exceptions for wartime situations. Neither should the US government.[75]

Attempts were made by the Public Citizen Health Research Group to prevent the military from using the experimental agents without informed consent, however the FDA 23(d) was already in the workings (Marwick).[76] George J. Annas of Boston University Schools of Medicine and Public health argued that "the most important thing we have learned is that informed consent is feasible in combat situations, and therefore rule 23 (d) was unnecessary... . General Norman Schwarzkopf, whose command did not seek or want the informed-consent waver, ordered that the botulinum-toxoid vaccine be given only with informed consent."[77]

More and more reports have been recently surfacing regarding Persian Gulf Syndrome (PGS). While no direct connections have been made between biological or chemical warfare and PGS, the potential exposure of the US troops is still a distinct possibility. A recent study suggested that the Persian Gulf Syndrome was caused by a beef allergy that developed as a result of an immunisation take prior to the war.[129] As reports of PGS grow stronger, it might be necessary to look at all aspects of the war, including the vaccinations made in preparation.

Ethical Analysis

Understanding the historical facts of the history of biological warfare is not enough. Several important ethical issues need to be discussed, and lessons need to be learned. Ethical considerations are inherent in biological warfare, yet biological warfare presents a situation where ethics can easily be overlooked. Individuals like Major Ishii Shiro were never taught the basic ethical values of life. Sheldon

Harris of California Sate University concluded his analysis of the Japanese biological warfare developments in the 1930s and 1940s by saying:

"It is evident that the moral issues relating to biological warfare have not affected many of the world's scientists as the twentieth century draws to a close. The frightening examples of the Japanese biological warfare experiments, as well as those of the Nazi doctors, have not deterred those who still seek fame and fortune, regardless of ethical considerations"[78]

Developments in biological warfare have reached a point where ethical considerations are a necessity. To what extent is research of biological warfare ethical at all? Is offensive biological warfare research different from defensive research? Where is the threshold between the two? Is the biological researcher completely responsible for the outcomes of his or her research? It is these questions that demand an ethical analysis of biological warfare.

In 1986, Doyle and lee from the University of Louisville published a paper entitled "Microbes, warfare, religion, and human institutions." The underlying premise of the paper was an attempt at educating microbiologists. It was stated that a significant number of microbiologists are unaware of the role biological warfare has played in the history of mankind. The paper discusses the tremendous effect biological warfare has played in shaping the outcomes of war, developments and success of religions, and success and downfalls of civilisations. "There is a need for microbiologists to have a historical perspective of some of the major ways in which a pathogen may influence civilised populations. Conditions may exist in contemporary society for a repeat of some of the kinds of plagues suffered by previous societies" (Doyle and Lee).[79]

Individual Efforts Towards Standards of Ethical Responsibility

Scientific developments are constantly under the ethical microscope. Several attempts have thus been made at establishing a universal standard of ethics in science, and social responsibility. However, few of these theories have directly assessed the role of biological warfare within the scientific paradigm. The social responsibility of the scientist is a well debated topic. Sir Solly Zuckerman notes the difficulty of creating universal standards, "When one talks about the social responsibility of scientists, it is thus carrying naiveté to the extreme to suppose that they speak with one voice and they share a common conscience when it comes to the application of scientific knowledge" (Chalk).[80]

Others, however, have made solid attempts at deriving ethical standards of social responsibility. In 1941, British historian J. G. Crowther suggested eight ethical standards for scientists, urging the consideration of possible social implications of their work, recommending political activity to help establish forces to regulate the conduct of research, and, in cases of war, advising the scientist to "consider which side was the less inimical to science, and then do what was possible to see that it was not defeated" (Chalk).[81]

In 1947, just after the bombings of Hiroshima and Nagasaki, a heated debate developed between Norbert Wiener and Louis Ridenour over Wiener's decision not to "publish any future work... which may do damage in the hands of irresponsible militarists." Wiener, a MIT mathematician, felt strongly that "to disseminate information about a weapon in the present state of our civilisation is to make it practically certain that the weapon will be used." Ridenour challenged Wiener, emphasising the problems with Wiener's interpretation of social responsibilities, and the implementation of such responsibilities. Ridenour felt that the social responsibility of the scientist is no different that of "every other thinking man." And, in war time situations, when the nation is preparing for war, any attempt to impede preparations for war should be avoided (Chalk).[82]

At the heart of the ethical analysis of scientific research as a whole, especially biological warfare research, is the extent to which a scientist can be held responsible for his or her research. Is the scientist completely responsible for all possible outcomes of the research? Newtol Press commented on this problem with an interesting analogy to Fritz Haber, A German scientist who was awarded the 1918 Nobel Prize for Chemistry for his work in ammonia synthesis. Two other French Nobel Prize winners declined their awards on the grounds of Haber being "morally unfit" for the honour, as his work played a major role in the development of poison gas:

"Ethicists sometimes say that spoons have been used to gouge out eyes and argue from this example that makers of tools have no responsibility for the ways their products are used. What holds true for spoons does not apply to poison gas. Haber had no illusions and was never publicly repentant; he believed that he was doing the right thing, that he was serving a noble end with an efficient means" (Press).[83]

In an article on ethical responsibilities in science, Daniel Callahan of the Hastings Center commented, "Life is a gamble...so, for that matter, is science... science can, directly and indirectly, produce useful knowledge and help improve the human lot. It can also, directly and indirectly, bring about great harm." Callahan continues by stating four moral propositions for scientists and researchers:

"1. Individuals and groups are ordinarily responsible only for the consequences of those actions that are voluntary and intentional on their part. However, they may also be held responsible for the unintended consequence of their actions if, through negligence, they failed to take into account such consequences;

2. Individuals and groups cannot be held responsible for these actions the consequences of which are totally unknown. However, if they voluntarily undertake such acts, they may be held responsible for the consequences unless there were serious reasons for the undertaking the action in the first place. One cannot, without serious reason, just "play around" in the unknown while simultaneously disclaiming responsibility for the results;

3. When others may be affected by our actions, they ordinarily have the right to demand that their wishes and values be respected. This is particularly the case when those actions may result in harm to them;

4. Individual scientists and scientific groups are subject to the same norms of ethical responsibility as those of all other individuals and groups in society. They have neither more responsibility for their actions nor less; there is no special ethic of responsibility applying to scientists that does not apply to others."[84]

In the context of biological warfare, Callahan explains that each individual scientist should base research decisions on his or her own moral principles. It is the obligation of the scientist to evaluate a proposed research assignment and decide whether the goal being sought is moral or immoral. If the scientist concludes that the goal is immoral, the scientist should withdraw from the work. Callahan is firmly against those who say "if I don't do it, then someone else will," or "I personally believe what I am doing is wrong, but I was not the one who made the decision to pursue this line of investigations."[85] Callahan admits that while his outlook on moral and social responsibilities may seem like a heavy burden, the scientist who evaluates his or her work and hypothesises on a worst-case-scenario basis will not act in a "morally irresponsible way.[86]

FIGURE 1

THE PLEDGE AGAINST THE MILITARY USE OF BIOLOGICAL RESEARCH

"We the undersigned biologists and chemists, oppose the use of our research for military purposes. Rapid advances in biotechnology have catalysed a growing interest by the military in many countries in chemical an biological weapons and in the possible development of new and novel chemical and biological warfare agents. We are concerned that his may lead to another arms race. We believe that biomedical research should support rather than threaten life. Therefore, WE PLEDGE not to engage knowingly in research and teaching that will further the development of chemical and biological warfare agents."

Leonard A. Cole of Rutgers University in his paper entitled "Ethics and Biological Warfare Research," agrees with Callahan's "imagination principle" of analysing all possible harmful outcomes of a scientists research. He adds, however, that in the case of biological weapons research, an additional element must be considered: the legal element. All biological researchers, Cole states, "should be familiar, for ethical as well as legal reasons, with applicable statutes and regulations. In the area of biological research, some work might be inappropriate while not being explicitly legal" (Cole).[87]

Efforts To Preserve Ethical Responsibility

Apart from individual attempts to standardise ethical and moral evaluations of science and biological warfare research, there have been several cases of individuals and groups playing the role activists against biological warfare research.

Newtol Press mentions a Soviet scientist who was investigating a mathematical model for the spread of influenza virus. After realising the potential for an "influenza pandemic," the scientists ordered an international review of this work to deter using if for the development of biological weapons.[88] The scientist, after evaluating the potential danger of his work, took an active role in preventing such work, using his own standards of ethical responsibility to come to such a realisation.

More and more scientific societies are acknowledging the need for open discussion of biological warfare related issues. Many other are establishing ethical committees to evaluate standards on ethical practice. Cassell, Miller, and Rest focused their article, entitled "Biological Warfare: Role of Scientific Societies" on the American Society for Mircobiology (ASM), the largest life sciences organisation. The ASM Committee on Ethical Practices is responsible for the ASM policy on biological warfare, and historically, has taken an active role in biological warfare awareness. The ASM Archives co-sponsored 1990 and 1991 conferences on biological warfare, and played a significant role in President Bush's May 1990 signing of the "Biological Weapons Anti-Terrorism Act of 1989" (Cassell, Miller, and Rest).[89]

In 1988, the Reagan administration attempted a fourfold increase in US military funding of biological research. In response to this proposed increase, the Council for Responsible Genetics developed a pledge for scientists against biological warfare (see Figure 1)[90]. By August of 1988, five hundred sixty researchers had signed the pledge (Norman).[91] By the August 1989 publication of *JAMA*, over eight hundred scientists had signed the petition (Jacobson and Rosenberg), [92] and by May 1990, more than one thousand (Shulman).[93] While the Council for Responsible Genetics, led by Jonathan King, has witnessed success as an activist organisation, it is not alone. Physicians for Social Responsibility has, for many years, opposed secrecy and unethical practices in biological warfare research, and has been a strong supporter of the 1972 BWC. Physicians for Social Responsibility feels that physicians share the responsibilities of the misuse of science. Their efforts concentrate on eliminating what Victor W. Sidel, MD labels "weapons of indiscriminate mass destruction" which pose "the greatest threats to the health of the people of the world" (Goldsmith).[94]

In 1984, planning began to develop a maximum containment lab at the US Army's Dungway Proving Ground, seventy miles southwest of Salt Lake City. The laboratory was to be the testing site for extremely hazardous biological agents, and would be one of five to possess P4 containment facilities. The plans were met with significant protest. Leading the way was Jeremy Rifkin, a leading critic of biotechnology, and the founder of the Foundation of Economic Trends, who brought the proposal to court. The court ordered a complete analysis of all possible environmental and public health side-effects (Norman).[95] Noted individuals like Moselio Schaechter, a molecular biologist at Tufts University, explained the fears felt by many scientists, "a lot better work can be performed in this lab, both for defensive as well as offensive purposes. By and large, there is no way to tell the difference. They are exactly the same. Richard Goldstein, professor of microbiology and molecular genetics at Harvard University agrees, "In my mind, the opening of this facility substantially escalates the biological arms race" (Smith).[96] The 1984 plans never passed Congress, as Senator James Sasser withdrew his approval after realising that the Pentagon "sought a reprogramming action under emergency fund statutes in order to avoid the regular authorisation and appropriation process of the Congress" (Smith).[97] In 1988, attempts to revamp efforts at building the facility began again. Immediately, opposition groups were formed in Utah, and a petition was signed in August by more than one hundred forty biological professionals protesting the Army's plans.[98]

Biological Weapons Defense Research The Ethical Debate

At heart of the Dungway Facility argument is the potential use of biological research. Jonathan King, who sponsored the pledge campaign for the Council for Responsible Genetics, argued that the major problem with biological research is that it is impossible to differentiate between defensive research to be used for completely defensive purposes, and defensive research that could be used for offensive developments. King's advice, "if it is really for civilian purposes, let's put the money into NIH (Norman), [99] This argument has been used by many critics of military biological research. What has resulted is a heated debate between those that favour military defensive biological research by the Department of Defense, and those, like King, who are diametrically opposed. Is biological research by the Department of Defense justified? Is it moral? Is it needed? An examination of the arguments of both sides will help clarify many significant ethical problems and questions.

Arguments in favour of military-led defensive biological research have mostly been developed by individuals within the military, especially from the US Army Medical Research Institute of Infectious Diseases (USAMRIID) at Fort Detrick, Maryland. There are several points that these authors emphasise.

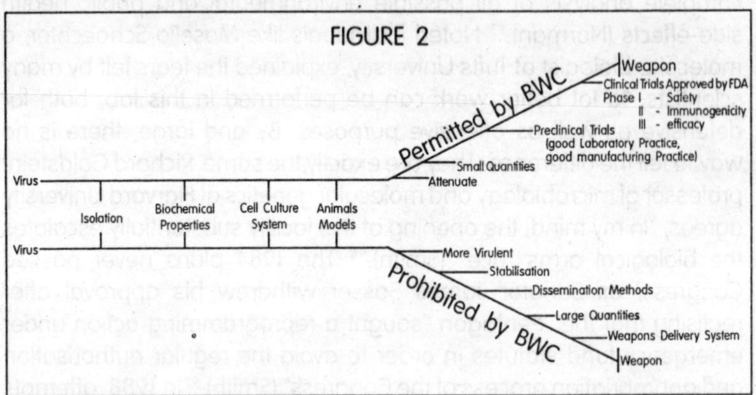

At the heart of most of the arguments concerning defensive research is the difficulty in distinguishing between what is considered defensive research and offensive research. Major Michael E. Frisina, assistant professor of philosophy at the US Military Academy, notes three major reasons that defensive research is distinguishable from offensive. Firstly, the BWC delineates the two types of research, allowing defensive research for the development of vaccines. Secondly, the ends of the two types of research are diametrically opposite. Defensive research attempts to develop a vaccine, while offensive research develop a weapon. The path to obtaining these two ends begins the same, but soon diverges intensely (see Figure 2). For a vaccine, only a small amount of experimental virus is needed, while in weapons production, the virus is developed into large quantities and placed into a delivery system. Finally, Frisina discusses his empirical research model. The empirical nature of defensive research differs from offensive. For offensive research, additional production of virus and testing of delivery systems are a necessity. These steps are missing from defensive research.[100] These three points point to a practical distinction between defensive and offensive research.

Huxsoll et al, reporting from the USAMRIID, notes political reasons for defensive biological research. It is important for the United States to continue to support the BWC and research against biological warfare. Many countries turn to the US for leadership in medical defense. Stopping defensive research would be a signal that the US is no longer supporting the BWC, and the US credibility and defense would be "seriously weakened."[102]

Proponents also cite moral justifications for defensive research. It is the moral reasons that are emphasised most and given the most importance by advocates of defensive research. Each country has a moral obligation to protect its soldiers and defenders. An additional moral obligation is placed on those associated with the healing profession, who have are morally obliged to promote health. Included within promoting health is the development of vaccines to protect both civilians and soldiers. As major Frisina put it, "thus defensive biological research to develop special vaccines for the military is a pragmatic and moral necessity."[103]

An interesting analysis was presented by Joseph Martino of the Research Institute at the University of Dayton. His paper discussed the role of military-sponsored research at universities. Martino discusses five major arguments against military biological research:

"1. All war is immoral; a university should have nothing to do with military activity;

2. Although defensive wars can be moral, weapons of mass destruction, such as nuclear weapons, are morally inadmissible, and a university should not conduct any research that might contribute to their development;

3. A university is devoted to the preservation and enhancement of culture. It should not support warlike activities because their practical result is the obliteration of culture;

4. Classified DOD research, which contradicts the demands of academic freedom, should not be conducted on a campus;

5. Although some amount of DOD research is acceptable, the current level of dependence of universities on DOD funding leads to militarisation of the university" (Martino)[104]

After refuting each of these arguments, Martino concludes that military biological research is moral and ethical, "to ask about the ethics of Department of Defense-sponsored research in the university is to get the question wrong. It implies either that there is something inherently unethical about the DOD... attempting to draw the line between the university and the DOD is to put it in the wrong place...

the proper place to draw the line is to between immoral research and any researcher, regardless of whether he works in a university, in industry, or in a government laboratory" (Martino).[105]

On the other end of the spectrum stand many of the activists who argue against defensive biological research. Historically, it is argued, defensive biological research has led "inevitably to offensive biological warfare research" (Harris)[106] Jacobson and Rosenberg site three major problems with defensive research[107]:

"1. The research may be highly dangerous to surrounding communities if virulent infectious organisms are released;
2. The research, even if truly defensive in intent, can be viewed by a potential military adversary as an attempt to develop protection for US military forces against an organism that the United States might wish to use for offensive purposes, thus permitting the United States to protect its own personnel in a biological "first strike";
3. The borderline offensive and defensive research on biological weapons is usually impossible to draw, since effective research on defenses may require testing and validation against potential offensive organisms. The fact that defensive medical research is being conducted by a branch of the US Army may appear ambiguous and may lead potential adversaries to conclude that the research is in part directed toward developing new materials for offensive use" (Sidel).

In response to arguments made by Huxsoll et al that there is a distinctly different path to developing vaccines than development of weapons (see Figure 2), Victor Sideal, chairman of the Working Group on Chemical and Biological Weapons for Physicians for Human Rights and Physicians for Social Responsibility, notes that the early steps in the development of vaccines leaves a large potential for the dissemination of diseases. As an example, Sidel describes an incident that developed at the University of Massachusetts in early 1989.[108] Microbiologist Curtis B. Thorne had been researching anthrax with funding from the US Army's biological defense programme. A large protest was held by nearly two hundred students, residents, and outside experts who feared that Thorne's research was a health risk to the entire community. Richard P. Novick, a molecular biologist and director of the Public Health Research Institute New York, supported the protests, claiming that the research "could not be construed as for peaceful purposes" (Shulman).[109] These same fears can be extended to the international community. Questions may soon arise whether countries involved in defensive research are actively developing biological weapons, possibly resulting in a biological arms race.

A strong argument against defensive biological research has been made by Jonathan King of the Committee for Responsible Genetics. This "no participation" argument, as Major Michael Frisina calls it, has several components. King argues that "the spread of disease is so unpredictable and the range of biological agents that could be used is so large that the very concept defending against biological warfare is misleading... offensive and defensive biological weapons programmes have the same components. The data gained form the defensive biological weapons testing is the same information needed to develop offensive capabilities." Research for vaccines and other defenses is thus essentially offensive, and a violation of traditional ethics of healing (Frisina).[110]

Others argue that vaccine research is pointless. Variations of biological weapons are so diverse that "vaccines against our home-grown bug would probably not work against a strain with slightly different surface properties" (Press).[111] Defensive research is thus forced at developing a large arsenal of possible agents, leaving the possibility of developing a specific agent in to an offensive weapon.

Several possible solutions to the debate have surfaced. For the most part, both sides agree that openness in research is the key. The major fears of the anti-research groups is the potential development of biological weapons. By releasing all aspects of the research, the public, especially special interest groups, would be able to keep a close eye on the developments. What the military cannot do is practice a policy of trust and secrecy. The suspicions will continue until a free flow of information is established.

In 1989, US Congressman Wayne Owens sponsored two resolutions regarding biological defense research. The first, HR5241, required that the National Institutes for Health perform "all federal research, development, testing, and evaluation of the medical aspects of the use of biological agents in the development of defenses against biological warfare." The second, HR 806, would require the Secretary of Defense to publish an annual *Federal Register* of the current stockpile of biological agents. The argument behind both of the resolutions was stated by Representative Owens, "there is nothing uniquely military about biomedical research, which is why the medical portion of the Biological Defense Research Programme belongs under the civilian control of the National Institutes of Health" (Gunby).[112]

While these two bills were never passed, an additional bill (HR 237 and S 993) that implements the BWC on a domestic level, adding criminal penalties for domestic violence, passed in May 1990. The passage was regarded by Jonathan King as, "a very important step [in] ensuring that biotechnology is developed for peaceful means" (Shulman).[113]

Still the debate continues. Doubts as to whether defensive biological research is moral or justified still cloud the mind of the researcher. At this point, it is up to the individual ethics of the scientist to decide whether to continue research or not. This is where comprehensive ethical decisions and moral judgements are an important necessity.

Future Possibilities for Prevention

Recent developments in technology have made biological weapons an even more dangerous option. Genetic engineering and biotechnology offer biological weapons a completely new dimension. Additionally, the appeal of biological weapons to terrorists and Third World nations has stirred even more fears. What has emerged is a situation that necessities awareness, precaution, and action. The BWC is the only current control measure for biological weapons. However, additional efforts have begun at developing alternative measures to help control the growing threat of biological weapons.

Much of the second review conference of the BWC was concentrated on the threat that genetic engineering brings to biological weapons development. At the conference, the US submitted a paper warning about the potential problems that modern technology brings:

"Verification of the Convention, always a difficult task, has been significantly complicated by the new technology... the ease and rapidity of genetic manipulation, the ready availability of a variety of production equipment, the proliferation of safety and environmental equipment and health procedures to numerous laboratories and production facilities throughout the world, are the signs of the growing role of biotechnology in the world's economy. But these very same signs also give concern for the possibility of misuse of this biotechnology to subvert the Convention" (Flowerree).[114]

This statement illustrates the potential problems that biotechnology brings to biological warfare. Concerns at the conference were met by a series of agreements known as "confidence building measures": exchange of information concerning all high-level containment facilities that house genetic engineering developments, the status of individual national biological defense programmes, and outbreaks of infectious diseases. Additionally, the second review conference encouraged the publication of scientific findings on vaccine development, and more contact between researchers (Freeman).[115] Despite efforts at controlling the threat added by biotechnology, concerns still exist.

As a weapon of mass destruction, biological weapons are much more tempting to a power-hungry militant leader than nuclear weapons. Nuclear weapons are significantly more difficult to develop, and are considerably more expensive. Biological weapons, known for many years as "poor man's atomic bomb," are easy to obtain, and much easier to develop. Newton Press reports that "it is actually possible to set up a laboratory in a house trailer and to equip a functional research facility for thousands of dollars rather than millions."[116] These traits leave biological weapons a potentially popular option to Third World powers or terrorist groups. As witnessed by Iraq during the 1991 Gulf War, Third World countries now see biological weapons as an easy route to world power. Third World nations learned an important lesson from the Gulf War: the superior technology of developed countries will outmatch the limited arsenals of the Third World. Aggressive Third World leaders are beginning to see biological weapons as the only way to contend with larger world powers. As we approach the 21st century, regarded by many as the Biological Age, biological weapons will become more and more tempting to modern warfare.

FIGURE 3

BIOLOGICAL WEAPONS CONVENTION

ARTICLE X

...[the signatory nations] undertake to facilitate, and have the right to participate in, the fullest possible exchange of equipment, materials, and scientific and technological information for the use of bacteriological (biological) agents and toxins for peaceful purposes. Parties to the Convention in a position to do so shall also co-operate in contributing individually or together with other states of international organisations to the further development and application of scientific discoveries in the field of bacteriology (biology) for the prevention of disease, or for other peaceful purposes.

What these threats amount to is a significant need for control and cooperation to develop the technologies of biological warfare for peaceful means. Several individuals have developed methods at aiding this effort.

In response to the 1991 Gulf War, a conference was held on April 4 and 5, 1991 entitled "The Microbiologist and Biological Defense Research: Ethics, Politics and International Security at the University of Maryland Baltimore Country. A book containing contributions from the meeting was subsequently published in a December 1992 edition of the Annals of the New York Academy of Sciences. During the meeting, a dinner talk was held discussing the outlook on biological weapons development and control. Carl-Goran Heden of the Biofocus Foundation Project of the World Academy of Art and Science published an article entitled "The 1991 Persian Gulf War: Implications for Biological Arms Control." The article stressed the lessons to be learned from the war, and discussed several alternative ways to develop biological research.

Firstly, Heden points out that the distinction between haves and the have-nots goes beyond access to resources, technology, food, and water. The distinction also is results from the availability of knowledge and lack of opportunities to use that knowledge. The gap between the haves and have-nots can be narrowed by sharing of knowledge and information. Cooperation in science and technology offers such an opportunity. He adds that bioengineers have an enormous potential to apply their skills to cooperation and peaceful developments. Bioengineers can help solve short-term environmental problems and make long-term contributions to solving ecological problems of their region. By sharing information and know-how resulting from peaceful developments in biotechnology, Heden feels that we can start the developing world on its way to a "biosociety."

Heden's most interesting suggestion is concerned with combating the proliferation of biological weapons. Heden emphasises that, in his opinion, the way to combat the rising threat of biological warfare is to develop Article X of the BWC (See Figure 3) which "enjoins signatory nations to cooperate in the peaceful development of applied microbiology. This action needs to be supported by an implementation programme that combines the resources of the United Nations system and appropriate scientific non-governmental organisations and brings them to bear on preventing development of biological weapons and fostering development of microbiology for humanitarian purposes" (Heden).[117] Heden's proposal is developed in detail by Raymond A. Zilinskas from the Center for Public Issues in Biotechnology at the University of Maryland Baltimore Country.

Zilinskas dedicates his article to a detailed description of a new technical agency—the Biological Hazards Early Warning Programme (BHP). The BHP would be an international agency dedicated to deterring the development of biological warfare. The BHP would, however, contain an additional agenda. It would be an international agency responsible for reporting and investigating any unusual outbreaks in diseases and attempt to determine their origin. In light of the fact that "international trust in the BWC is not being enhanced," the BWC would help strengthen the confidence-building measures set forth in the BWC and its subsequent review conferences.

The focal point of Zilinskas' plan to strengthen the BWC is placing more emphasis on Article X of the BWC. In fact, Zilinskas and Heden advised the second BWC review conference to take such action. They advised the review conference to raise necessary funds and work with the International Council of Scientific Unions to develop an international programme that would include "training, short-term exchange visits, joint research between scientists and between laboratories, and linking institutes via computerised communication networks. Participating scientists would have a moral and professional organisation as they carry out their normal activities of research, teaching, and training." Although RC3 agreed to carry out the advice, funding was never provided (Zilinskas).[118]

Details of the BHP are developed in an article published in the *Annals of the New York Academy of Sciences* article entitled "Confronting Biological Threats to International Security: A Biological Hazards Early Warning Programme." Zilinskas discusses in detail the BHP purpose, structure, legal justification, and funding. One of the most significant aspect of the BHP would be its ability to enhance international arms control and security from biological threats through early detection, monitoring, and suppression of diseases.[119]

Attempts to prevent the spread of biological weapons must go beyond the BWC, as the ethical dilemmas of biological weapons span beyond the collective level. Individual efforts of scientists are the source of biological weapons development. What results is a question of social responsibilities, morals, and ethics.

In attempting to derive possible solutions to the problem of moral and social responsibilities, Rosemary Chalk of the Institute of Medicine at the National Academy of Sciences in Washington, DC, discusses establishing a reference of "unlawful science." She suggests to "develop a list of weapons that are banned by international treaty and to formulate an international protocol urging scientists of very nation to refuse assignments that directly contribute to the development of such weapons" (Chalk).[120]

Leonard Cole offers an additional policy approach to solving the ethical problems of biological warfare. He suggests establishing a consultative agency to review biological research, and answer any ethical questions or doubts. He recommends the Institutional Review Boards, who currently oversee experimentation on human subjects, as a possible agency to fulfill this role.[121]

The role of the physician in the biological field can not be underemphasised. After all, it is these professionals who develop biological weapons and their defense. As Alan H. Lockwood, MD of SUNY at Buffalo School of Medicine and Biomedical Science explains, "Our professional responsibility should compel us to be leaders in the process that will prevent [chemical and biological weapons] proliferation and lead to their abolition."[122] Unfortunately, most medical training does not emphasise awareness, preparation or treatment of biological threats. It is this subject that Surgeon Lieutenant Commander a G Robertson of the Royal Australian Navy and Major Morgan-Jones of the Royal Army Medical Corps discuss in their article "First

Line Nuclear, Biological and Chemical Defense Training—The Way Ahead." The discussion emphasises the need for a protocol for medical professional training, citing a need for "training at paramedic, first line medical officer, medical specialist, nursing officer and medical services officer level." They mention six aims of this protocol: "First, it should complement the single service general NBC training given to all personnel and provide a good foundation from which the student could progress onto other more advanced courses such as the Chemical Defense Science Course and the DRPS Senior medical Officer's Radiation Protection Course. Second, the training should provide treatment protocols to support the medical person in the field. Third, as the principles of treatment are applicable to all three services, it should be organised on a tri-service basis utilising tri-service resources. Fourth, the training should be based on the basic resuscitative principles of the Battle Acute Trauma Life Support (BATLS) concept. Once developed, these concepts can be applied to other toxicological emergencies, e.g. in industry, and could be usefully applied in the training of civilian physicians. Fifth, training should initially be developed for junior medical and dental officers as they are most likely to be required to put these concepts into action. Spealised courses for medical specialists providing third and fourth line care, paramedics, nursing staff and other health service personnel should then follow. The medical specialists would then form a nucleus for development of coherent responses by National Health Service staff. Finally, there should be revision courses every three years to ensure that personnel remain current" (Robertson and Morgan-Jones).[123]

Commentary

Tremendous lessons can be learned by understanding biological warfare and biological weapons development. Biological warfare has been present for thousands of years, yet only recently has the potential threat been understood. Nonetheless, this threat is growing. Although great efforts have been made at curtailing the proliferation of biological weapons, development continues. A 1994 Pentagon report noted that as many as twenty-five nations including North Korea, Iran and Iraqi, are currently developing biological weapons. Many already possess biological stockpilers.[124] If successful deterrent does not come soon many other countries will join the list. The Biological Weapons Convention of 1972 provides the basis for such deterrent. It contains the guidelines for successful control of the biological threat. However, the document alone is not enough. What is needed is international cooperation by all countries to adhere to the regulations of the Convention. History has taught us that biological research and weapons development can be abused. One only needs to look back as far as World War II to discover the lack of morals and human decency that took place in prewar Japan. Leaders like Ishii Shiro are not alone. More and more aggressive and unscrupulous leaders are emerging in the modern world; leaders who see little need to consider ethical and moral values in devising their plans to become world powers. It is these leaders and terrorists that pose the most intense threats. Biological weapons are easy to create, easy to hide, and difficult to detect. The propensity of these individuals to take advantage of the "poor man's atomic bomb" becomes more and more realistic as time progresses and development continues.

From an ethical standpoint, this creates a threatening problem. New moral obligations must now be considered in performing biological research. The scientist must be fully aware of what he or she is about to embark upon. The potential of biological technology falling into the wrong hands grows larger with time. Lists of ethical standards, moral obligations, and social responsibilities can only go so far. Ultimately, the decision can only be made by the individual scientist. Yet the scientist has the added obligation of awareness. An obligation of understanding the project at hand and the potential abuse that may develop from it. An obligation that attempts to cover all corners, predict the future, and protect for the future.

The time has come to learn from the past. Any miscalculation in the area of biological warfare can be disastrous. Had Saddam Hussein chosen to use his biological arsenal during the Gulf War, devastation would have resulted. Matthew Meselson estimated that had 200 kilograms of anthrax spores (the amount contained in one Iraqi Scud) been detonated a few feet above ground, it "would kill every living thing—at least all humans and bovids for hundreds of kilometers in the downwind plume" (Dickey).[125] Fortunately, Iraq's biological arsenal was untouched. Yet, the 1991 Gulf War taught the world a valuable lesson about biological warfare; a lesson filled with a need for awareness and precaution; a lesson filled with ethical and moral responsibilities and considerations; a lesson about the importance of understanding the growing threat of biological warfare.

We are living in a time period where peaceful negotiations are rising. Concepts of peace are beginning to fill the "minds of men." We are beginning to understand the root of violence, and are just starting to use the root of violence to combat violence. In 1986, a diverse group of scientists met in Seville to ask an important question, "Does modern biology and social sciences know of any biological factors that are an insurmountable or serious obstacle to the goal or world peace?" The answer was precise, "biology does not condemn humanity to war, and that humanity can be freed from the bondage of biological pessimism and empowered with the confidence to undertake the transformative tasks needed in this International Year of Peace and in the years to come.... The same species who invented war is capable of inventing peace. The responsibility lies with each of us."[131]

Violence is thus not genetic. Rather, it is a learned behaviour that develops out of environment. Man learns violence. If violent behaviour and violent thoughts may be "unlearned." Take for example Western Europe. One would only have to look back to the beginning of this century to find a warring Europe filled with violence. Two of the largest and bloodiest wars in the history of man were fought on this land. The entire world was shocked at the depths of man's inhumanity illustrated by the two world wars. Yet, soon after World War II, Western Europe began to use the war as an impetus for science. Science and technology flourished. What has manifested itself in the European Community are a group of countries united together through cooperation. The transition is almost complete for Western Europe.

Presently, the social, economical, and cultural attractiveness of weapons has resulted in a unified desire to increase the prestige of a country. Lesser developed countries see weapons development as an easy ticket to stardom; a backstage pass to world reverence. What if the prestige associated with weapons development were to be reversed. What if weapons lost their prestigious value? What if countries began working together on positive developments in science and technology instead of concentrating efforts to the growing technologies of weapons development and war technologies? The case of biological warfare gives a perfect opportunity for such a reversal. What if scientists concentrated on coordinating their efforts at developing vaccines instead of biological weapons on a global level, with full dissemination of all developed technologies to all countries? When these questions are answered with large scale cooperation, scientists will begin to work together on an international level; development of technologies of war will be replaced with the development of technologies of peace. Scientists will be united by a common desire—preservation of peace through science.

How does such a large scale cooperation develop? As an answer to this question, it is necessary to turn to two overwhelmingly influential systems of society: education and religion. It is through these two mechanisms that international cooperation can begin to see success. Merging these two disciplines together may be the answer.

Education will always be the primary means of generating awareness. The effectiveness of education, however, depends on the method of communication. The world needs a lesson on ethics and morals. The educations systems need to begin emphasising ethical and moral standards at all levels of education. In the scientific community, this emphasis becomes essential. Scientists involved in technological and scientific developments are constantly putting ethics and morals to the test. Yet scientists are often the last to be given the proper morals and ethics courses they need. The example of biological weapons development is a clear example.

The ethical systems developed within religion can provide the stepping stone to such an educational system. A system that has the potential to develop the needed international awareness. The key will be utilisation of the two uniting principals of all religions: non-aggression and cooperation. The solution thus rests in the delivery of such an ethical system. Who better to perform such an act than the religious and spiritual leaders! There is a profound need to integrate the role of spiritual and religious leaders in the major decisions of biological warfare awareness.

The importance of religion in fostering international awareness is nothing new. The UNESCO Catalunya Center in Barcelona has organised several meetings on "The Contribution of Religions to the Culture of Peace." At the 1994 meeting, participants from all major religious groups issue the "Declaration on the Role of Religion in the Promotion of a Culture of Peace," identifying the need to unite all religions and cultural traditions.[132]

The recent stress given to bioethics committees is not unjustified. Bioethics committees are able to analyse scientific decisions from a multi-faceted approach, emphasising the important ethical standards in question and give the necessary advice on these issues. These committees are traditionally made up of physicians, lawyers, scientists, and theologians. Emphasis on the development of bioethics committees dedicated to the threat of biological warfare may foster such a solution. Within the context of bioethics committees religious and spiritual leaders will be able to take local awareness to an international level, emphasising two of the most significant moral values that have been preserved throughout all religions: non-aggression and cooperation.

Efforts by international organisations to develop and preserve peace are increasing. The United Nations Educational, Scientific and Cultural Organisation (UNESCO) has played an important role in formulating ethical principles, promoting ethical behaviour, and establishing peaceful developments within the United Nations system. As a statement made at the 26th Session of the General Conference of UNESCO in 1991 explains, "Within the UN system, UNESCO has been entrusted with a special ethical mission in the promotion of a democratic culture that is conductive to the effective application of human rights and the establishment of a culture of peace"[126] UNESCO's Culture of Peace Programme is now gaining international recognition and support for their attempts to "ensure that the conflicts inherent in human relationships be resolved non-violently, based on the traditional values of peace."[127] The Culture of Peace Programme is currently involved in national peacemaking programmes throughout the world.

The recently established UNESCO Hebrew University International School for Molecular Biology and Microbiology (ISMBM), has combined the emphasis of peaceful negotiation with developments in science. The International School centers efforts on the implementation of "Science for Peace" through international cooperation in science. As Leah Boehm, chief scientist at Israel's Science Ministry put it, "Science is less political than other issues, it's a bridge for peace."[128] The central issues of biological warfare fit perfectly into this bridge. The future of biological weapons development and research will depend on the outcome of ethical and moral responsibilities. An outcome that will be essential to both peace and science.

References

1. Poupard, James A and Linda A Miller. "History of biological ware: catapults to capsomeres." *Annals NY Acad Sci.* Volume 666. December 31, 1992, p. 9-20 (10).
2. See note 1; p. 15.
3. Cole, Leonard A.. "Ethics and biological warfare research." *Ann NY Acad Sci,* Volume 577. December 29, 1989, p. 154-163 (154).
4. See note 1; p. 13.
5. Mobley, James A. "Biological Warfare in the Twenteth Centur: Lessons from the Past, Challenges for the Future. *Military Medicine.* Volume 160, Number 11, p. 547-553 (547).
6. Goldsmith, Marsha F. "Often thwarted treaty efforts leave chemical, biological weapons a still potential threat." *JAMA.* Volume 265. Number 6. February 13, 1991, p. 705.
7. Flowerree, Charles C. "The Biological Weapons Convention and the researcher." *Ann NY Acad Sci.* Volume 666. December 31, 1992, p. 113-128 (113-4).
8. Harris, Sheldon. Japanese biological warfare research on humans: A case study of microbes and ethics." *Ann NY Acad Sci.* Volume 666. December 31, 1992, p. 21-49 (21).
9. See note 8; p. 23-4.
10. See note 8; p. 24-5.
11. See note 8; p. 26-7.
12. See note 8; p. 27-9.
13. See note 8; p. 30.
14. See note 5; p. 548.
15. See note 8; p. 33-4.
16. Freeman, Shirley. Rev. *Biological Warfare in the 21st Century* by Malcolm Dando. *Medicine and War.* Volume 11. Number 3. July-Sept 1995, p. 112-4 (113).
17. See note 8; p. 41-2.
18. "Chemical and bacteriological weapons in the 1980's." *The Lancet.* July 21, 1984, p. 141-3 (141).
19. Aldhous, Peter. "Gurinard Island handed back." *Nature.* Volume 344. April 26, 1990, p. 801.
20. See note 19; p. 801.
21. Doyle, RJ and Nancy C. Lee. "Microbes, warfare, religion, and human institutions." *Can J Microbiol.* Volume 32. March 1986, p. 193-200 (195).
22. See note 5; p. 548.
23. Bernstein, Barnstein, Barton J. "The birth of the US biological warfare programme." Scientific American, June 1987, p. 94-9 (97).

24. Press, Newtol. "Haber's choice, Hobson's choice, and biological *Persp in Biology and Medicine*. Volume 29. Number 1. Autumn 1985, p. 92-108 (95).

25. See note 5; p. 552 (n. 18).

26. See note 23; p. 97.

27. See note 23; p. 94-6.

28. See note 1; p. 14.

29. See note 1; p. 14.

30. Stanier, Roger Y. "The Journey, not the arrival matters." *Ann Rev Microbio.* Volume 34. 1980, p. 1-48 (18).

31. *Facts on File World News Digest.* December 21, 1984, p. 949 D3.

32. Huxsoll, David L, Cheryl D Parrott, and William C Patrick III. "Medicine in defense against biological warfare." *JAMA*. Volume 262. Number 5. August 4, 1989, p. 677-9 (678).

33. Moon, John Ellis Van Courtland. "Biological warfare allegations: The Korean War case." *Ann NY Acad Sci.* Volume 666. December 31, 1992, p. 53-81 (54).

34. See note 1; p. 13.

35. Robertson, Andrew G. "From Asps to Allegations: Biological Warfare in History." *Military Medicine.*Volume 160. Number 8. August 1995, p. 369-72 (371).

36. 36. See note 8; p. 33.

37. Rolicka, Mary. "New studies disputing allegations of bacteriological warfare during Korean conflict." *Military Medicine.* March 1995, 1995, p. 97-100 (97).

38. See note 37; p. 98.

39. See note 33; p. 58

40. See note 33; p. 58-60.

41. See note 5; p. 552 (n. 19).

42. See note 37; p. 98.

43. Cowdrey, Albert e. "Germ warfare and public health in the Korean conflict." *Journ of Hist Med.* Volume 39. April 1984, p. 153-72 (157).

44. See note 43; p. 166.

45. See note 33; p. 66.

46. See note 33; p. 66.

47. Seale, John. "AIDS virus infection: A Soviet view of its origin." *Journal of the Royal Society of Medicine.* Volume 79. August 1986, P. 494-5 (494).

48. Medvedev, Zhores A. "AIDS virus infection: A Soviet view of its origin." *Journal of the Royal Society of Medicine.* Volume 79. August 1986, p. 494.

49. See note 24; p. 93.

50. See note 16; p. 112-3.

51. See note 1; p. 15.

52. See note 32; p. 677-9 (677).
53. Dickson, David. "Gene splicing dominates review of weapons pact." *Science.* Volume 234. October 10, 1986, p. 143-45 (143).
54. See note 7; p. 114.
55. Orient, Jane M. "Chemical and biological warfare: should defenses be researched and deployed? *JAMA.* Volume 262. Number 5. August 4, 1989, p. 644-648 (646).
56. Meselson, Matthew, et al., "Yellow Rain." *Scientific American.* September 1985, p. 122-31 (122).
57. Rosen, Robert T and Joseph D Rosen. "Presence of 4 fusarium mycotoxins and synthetic material in "Yellow Rain": Evidence for the use of chemical weapons in Laos." *Biomedical Mass Spectrometry.* Volume 9. Number 10. 1982, p. 443-50 (450).
58. See note 56 p. 131.
59. Rosen, Joseph D. Letters. "Yellow Rain." *Science.* August 19, 1983, p. 698.
60. Meselson, Matthew et al. "The Sverdlovsk anthrax outbreak of 1979." *Science.* Volume 266. November 18, 1984, p. 1202-1207.
61. See note 55; p. 647.
62. See note 18; p. 142.
63. See note 60; p. 1203.
64. Dickey, Christopher. *Newsweek*: September 4, 1995, p. 14-15 (14).
65. Knudson, Gregory B. "Operation Desert Shield: Medical aspects of weapons of mass destruction." *Military Medicine.* Volume 156. June 1991, p. 267-271 (268).
66. See note 65; p. 268.
67. Marwick, Charles. "Defense appears to have advantage over offense presently in biological weapons." *JAMA.* Volume 265. Number 6. February 13, 1991, p. 700.
68. See note 65; p. 271.
69. Horgan, John. "Biowarfare wars." *Scientific American.* January 1994; p. 12.
70. Se note 64; p. 14.
71. See note 64; p. 14.
72. Goldsmith, Marsha F. "Defensive biological warfare researchers prepare to counteract 'natural' enemies in battle, at home." *JAMA.* Volume 226. Number 18. November 13, 1991, p. 2522-3 (2522).
73. 10 U.S.C. sec. 980.
74. Annas, George J. "Changing the consent rules for Desert Storm "*The New England Journ Med.* March 12, 1992, p. 770-3 (770).
75. See note 74; p. 770-3.
76. See note 65; p. 700.

77. Annas, George J. Correspondence. "Medicine and War." *The New England Journ Med.* October 8, 1992, p. 1098.
78. See note 8; p. 43.
79. See note 21; p. 193.
80. Chalk Rosemary. "Drawing the line: An examination of conscientious objection in science." Volume 577. December 29, 1989, p. 61-74 (68).
81. See 80; p. 69.
82. See 80; p. 69-70.
83. See 24; p. 107.
84. Callahan, Daniel. "Ethical responsibility in science in the face of uncertain consequences." Ann NY Acad Sci. Volume 265. January 23, 1976, p. 1-12 (2).
85. See note 84; p. 4.
86. See note 84; p. 8.
87. See note 3; p. 158.
88. See note 24; p. 100.
89. Cassell, Gail H, Linda A Miller, and Richard F. Rest. "Biological warfare: Rule of scientific societies." *Ann Ny Acad Sci.* Volume 666. December 31, 1992, p. 230-238 (234).
90. Frisina, Michael E. "The offensive-defensive distinction in military biological research." *Hastings Center Report.* May - June 1990; p. 19-22 (21).
91. Norman, Collin. "Biologists eschew weapons research." *Science*, Volume 241. August 5, 1988.
92. Jacobson, Jay A. "Biological defense research: Charting a safer course." *JAMA.* Volume 262. Number 5. August 4, 1989, p. 675-6 (676).
93. Shulman, Seth. "International treaty made domestic law." *Nature.* Volume 345. May 17, 1990; p. 192.
94. See note 6; p. 705.
95. Norman, Colin. "Biowarfare lab faces mounting opposition. "*Science.* April 8, 1988; p. 135.
96. Smith, R. Jeffrey. "New army biowarfare lab raises concerns." *Science.* Volume 226. December 7, 1984; p. 1176-8 (1178).
97. See 96, p. 1176.
98. Franklin, Naomi C. "Petition on Dungway facility." *Science.* January 6, 1989; p. 11-12.
99. See note 95.
100. Frisina, Michael E. "The offensive-defensive distinction in military biological research. "*Hastings Center Report.* May-June 1990; p. 19-22 (21).
101. Frisina, Michael E. "A healing-killing conflict in military research" Hastings Center Report. September-October 1989; p. 2.

102. Huxsoll, David L, Cheryl D Parrott, and William C Patrick III. " Medicine in defense against biological warfare." *JAMA*. Volume 262. Number 5, August 4, 1989; p. 677-79 (677).

103. Frisina, Michael E. "The offensive-defensive distinction in military biological research." Hastings Center Report. May-June 1990; p. 21.

104. Martino, Joseph P. "The place of department of defense-sponsored research at the university." *Ann NY.Acad Sci.* Volume 577. December 29, 1989; p. 172-83 (173).

105. See note 104; p. 182-3.

106. See note 8; p. 43.

107. This summary appears in an editorial by Victor Sidel:Sidel, Victor W. Weapons of mass destruction: The greatest threat to public health." *JAMA*. Volume 262. Number 5. August 4, 1989; p. 680-82 (681).

108. Sidel, Victor W. "Weapons of mass destruction: The greatest threat to public health." *JAMA*. Volume 262. Number 5. August 4, 1989; p. 680-82 (681).

109. Shulman, Seth, "Microbiologists butt of protests." *Nature*. Volume 339. May 4, 1989; p. 6.

110. See note 90; p. 20.

111. See note 24; p. 106.

112. Gunby, Phil, "Biological weapons proliferation arouses US and international concern." *JAMA*. Volume 262. Number 5. August 4, 1989; p. 605-6 (605).

113. See note 93; p. 192.

114. See note 7; p. 117.

115. Freeman, Shirley E. "The Biological Weapons Convention: The 1991 Review Conference." *Medicine and War*. Volume 8. 1992, p. 128-30 (128).

116. See note 24; p. 101.

117. Heden, Carl-Goran. "The 1991 Persian Gulf War: Implications for biological arms control." *Ann NY Acad Sci.* Volume 666. December 31, 1992, p.1-8 (5).

118. Zilinskas, Raymond A. "Confronting biological threats to international security: A biological hazards early warning programme." Volume 666. December 31, 1992, p. 146-76 (156).

119. See note 118; p. 146-76.

120. See note 80; p. 72.

121. See note 3; p. 162.

122. Lockwood, Alan H. Letter. "Chemical and biological weapons." *JAMA* Volume 226. Number 5. August 7, 1991, p. 652.

123. Robertson, a G and D J Morgan-Jones. "First line nuclear, biological, and chemical defense training—the way ahead." *J Roy Nav Med Serv.* Volume 80. 1984, p. 90-94 (93).

124. See note 69; p. 12.

125. See note 64; p. 14.
126. Poteliakhoff, Alex. "The UN and Global Ethics." *Medicine, Conflict, and Survival,* Volume 12. 1996, p. 4-13 (5).
127. International Forum on the Culture of Peace, San Salvador, February 1994 taken from UNESCO's Culture of Peace programme brochure, 1995.
128. "Science in an era of political change." *Nature.* Volume 375. June 29, 1995, p. 717-720 (719).
129. Nowell-Smith, Patrick H. "Religion." *The Encyclopedia of Philosophy.* Paul Edwards, Ed., Volume 7. Macmillan Publishing Co., Inc & The Free Press, New York, p. 153.
130. "Religion." *The Encyclopedia Americana.* Volume 23. Americana Corporation, Connecticut, 1980, p. 361-2.
131. Adams, David, Ed., *Culture of Peace: Promoting a Global Movement.* Culture of Peace Programme, France, 1995, p. 41.
132. See note 131; p. 54.

9

Conversion of Biological Weapon Research and Development

—M. *Leitenberg, United States of America*

The Biological Weapons Convention (BWC) was signed on April 10, 1972. The United States, the USSR, and United Kingdom deposited their instruments of Ratification of the Convention on March 26, 1975, and the Treaty came into force. It was the first—and for a long time only—post-World War II disarmament treaty in which an entire class of weapons of mass destruction was done away with. Contrary to the Nuclear Non-proliferation Treaty of 1968 (NPT), their was to be no preferred group of countries that would continue to retain the weapons. Biological weapons were to be prohibited to all, into the future. This was the first major and unique distinction of the subject.

The second unique aspect was that one of the two superpowers—the United States—that did possess biological weapons, gave them up and destroyed them, even before the Treaty came into being. Instead of first negotiating a treaty and then implementing its provisions, an entire class of weapons was renounced by a major possessor without any prior international agreement.

There was also a third major and unique distinction of the BWC: in 1992, Russia admitted that the former USSR had been in gross, generic violation of the Treaty, the only instance in which one of the superpowers admitted having been in total violation of a post-world War II arms control treaty.

Before looking at the technical aspects of BW R & D conversion, one must know who and what there is to convert, that is who currently has offensive BW R & D programmes and, in the case of Russia, the very large and more or less indeterminate remaining residue of institutions, researchers and programmes that still need to be converted. According to official US Government estimates, four nations had offensive BW programmes in 1972 while ten nations — and apparently now 12—have had them in the 1990s. Some of the ten countries are signatories of the Biological Weapons Convention, and two of them may be producing and stockpiling agents.

The biggest problem is the magnitude of institutions in Russia still requiring conversion. The second largest problem is the ongoing situation in Iraq—ongoing since 1991—in which the United Nations and the United Nations Security Council in particular have not been able to convince or force Iraq to give up and to cease its BW programme as required by UN resolutions. Iraq committed itself to abide by those resolutions in 1991, however it has strenuously and actively avoided doing so, and it has continually been in material breach of the resolutions. Compounding this problem is the clear and repeated failure of three of the five permanent members of the UN Security Council to hold Iraq to those commitments. That has undercut the ability of the international community to see an end to Iraq's offensive BW programme, and obviously that must precede conversion.

As regards the technical aspects of conversion and its feasibility, there is probably no area of defense industrial production or development that should be easier to convert than BW R & D. A microbiologist or virologist uses the same knowledge base and laboratory technology no matter what applied aspect he or she works on. The alternative research areas and products are numerous, well understood, required by societies, and will be described in detail.

Biological weapons (BW), along with chemical and nuclear weapons are the three systems designated as Weapons of Mass Destruction. The Biological Weapons Convention (BWC), banning the development, production and stockpiling of BW, as well as BW research for offensive purposes, was signed on April 10, 1972. The treaty came into force on March 16, 1975, when the United States, the USSR, and United Kingdom deposited their instruments of ratification for the convention. It was the first - and for a long time the only—post-World War II disarmament treaty in which an entire class of weapons of mass destruction was done away with—or so it was widely assumed at the time—and the arms control community by and large thought biological warfare had been removed from the scene. Contrary to the nuclear Nonproliferation Treaty of 1968 (NPT), there was to be no preferred group of countries that would continue to retain the weapons. Biological weapons were to be prohibited to all, into the future. This was to be the first major and unique distinction of the subject.

The second was that, the United States, one of the two superpowers that did possess biological weapons, gave them up and destroyed them in 1969, even before the Treaty came into being. The United States chose this policy at the time to dissociate biological from chemical weapons, the combined and historical framework under which arms control deliberations on them had been carried on prior to 1969 in Geneva. Article 9 of the BWC was an undertaking to continue negotiations to achieve a chemical weapons disarmament treaty—but an additional 22 years would pass before that would be achieved. The BWC additionally carried *no* verification provisions; on-site verification was not something that the USSR would consider or accept before the Stockholm Conference in 1986 and the Intermediate Range Nuclear Forces (INF) Treaty in December 1987. Nonetheless, the BWC does address the question of non-compliance.

There was, however, a third major and unique distinction of the BWC: in 1992 Russia admitted that the former USSR had been in gross, generic violation of the Treaty, the only instance in which one of the superpowers admitted to having been in total violation of a post-World War II arms control treaty. By the end of the 1980s, it had also become clear that a half dozen or more countries had decided to develop biological weapons in the intervening twenty years. One official U.S. Government estimate is that "The number of nations having or suspected of having offensive biological or toxin warfare programmes has increased from four to ten since 1972." And as the same statement noted, some of the 10 nations in question"... are signatories of the BWC." Thus the assumed achievement of the 1970s had been, at least in part, reversed. Chemical weapons had been used in the war between Iraq and Iran in the 1980s, and allied troops that fought Iraq in the gulf War in 1991 ran a risk of being attacked by both chemical and biological weapons. From the mid- 1980s on, there had also been movement to strengthen the BWC and add some kind of verification provisions to it, particularly once the Chemical Weapons Convention was signed in January 1993.

In short, the two international legal regimes that prohibit and control the development, production and use of biological weapons —the Geneva Protocol of 1925 (prohibiting use) and the Biological and Toxin Weapons Convention of 1975 (prohibiting development and production)—have been under clear pressure:

- by the increase in the number of states that have undertaken offensive BW programmes in the past two decades since the treaty came into force;
- by the admission by Russia that the USSR had been in generic violation of the BTWC between 1975 and 1992, and by US and British suspicions that not all the remnants of that programme have yet been entirely curtailed since 1992;
- and by the evidence disclosed by the United Nations Special Commission (UNSCOM) investigatory process in Iraq, that Iraq had produced and deployed a stockpile of biological munitions by 1990 and was prepared to use them;
- and by the continuing inability of the United Nations security Council, over a period of nearly seven years, to force Iraq to entirely give up its BW programme and hidden materials.

The Proliferation of Biological Weapons

The first question to be addressed then is which nations currently have offensive BW programmes, and therefore facilities and researchers that require conversion. The years since 1972 and 1975—when the Biological Weapons Convention was signed and then entered into force—have been a severe disappointment for arms control in the biological field. One official U.S. estimate is that "The number of nations having or suspected of having offensive biological and toxin warfare programmes has increased from four to ten since 1972."[1] These numbers were fist presented in U.S. Government testimony to Congress the year before, in 1988, by Dr. Thomas J. Welch of the U.S. Department of Defense, to the House committee on Armed Services. See also, John H. Cushman Jr., "US. Cites Increase in Biological Arms," *New York Times,* May 4, 1988.

Another version of this estimate reads, "During the 20 years the BW Convention has existed, the number of countries considered to be developing or recently engaged in offensive BW programmes has risen from 4 in 1972 to 10 in 1992 - some of which are members of the convention." U.S. and *International Efforts to Ban Biological Weapons,* U.S. General Accounting Office, GAO/NSIAD- 93-113, December 1992, pp. 2-3, 16. And as the same statement noted, some of the 10 nations in question"...are signatories of the BWC." A statement by the Director of the US Arms Control and Disarmament Agency in November 1996 raised that number to 12.[2] A substantial number of these countries are in the Middle East, and these have either not signed not ratified the BWC. In 1992, the Bush Administration made a concerted effort, but failed, in the attempt to convince several of the major Middle East antagonists to either sign and /or ratify the BW Convention.[3]

With the exception of both the former USSR—now Russia—and Iraq, (as a result of the Gulf War and the United Nations Special Commission (UNSCOM) process which followed it), there has however been *no* international pressure or penalty applied against any of the suspected BW states. Until around 1988, no national or international spokesperson even made reference to the development; and since then, it has been virtually only U.S. Spokepeople who have done so. The statements have been constantly plagued, however, with ambiguities in their descriptive terminology, such as the words" ...or suspected of having" in the statement quoted above. In 1990, Admiral Trost, the Chief of Naval Operations, told Congress that "three countries worldwide now have bacteriological weapons," and that 15 others were *suspected* of *developing* them.[4] Three weeks later the Director of naval Intelligence identified iraq, Syria, and the former USSR as the three "assessed to have (BW) *capability*.[5] In 1988, his predecessor, Admiral Studeman, had also identified China, Taiwan, and North Korea by name. But what the U.S. Government's criteria were for the categories of suspected," "developing," and " capability" were never specified, although in this particular pair of statements "capability" apparently meant weapons' possession.

A statement in the 1992 British Defense White Paper uses the same pattern of ambiguous phrasing, nothing that "about ten (nations) have *or are seeking* biological weapons." What was worse is that the number of nations "developing" or with "capability" were frequently aggregated with those doing the same for chemical weapons. The facts that one wanted to know explicitly were which nations had BW R&D programmes, which nations had gone into weapons development, and which into production and stockpiling of weapons and the BW agents to fill them. That information was unavailable publicly.[6] In 1993, the Russian Government released a report which identified some nations that had biological weapons programmes; this report was somewhat more explicit in categorising their relative stages of development.[7] A larger study on *Biological Weapons Proliferation* released by two U.S. Government agencies in April 1994 contained only three-and-a-half pages out of 90 with information on specific BW- proliferating nations, and contained little that was not already in the public domain.[8] Notably, in 1994, two senior U.S. Government officials stated in private meetings that no nation was then known to be producing and stockpiling BW agents. Iraq had been doing so, but as a result of the Gulf War in 1991, presumably it had not been doing so since then. Some countries apparently have BW production and assembly facilities, but they have been maintaining them in a standby capacity, and-at least in 1994—were not actually producing BW agents in them. Whether such countries had tested weapons and tested the production lines, etc., again remained unstated, but one would have to presume that they had. However, the official US Arms Control Non-Compliance Statements issued for the years 1994 and 1995 both stated that Iran was *producing* biological agents and had "...Weaponised a small quantity of those agents."[9] "Adherence to and Compliance with Arms Control Agreements," May 30, 1995, US Arms Control and Disarmament Agency, p. 16.

Of those countries that developed BW after World War II to the stage or weapons acquisition, virtually all either acquired all three categories of weapons of mass destruction (nuclear, chemical, and biological), or at least two, and have made attempts at a third:

- The United States, USSR, France, the UK, China, and South Africa procured all three;
- Iraq had chemical and biological and was in advanced development of nuclear;
- Israel has nuclear and chemical; and an offensive BW programme;
- Iran has chemical and biological; seeks nuclear;
- Libya, has chemical; has sought nuclear, for decades, and is seeking biological;
- Syria has chemical and biological;

- North Korea has chemical; sought nuclear, and accepting the Russian assessment, apparently has biological;
- India and Pakistan have nuclear; India.has chemical, and biological are unknown,[10]
- Taiwan has chemical, South Korean chemical is ambiguous, and both had incipient nuclear programmes in the late 1970s.

Israel is omitted from annual U.S. arms control non-compliance statements because it has neither signed nor ratified any of the non-proliferation treaties, including the Biological Weapons Convention. It is also omitted entirely in the U.S. Department of Defense's annual report on proliferation of weapons of mass destruction, *Proliferation: threat and Response.* No mention whatsoever is made of Israel; in fact the country is not even listed in the geographical section of countries in "The Middle East".

According to a statement by former CIA Director Woolsey in 1994, nations developing and procuring BW have usually done so following their procurement of CW, and it has frequently been stated that various Arab states in the Middle East developed chemical weapons because of Israel's possession of nuclear weapons. There are no statements or analyses that have extended this rationale specifically to their development of biological weapons as well, although it is an easy, logical extension to make. Anthony Cordesman commented that, "Nations that are interested in biological weapons are already interested because they offer an alternative to nuclear weapons..." It would not be altogether surprising if one learned that some governmental policy group in these states that had considered or was urging the acquisition of nuclear weapons had spun off the suggestion to develop biological weapons. Nevertheless, nothing is publicly known regarding the policy decisions in these states regarding BW development.

It is interesting to look for a moment at the historical record of allegations regarding national BW programmes, and their eventual resolution:

US allegations between 1976 and the early 1990s of a Soviet programme—and even of a continuing Russian one—proved to have been correct. The Soviet denials were false.

Israeli and other allegations in the late 1980s regarding the Iraqi programme proved to have been correct, and the years of Iraqi denials —both before and after 1990—were false[11]

It is also important to note that the US noncompliance statements, including the classified versions that the administration supplies to US Congressional Committees, have never included a discussion of either South Africa or Israel, although it seems patently clear that the US Government considered the South African BW programme to be an offensive one.[12] Since Israel is not a signatory of the BWC, it is not discussed at all in the US Non compliance Reports. A recent study attempted to provide the maximum information available on those nations alleged to have offensive BW programmes, and a summary table is provided here[13] (see table 1).

TABLE 1: NATIONS HAVING BW PROGRAMMES AT LEAST APPROACHING WEAPONISATION

	U.S. Gov't Arms Control Compliance Reports to Congress (1993, 1995)	Admirals Brooks,[1] Studeman, Trost (1988, 1990,1991); Sec. Cheney, 1990	U.S. and UK Governments (1995)[2]	Russian Federation Foreign Intelligence Report, 1993
Middle East				
Iraq	X	X		
Libya	X	X		X
Syria	X	X		
Iran	X	X		X
Egypt	X			
South/East Asia				
China	X	X		
North Korea		X		X
Taiwan	?	X		
India[4]				?
South Korea				?
Africa				
South Africa			X	
Russia	Ambiguity regarding continuation of offensive programme			

1. "Statement of "Rear Admiral Thomas A. Brooks, USN, Director of Naval Intelligence, before the Seapower, Strategic, and Critical Materials Subcommittee of the House Armed Services Committee on Intelligence Issues," March 14, 1990, p. 54;

"Statement of Rear Admiral William O. Studeman, USN, Director of naval Intelligence, before the Seapower, Strategic, and Critical Materials Subcommittee of the House Armed Services Committee on Intelligence Issues," March 1, 1988, p. 48; "Statement of Admiral C.A.H. Trost, USN, Chief of Naval Operations, before the Senate Armed Services Committee on the Posture and Fiscal Year 1991 Budget of the United States Navy," February 28, 1990; "Remarks Prepared for Delivery by the Honorable Dick Cheney, Secretary of Defense, American Israel Public Affairs Committee, Washington, D.C., June 11, 1990," News Release, No. 294-90, p. 4.

2. The South African Government claims that its programme was disbanded in 1992. Official UK Government statements refer only to "around 10" nations with "or seeking" BW, but do not name any countries aside from the separate identification of South Africa in 1995.
3. *Proliferation Issues: A New Challenge After the Cold War, Proliferation of Weapons of Mass Destruction,* Russian Federation Foreign Intelligence Report, (translation), JPRS-TND-93-007, March 5, 1993.

The Case of Iraq and UNSCOM

During the war between Iraq and Iran there was no serious international response to Iraq's use of chemical weapons against Iran in 1984. Not only did Iraq's use of CW take place penalty-free, but it is widely assumed that the lack of a serious international response served as the stimulus for Iran's development and production of chemical weapons, as well as for the initiation of the Iranian biological weapons programme. The UNSCOM process disclosed that Iraq had been prepared to use BW in 1990-91. Weapons had been filled with agent and had been forward deployed both for aircraft and missile delivery. Iraqi field commanders had been given authority to use the weapons, under particular conditions, in the Gulf war. The targets are assumed to have included Riyadh and Tel Aviv, that is, major civilian population centers, as well as the coalition forces in the northern part of Saudi Arabia.[14] In 1990, Iraq had signed but not yet ratified the BWC. Iraq's BW developments programme had been substantial: investigations had been carried out on 5 bacterial strains, 1 fungal agent, 5 viruses, 4 toxins, and two bacterial strains to use in simulation testing.[15] One estimate of its cost, from UNSCOM sources, was $ 100 million. The sanctions that have been applied to Iraq following the Gulf War occurred *only* due to the circumstances of its partial defeat in that war. Iraq was forced to agree to the terms of UN Security Council Resolution 687 and to the establishment of UNSCOM, which was allotted powers unique in the history of the UN Security Council.

Nevertheless, although UNSCOM's resolve and performance have been superb, members of the UN Security Council—in particular, France, China and Russia, three of the five members of the "Permanent 5"—have wavered on continuing the original requirement that the sanctions be maintained until Iraq met the conditions mandated by the Security Council under Resolution 687. That resolution required Iraq to give up all its weapons of mass destruction, missile delivery systems, the production machinery and facilities for these, and attendant government records and documentation. Iraq was to provide a "Full, Final and Complete Disclosure" regarding its WMD programmes within 120 days of April 1991. In the intervening six and a half years Iraq has successively submitted eight of these documents to UNSCOM; all have been rejected as inadequate, insufficient and openly deceptive. Nevertheless until the altogether damning disclosures in August 1995, following the defection of General Kamel, France, Russia and China and been pressing for an end to the sanctions, and they reinitiated those efforts in 1996 and 1997. Iraq has continued to violate the terms of the UN resolutions requiring it to end and to fully disclose the record of its weapons of mass destruction programmes, but there is little indication of interest to pressure Iraq by these governments.[16] The commitments that Iraq made in June 1996—in fact, no more than re-commitments to its originally agreed conditions—to permit UNSCOM inspectors "unimpeded access" to all sites, proved within one short month to be as worthless as all the "Full, Final, and Complete Disclosures" that Iraq had submitted for five previous years. Once Iraq obtained UN authority to resume limited oil exports, there was every likelihood that it would carry out a programme of tactical delays, repeated negotiations, promises of better behaviour, and almost immediate violation of those "promises," each cycle eating up four to six months of time—and permitting Iraq to continue evading the UN resolutions for a fifth and sixth year and, in all likelihood, indefinitely.[17] Iraq is emboldened to continue in violation of the UN Security Council resolutions as long as it sees support among the Perm 5 despite its unquestioned intransigence.

When the Biological Weapons Convention Review Conference, an event that takes place every five years, was held in the fall of 1996, Iraq was present by virtue of having had to join the treaty regime after 1991. Due to that circumstance, and following United Nations procedures, Iraq was able to prevent agreement by consensus that it be stated that Iraq was a violator of the BWC, and it was able to prevent the presentation of UNSCOM's report to the conference as a formal document. Iraq was able to defend its own violations, and it was further supported by a statement from the Russian delegation to the conference. No other country condemned the nature of those proceedings. It was an egregious example of the counter productive nature of consensus proceedings in the United Nations.

What was altogether predictable continued through 1997. It begân with the UN Security Council's decisions of June 22, 1997. An UNSCOM report in April 1997 was unequivocal in stating that Iraq remained in conscious and intentional violation of the UN resolutions demanding that it end its WMD programmes. Nevertheless, the Iraqi Government continued to intermittently impede UNSCOM inspections in the spring of 1997. Finally, after UNSCOM were denied access to two sites, the UNSCOM Executive Chairman presented a record of Iraq's recent refusals to permit access to its facilities to the UN Security Council on June 19, 1997. He stated that these were "...a clear violation of UN Security Council resolutions" and he requested that the Security Council mandate a "firm reaction" to force Iraq to accede to the UN resolutions. However, China and Russia threatened to veto the most limited sanctions suggested.[18] A "compromise" was negotiated between the US, Russian and French foreign ministers: Sanctions could be placed on the international travel of "categories of Iraqi officials responsible for noncompliance"—in effect, senior Iraqi military and intelligence officers—however the sanctions would not into effect until after a period of four months, on October 11, 1997.[19] Since one can scarcely imagine much international travel at the present time by the individuals targeted, *a trivial and extremely limited sanction was selected, and then its implementation delayed.* What this meant in effect was that Iraq was given roughly ten months in 1997 in which it either had, or could continue to impede UNSCOM inspections without any penalty and during which the partial oil exports permitted by the UN were allowed to continue.

When October 1997 arrived, the Security Council could not even carry out the minimalist pressures that it had ostensibly reached a compromise agreement on in June. After three weeks of deadlocked debate, China, France, Russia, Kenya and Egypt abstained from a vote to invoke the additional sanctions. This was despite the fact the UNSCOM had released a new and devastating report on October 6, documenting Iraq's obstructionist policies, that its latest "Full, Final and Complete Declaration", its eighth, was totally inadequate (in regard to its portions on biological weapons, that it "...was not even remotely credible"), that Iraq had been operating a deception strategy and organisation for years, and the fact that Iraq was hiding and moving about the country the very materials that UNSCOM was trying to track down.[20] On top of this, during the weeks just prior to, as well as during, the Security Council's debate, Iraq forcibly prevented inspectors from visiting several sites. With the indication of his continued protection by 3 of the 5 permanent members of the Security Council, Saddam Hussein waited only 24 hours after the vote on October 23, 1997 to take his next step. Iraq demanded:

- a specified timetable for the end of UNSCOM inspections;
- a scheduled end to the sanctions;
- the right of Iraq to specify locations that UNSCOM would be prohibited from inspecting, which were precisely the sites that UNSCOM was attempting to inspect and that Iraq was forcibly preventing it from seeing;
- an end to U-2 reconnaissance flights, flown on behalf of UNSCOM inspections by the United States;
- the immediate expulsion of US members of UNSCOM inspection teams.

In the month long halt to UNSCOM operations that ensued, UNSCOM was able to record that Iraqi personnel tampered with monitoring devices and relocated equipment that was not to have been moved.[21] All inspections were halted on October 29, US inspectors were expelled on November 13, and all UNSCOM inspectors left Iraq on November 14. Russian Government intercession produced an agreement on November 21 to permit the UNSCOM inspectors to return. However the situation at the end of November 1997 was frozen exactly as it had been a month before, and it was clear that Russian and French Government pressures in the UN Security Council would follow to modify the sanction conditions and revise Resolution 687.

Suggestions already made by France and Russia in the proceeding months had varied from removing the sanctions entirely, removing them in part, or altering the parameters of UNSCOM inspections to reduce their frequency or powers, or ending them in regard to particular weapon categories, that is, impeding them operationally. All were versions of Iraq's demands. France and Russia had been negotiating oil exploration and export contracts with Iraq for most of 1997, and both also hoped to sell major weapons systems to Iraq once the sanctions were lifted. This appeared to be their main priority, rather than the maintenance of the prohibitions against Iraq's weapons of mass destruction, or even the very integrity of the UN Security Council's Resolutions and the ability of the United Nations to maintain its decisions in the face of an unquestioned and persistent violator.

It is certain that nothing at all approximating either the UN sanctions or the UNSCOM process would have come to pass in any other circumstances than in the aftermatch of the war. It is astonishing to realise that the Iraqi Government, although it nominally committed itself to the total destruction of all of its weapons of mass destruction by agreeing to the terms of the UN resolutions, has *willingly* given up over $130 billion in oil export earnings between April 1991 and November 1997 in order to retain the products of its BW programme, instead of providing the conditions which would have led to the withdrawal of the sanctions. Reading the detailed record of Iraq's lying over a period of six years demonstrates an unprecedented display of diplomatic obstruction and deception, on top of considering the significance of foregoing $130 billion in earnings.[22] It is clear that Iraq had intended to see the sanctions lifted while maintaining its BW programme and without complying with the UN resolutions, to the point of threatening the UN with an ultimatum in mid-July 1995, only three weeks before the defection of General Kamel. In June 1996 following the crisis that Iraq had precipitated at that time, Iraq promised "unimpeded access" to *all* sites, as the original UN resolutions required, without exception. But at the end of November 1997, Iraq was insisting that some 80 sites were "sovereign from the beginning", and immune to UNSCOM inspection.

The Magnitude of the Biological Weapons Programme of the Former USSR: Facilities and Personnel

For the past half dozen years or a bit more, the overwhelming practical issue in the area of the conversion of former BW R & D facilities has been the BW R & D institutes of the former USSR. The former USSR's BW R & D infrastructure, as in most other areas of its defense sector, was quite large, with overcapacity in both R & D and production facilities. The exact number of institutes in question is actually unknown, at least to the international arms control community, even now, five years after the dissolution of the USSR. In October 1987, in its first submission of information under the confidence building measures (CBM's) agreed to a year earlier at a Review Conference of the BWC, the USSR reported five BW laboratories under Ministry of Defense control that met the reporting requirements: (1) Leningrad (now St. Petersburg); (2) Kirov; (3) Sverdlovsk (now Ekaterinburg); (4) Zagorsk (subsequently referred to under its pre-1917 name, Sergiyev-Posad), Moscow oblast; and (5) Aralsk, Kzyl-Ordinsky oblast. The Leningrad site was under the jurisdiction of the USSR Ministry of Defense Scientific Research Institute for Military Medicine, the remaining four were under the USSR Ministry of Defense Scientific Research Institute for Microbiology. The documentation also listed fourteen civilian laboratories, for a total of nineteen sites that fit the reporting criteria, including those in Ukraine and what was then Byelorussia. However these fourteen were civilian laboratories, they were *not* reported as BW facilities, and some of them, or portions of them, may not have been. However, the 1987 Soviet documentation was *very* incomplete. Obolensk was not mentioned in 1987, neither among the military nor the civilian institutions, but appeared in the 1989 submission as the All-Union Scientific Institute of Applied Microbiology under the jurisdiction of the USSR Ministry of the Medical and Microbiological Industry. (It presumably met the criteria which should have mandated its inclusion in 1987.) Then in 1992, all the disclosures about the Biopreparat system appeared, both in Russian press sources and in the West, with the result that the number of former Soviet institutions that were involved in BW R & D and production appeared to be between twenty and twenty-five.[23]

In an article published in *Izvestiya* on October 15, 1997, a former Soviet scientist who was an advisor to this organisation for several decades writes of the "birth of a system that included *dozens* of institutes and test ranges, institutions with the most diverse names and completely anonymous institutions, and scientific cities which denied access to outsiders." He referred to it as the "Organisation

P/YA A-1063" (P.O. Box A-1063), which was its early Soviet code name.[24] Later on this organisation was also named Glavmikrobioprom. This was well over an order of magnitude larger than the US programme had been at its height prior to 1969, and when it still included an offensive component. As troubling as the magnitude of the programme was the fact that it was established by the Soviet Government in 1973, and built up following that, precisely in the years during which the USSR had signed and ratified the BWC. In September 1992, then Deputy Foreign Minister Berdennikov called the Bio-preparat system "one of the best-guarded secrets in the old Soviet Union," and one whose operations the Foreign Ministry professed to know nothing about as late as the end of 1992. Its management staff was apparently taken from the Ministry of Medium Machine Building, one of the former USSR's eight defense industrial ministries, and the one responsible for producing the USSR's nuclear weapons. The overall management of the entire BW R & D system appears to have been in the hands of "the 15th Directorate" of the Soviet General Staff.

However, the identified facilities and their total number still appears to have been incomplete, and would apparently have included the following:

(a) the five Ministry of Defense institutes;

(b) some portion of the Bio-preparat laboratories and production sites. These included the "mobilisation capacity", fully outfitted, tested, and ready BW production sites, with weapon filling lines. In Russia's 1992 "Form F" submission, two of the Ministry of Defense facilities are named as production sites: Zagorsk and Sverdlovsk, and it is additionally stated that an unidentified number of additional"...industrial facilities with storage capabilities were created in Glavmikrobioprom... that might also produce biological agents during a crisis." One of these, an extremely large facility at Stepnogorsk, in Kazakhstan, was identified by the United States in 1996. At least three other sites in Russia are also suspected of having had the same function, which would mean a minimum of six BW production facilities. These stood ready to begin production should that be ordered, until late 1991, when the weapon filling lines were removed from these locations, but the rest of the production and processing portions left intact;

(c) some or all of the 12 "Plague Institutes";

(d) some of the Academy of Sciences institutes and the 14 "civilian" R & D institutes listed in 1987;

(e) and perhaps others, some of which have recently been alleged in Kazakstan.

It is possible that the majority of these institutes were primarily involved with research—setting aside for the moment the problem of distinguishing between offensively intended research, and defensive—and that weapon development was relegated to the Ministry of Defense laboratories and to the Bio-preparat system. If that were the case, it would also be important to understand what the organisational and production status was prior to 1973? The answers to these questions are unknown.

In the 1996 Russian submission to the United Nations in accordance with the annual BWTC CBMs, the "Organisation of Russia's programme for R & D on Medical and Technical Measures for Defense Against Biological Weapons" shows the Russian Ministry of Defense having undefined relations with *fifteen* institutions: four directly under the Ministry, four under Bio-preparat, and seven others, including institutes primarily affiliated with the Russian Academy of Sciences, the Russian Academy of Medical Sciences, the Ministry of Health and the Ministry of Medical Products. (In its 1993 declaration, these same institutions were shown under joint organisational matrices of both the Ministry of Defense and the Ministry of Health. Therefore, exactly what these relationships are: management, funding, or contractual is not clear). In the 1993 declaration, Russia had listed five primary facilities, presumably the five under Ministry of Defense control, and twelve others, with a total staff of at least 6,000. As to the total number of personnel that were working in the Soviet BW R&D programme at its peak, the number is unkown. Zilinskas refers to"...more than 25,000 people",[25] While an article in *Science* in November 1997 estimated " ...about 10,000."[26] Thus, the present Russian programme, in 1997, still appears to be several times larger than the US programme ever was, but this may be misleading as it would depend entirely on how many people within each institute were occupied wholly or inpart with BW work.

At the same time, there has presumably been some overall retrenchment—perhaps substantial—in this former system through a combination of intentional and coincidental processes. There was allegedly reported to have been some consolidation in the scope of the USSR's BW effort in the five years between 1987 and 1992. Early in 1992, General Volkogonov, at the time a spokesman for President Yeltsin, declared that "a number of centers dealing with this issue have been closed."[27] If this was in fact the case, exactly which they were, other than the Vozrozhdeniye island test site in the Aral Sea, has never been specified. In September 1992, in conjunction with the announcement of the US-UK-Russian Trilateral Agreement, Russian officials stated that the number of personnel working in military biological programmes would be cut in half, and that funding for the programme would be reduced by 30 per cent[28]. Aside from the fact that there is no way to document that such reductions took place, that was not quite so great a reduction given the available estimates of the number of research personnel in these former Soviet BW facilities.[29]

Nevertheless, due to the sharp drop in R&D funding by the Russian state since 1992,[30] as well as by the other independent states of the former USSR, the overall number of working scientists has dropped substantially, and that has undoubtedly occurred to varying degrees in the former BW R & D institutes as well. Even here, available estimates and numbers vary substantially. In 1995, Deni and Harrington quoted "the number of research workers and science instructors in Russia—initially around 900,000 in 1991..." and that the "Russian Federation State Committee for Statistics... [noted] that between 1990 and 1993, the number of Russians employed in science and technology declined by 27 per cent."[31] A 1997 report in *Science News* quoted a "1991... Soviet...scientific work force of 1.5 million people," with a loss of approximately 600,000, or 40 per cent, of that work force by 1997.[32] The report added that as much as 80 per cent of the 1991 Soviet budget for R & D was "for military projects." Also in 1997, Maurizio Martellini quoted figures relatively close to these: quoting Goscomstat, " ...the total number of employees in the R & D sphere in Russia... shrinks from about 1.7 millions in 1991 to about 1 million in 1996[33]. A Russian researcher recently used still higher figures, stating that from a "scientific workforce" of "about three million" in 1991, "about half" (1.5 million) remained in 1996."[34] This shows a greater percentage drop, but leaves as high a number in 1996 as the previous estimate began with in 1991. A final statistic, again from a semi-official Russian source, and showing the continued disparity of estimates, came from the Center for Science Research and Statistics in Moscow in March 1996. It reported that employment in the science sector fell by one half between 1989 and 1995: the number of full-time personnel in the R & D sector dropping from 2.2 million to 1.1 million (and that budget appropriations for 1995 were 20 per cent of the amount in 1989).

Exactly how the personnel reductions, whatever their actual level may have been, is specifically reflected at the former BW institutions, is not known. A Swedish scientific team visited eight bio-technology-capable facilities in Russia in 1996, some of which had been part of the USSR's BW R & D infrastructure, and some of which had not. They were provided with pre-1990 employment levels of these facilities as well as with their present employment levels. These showed great variations, from virtually no change in employment levels to near-total extinction.[35] How representative the sample is, however, is impossible to say, and how representative it is of the former BW-related institutes, those that were directly affiliated with the Ministry of Defense, or those in the former Bio-preparat system, or under any other jurisdiction, is also impossible to say. There are no known official comparative personnel levels, pre-1990 or pre-1992 and presently, for the All Union Research Institute for Molecular Biology (Koltsovo), the All Union Scientific Institute of Applied Microbiology (Obolensk), the Institute of High Purity Preparations (Leningrad), the Institute of Immunology (Lyubuchany) and others. There are also only isolated and fragmentary reports regarding the numbers of scientists emigrating. Deni and Harrington quote the figure of 9,200 former-Soviet scientists in all disciplines who found jobs overseas in 1992 alone.[36] Most of these went to Western Europe, the United States, and Israel. The vast majority of the reduction in the number of working scientists has been due to what is termed "internal brain drain," the loss of scientists to other professions within Russia. In 1996 testimony to the US Congress, the US Defense Intelligence Agency stated that "Last year, reportedly 20-25 per cent of scientists younger than 45 years old left, from one Russian BW institute alone."[37] However whether the "20-25 per cent" were 5,10, 50, or 100 scientists was not stated, which would make a great deal of difference and it is apparently the lower numbers. More significantly for our discussion, reliable information regarding the emigration of former Soviet-BW researchers to countries of current BW proliferation concern is virtually non-existent. It appears that the number of scientists from the former USSR's BW R & D system that have left Russia—in contrast to the thousands in all disciplines referred to above for a single year—has been only several hundred since 1990, and of these approximately 90 per cent went to western countries.[38]

Conversion: The Conversion of the US BW R & D and Production Facilities: Fort Detrick and Pine Bluff Arsenal

The US Government decision in November 1969 to unilaterally end its offensive BW programme required the conversion of two major facilities, the US Army's Biological Defense Research Center at Fort Detrick, Maryland, and the Division of Biological Operations, the production facility at Pine Bluff Arsenal, Arkansas. It also required the distruction of stored agents and weapons. 300 people were employed at Pine Bluff, at all levels, and 1,800 at Fort Detrick, at all levels, professional, technical, and support.

I have included an appendix which supplies a more detailed examination of the US process, as this was published in an earlier study in 1994, and will therefore only include several points here by way of summary. (See Appendix I) On May 1, 1972, the former Pine Bluff Arsenal was turned over to the Food and Drug Administration (FDA) of the Department of Health, Education and Welfare as a new National Center for Toxicological Research. The entire destruction programme of stocks and munitions was carried out on schedule. Fort Detrick was divided into several major components, all but one of these being transferred to different civilian government agencies, such as the National Cancer Institute, a branch of the US National Institutes of Health. One section remained as USAMRIID, the United States Army Medical Research Institute for Infectious Deseases, which became the US research institute for defense against BW. The entire conversion process took place quite quickly, at little expense, and with minimal problems for personnel to find new professional civilian employment. Many found positions in their relevant disciplines in the facilities that took over parts of Fort Detrick and Pine Bluff, others in commercial or academic institutions. However, it is important to note major differences in the employment and economic context in the United States in 1970 to 1973 in comparison to that in the USSR and Russia from 1990 to the present time:

- The US economy was not suffering the major economic and industrial decline that the USSR and then Russia has undergone since 1990.
- The US Government continued very generous funding to scientific research in general, as well as to the civilian government agencies that took over the physical plant of the converted BW installations. In contrast, the USSR and then Russia have drastically slashed R & D funding to most if not all R & D institutions in the FSU.
- Finally, US researchers were routinely accustomed to moving between positions in government, commercial or academic institutions.

Britain also converted its single biological warfare R & D facility, but its size was much smaller and it had not been doing offensive work for many years. After the end of World War II, Britain's biological warfare laboratory was renamed the Micro biological Research Establishment (MRE). In 1979, four years after the BWC had come into force, the MRS was transferred from the Ministry of Defence to the Department of Health, and became the Centre for Applied Microbiology and Research within the Public Health Laboratory Service. At the same time Britain's Chemical Defence Establishment (CDE) assumed responsibility for work on biological defense as well as setting up a new Defence Microbiology Division, and eventually the entire facility at Porton Down was renamed the Chemical and Biological Defence Establishment (CBDE).

Conversion: 1990 to 1994 in the USSR and Russia. The Beginning of Consideration Within BW R & D Institutes

As a description of this period has also been published previously, I have included it as Appendix II. I will include only one example here. In a press conference given in the spring of 1990 by the Ministry of Defense BW facility in Ekaterinburg, now renamed the Sector for Military Epidemiology of the Microbiology Research Institute, the spokesman listed the following capabilities of the laboratory that could be applied to civilian problems:

- production of nutrients for bacteriological cultures;
- chemical analysis;

- monitoring of urban and industrial air quality;
- assaying and combating damage to metal structures by bacteriological organisms;
- detection of harmful substances in the environment, and other ecological research.

In June 1992 the Deputy Chief of the Center for Science at the same facility listed a slightly different agenda of projects in the civilian sector that could be performed by their staff.

They were reportedly preparing to do the following:

- cleanse maternity hospitals of *Staphylococcus*;
- disinfect poultry farms and railroad cars, removing every kind of fungus or mold;
- destroy chemical weapons and highly toxic waste with their own bacteria, clean up spills of oil and kerosene with micro-organisms that will reprocess the oil into protein that will go in food for fish, [and]
- set up production of the very antibiotics that hospitals are so catastrophically short of.

Two of these proposals made use of the facility's clean-up capacity, turned to commercial use, two had been proposed earlier, and CW detoxification was newly added. In March 1992 it was claimed that work was under way on two of these technologies. Scientists at the Ekaterinburg laboratory had developed a method of eliminating water and soil contamination by fuel oil and other oil products, using micro-organisms that metabolise the fuel oil, and producing a protein product that could be used as a fish food. The process was developed jointly with researchers at the Tyumen Petroleum and Gas Institute. The organisms that degrade petroleum residues have long been known and have been commercialised in the West, and given the degree of oil pollution being reported in the former USSR and in Eastern Europe this should certainly be a viable project.

It clearly is not a great task to arrive at research menus for former BW R & D institutes. In the period of political "detente" between the US and USSR, in the early 1970's, a wide range of scientific exchange programmes were established between the US and USSR Several pertained particularly to microbiology, virology and biotechnology. This agreement led to the *US-USSR Joint Working Group on Production of Substances by Microbiological Means.* Following the meeting of the US-USSR Joint Commission on Science and Technology, in March 1973 in Washington, DC, it was agreed to develop cooperative projects in six broad areas, one of which was "Microbiological Synthesis", which in turn established six project areas:

1) Development of Technology for Industrial Production of Food and Feed Proteins by Mircobiological means;
2) Engineering Research and Development of Instrumentation and Methods for the Computerised Simulation, Design, and Control of Processes of Microbial Technology;
3) Molecular Biology of Industrial Microorganisms;
4) Immobilised Enzymes (Development of Methods of Producing and Using Enzymes and Other Biologically Active Substances for Agricultural, Industrial and Analytical Purposes);
5) Microbial Control of Pests of Agricultural Crops;
6) Geomicrobiological Research.

The US side was interested in basic R & D, while the Soviet counterparts—the State Committee for the Microbiological Industry (later to become the Ministry of the Microbiological Industry—were interested in the syntheses and applied work.

Conversion: Russia, 1994 to the Present

There is no systematic record of what is occurring in each and all of the former USSR's BW R & D institutes. One could additionally attempt to survey what has happened in various of these institutions in the past half dozen years through:

- local production initiatives;
- commercial joint ventures with foreign pharmaceutical corporations;
- funding for research provided through external grant funders:' The International Science and Technology Center (ISTC) and the Science and Technology Center in Ukraine (STCU), and the International Science Foundation (Soros).

Two tables of summary data regarding the last of these alternatives are included below:

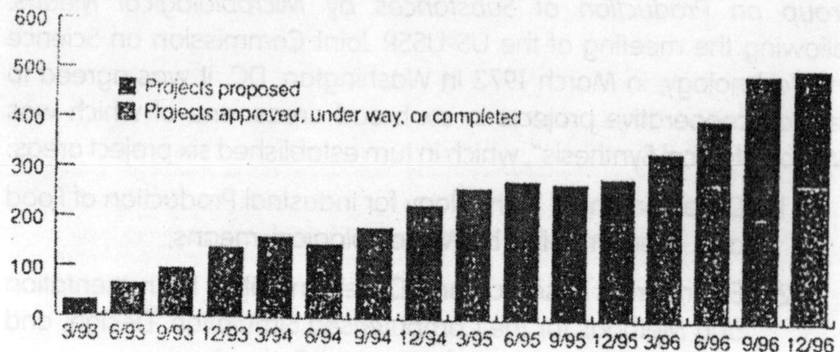

The numbers of U.S.-FSU collaborations funded by the ISTC

EAST-WEST R & D COLLABORATIONS

Programme	Funds for Projects ($ millions)	FSU Scientists Active Projects	Supported
ISTC	121	327	15,400
IPP	109	150	2,200
INTAS	69	1204	15,000
ISTC-Ukraine	10.4	87	1,650
CRDF	8.2	257	1,400

ISTC: International Science and Technology Center (E.U., Japan, Russia Sweden, U.S.)

IPP: Initiatives for Proliferation Prevention (U.S.)

INTAS: International Association for the Promotion of Cooperation with Scientists from the New Independent; states of the FSU (E.U.)

ISTC: Ukraine (Canada, Sweden, Ukraine, U.S.)

CRDF: Civil and R & D foundation for the Independent States of the FSU (U.S.)

Conversion of BW R & D Facilities, General Considerations

When the United States was preparing to convert Pine Bluff Arsenal in 1969-70, a study paper proposed a large group of research projects that might be undertaken in the future. These fell into two groupings. One group was clearly based on new priorities and needs, and concerned pollution and pesticide problems, recycling and food production. These included; (a) selective biological destruction of undesirable plants; (b) viral control of algae; (c) disposal of toxic wastes by micro-organisms; (d) destruction of pesticides by micro-organisms; (e) control of oil spills with micro-organisms; (f) utilisation of waste material; (g) production of single-cell protein food supplements, amino acids and growth factors; and (h) production of cellulose and other enzymes.[39]

A second group of research topics, however, was either much more closely related to and/ or made more particular use of Pine Bluff's existing specialised facilities: (a) microbial insecticides; (b) viral and rickettsial vaccine production; (c) production of bacterial and fungal vaccines and toxoids; (d) aerosol immunisation; (e) development of decontamination system; (f) aerosol applications of air pollution control; and (g) development of rapid, automated bioanalysis techniques.

Carl Göran Heden pointed out that these topics "belong to the areas... where the reciprocity of international cooperation and fellowship programmes might serve a verification purpose", as it is this group that comes closest to classical biological warfare research. He indicated that "it would be unwise to initiate one of the projects: 'production of toxins, specific bacterial pathogens, viruses and rickettsiae', before such international cooperation had started".[40] Twenty-five years later, Heden again commented on the conversion of BW R & D capabilities ("redeployment" in his phrase) in a remarkable paper of nearly prescient insights. He specified areas of work beneficial to developing nations as "...methods to upgrade foods, practice biological pest control, recycle organic materials, improve soil structures, detoxify chemicals, and improve recovery of mineral resources."[41] The clear and absolute avoidance of anything having to do with "dual threat" agents is obvious.

The important context is the intent to convert facilities in which military R & D has been carried out, after weapon development of that type is renounced. If existing biological warfare facilities are converted, it is clearly beneficial from the disarmament and arms control standpoint the more "distance" there is in the projects chosen for the converted facility from those that might have biological warfare relevance as well. Thus *from this standpoint* research in production of cattle feed from paper waste, or of fertiliser, cattle feed or single—cell protein from mixed waste—all processes in which there already existed research in 1972—would clearly have been preferable to projects in biological methods of pest control or in aerobiological aspects of disease spread or control, and even more so to viral and rickettsial production, however useful knowledge or products in the latter areas might be. It would simply allay suspicion of intent. This would be so despite the fact that research in biological methods of pest control might already be under way in the laboratories of an agriculture ministry in the same country, and that such research would continue. It would equally be so even if a military agency possessed the ability to monitor and attempted to adapt such research results in the absence of any other analogous research of its own. This example in fact points to the explicit formal relationships that underlie judgements of intent in areas of prospective weapon development.

It is for these generic reasons that recommending projects on "dual threat agents", for BW R & D institutes seeking conversion pathways is extremely undesirable. If one rationale for these proposals was, as it was claimed, their intended benefit to international public health, and most particularly in developing countries, then these were the wrong agents and the wrong diseases to focus on. Instead, the diseases listed by the WHO as the largest killers—malaria, schistosomiasis, leprosy, tuberculosis, etc.—are the ones to which work should be directed. There is literally *no* overlap in the two lists of agents—"dual threat agents" and international public health killers. One has only to look at two tables produced each year by the World Health Organisation (WHO): (1) *The World's Most Prevalent Diseases*, which tabulates the number of people infected globally, and (2) *The World's Most Common Cause of Death from Disease*. Each table begins with a category of "vaccine preventable diseases." Other statistical compilations which facilitate the same selection are the WHO's "Infection-Associated Child Mortality," and data on the incidence of tropical and parasitic diseases. In 1993, 267 million people had malaria. (In 1994, the WHO indicated that between 275 and 490 million people suffered from malaria.) Two hundred million had schistosomiasis, about 100 million had leprosy, 89 million lymphatic filariasis, 17 million river blindness, 17 million Chagas disease. In 1990, 1.7 billion people were infected with tuberculosis, 49 million with measles, and 46 million with pertussis. Nearly 40 per cent of the world's population lives in regions where malaria is endemic, and the disease is currently spreading into areas that have been malaria-free for decades. Mosquitoes transmit disease to at least 700 million people each year, about one-eighth of the world's population, and cause about 3 million fatalities per year. Malaria alone is responsible for 25 per cent of childhood deaths in the third world. A consortium of the world's leading pharmaceutical companies recently rejected a proposal to join public and private organisations in research on malaria and other tropical diseases.[42] At the same time, a major portion of present US Government research for a potential malaria vaccine is funded through the medical laboratories of the US Navy and the US Army.[43] Obviously, this should therefore be a prime candidate for US Government funding for FSU BW R & D institutions.

If the second rationale is to "convert", end, or reduce the BW R & D capacity of present-day Russia, or of any other country, then work on "dual threat agents" should be the last thing in the world that anyone should want to sponsor work on. Simply put, it is the most certain way to keep the maximum research personnel working on exactly the same BW agents that they had been working on previously.

More recently, and for exactly the same reasons, a proposal in November 1997 by a committee of the US National Academy of Sciences (NAS), suggesting that projects dealing in particular with anthrax, plague, viral hemorrhagic fevers, and two other directly BW-relevant agents be funded in former BW R & D institutes in Russia is an extremely unfortunate mistake.[44] A simple question to the Director and the Deputy Director, respectively, of two of the relevant institutes in Russia immediately obtained the response that of course they would have *no* objection to obtaining research funds to work on malaria, or other third world parasitical diseases that kill millions annually. But in any case a *large* number of research problems in molecular biology that do not deal with disease agents at all could have been specified, problems which require the most advanced competence in molecular genetics, as well as having direct applied and commercial possibilities.

The remark of Dr. Peter Jahrling, as a justification for the NAS panel's choices, that "Any one of these projects that doesn't entail looking at Paramecia in pond water has a certain level of risk" is simply not true. The November 1997 issue of *Scientific American* contained in the single issue articles on:

- Using genetic engineering to protect rice from disease. The author pointed out that when she began her work on this problem in 1990, no one had ever succeeded in cloning a disease-resistant gene from any plant. The blight-resistant segment of the rice DNA was cloned in bacteria.[45]
- The beginning of field trials of human antibodies produced by field crops—corn in this case—against tumor cells; it is hoped that the plant antibody will avoid human rejection caused by cloning done in mouse cells.[46]

If a single issue of *Scientific American* can provide two such examples of the highest caliber molecular genetics—both with obvious commercial application—then there should be no problem for the members of an NAS panel or the molecular geneticists at any of Russia's former BW R & D institutes to come up with equally valid research topics to make use of their talent that are very fare from "Paramecia in pond water," and equally far from BW relevant organisms. As if to emphasise this with pointed irony, a report in the general press in December 1997 described the advanced DNA research being carried out on Oxytricha, a protozoan cousin of "Paramecia in pond water."[47] Far more to the point, research on developing a malaria vaccine has recently focussed on DNA vaccines, one of the most advanced techniques available in the field.[48]

In the table below I have combined various subject areas that were suggested in the US in 1970, and in the USSR/Russia in 1990-1994:

(1) nitrogen fixation of non-leguminous food crops, particularly grains and corn (Success with bacterial inoculation of non leguminous crops was reported in the USSR by the mid 1960's. These reports may not have been valid, nevertheless this would be an area of research of inestimable value to global agricultural production, agricultural economics, and the continued viability of both agriculture and the earth which sustains it.)

(2) drugs and pharmaceutical products

(3) remediation of oil pollution

(4) remediation of other environmental contamination (heavy metals, etc.)

(5) microbial insecticides

(6) microbial herbicides

(7) anti-food spoilage

(8) anti-industrial and raw material microbial destruction

(9) malarial, TB, or other vaccines for diseases on the WHO list.

Conversion is entirely feasible. There is probably no other area of defense-related R & D or defense industrial production in which conversion should be as easy. Facilities require little modification and personnel virtually no retraining. Microbiologists, virologists, molecular geneticists, biochemists do not need to be retrained when they turn to a new research problem or work with a new organism; that is part of what they routinely do throughout their careers. Finally, the national need for the converted personnel and products is supremely manifest in Russia and in all of the former Soviet Union, unquestionably more so than anywhere else in the industrialised world. There is no question that researchers and institutes seeking conversion possibilities have other constructive and economically viable alternatives to which to turn.

Conversion: National Policy and Governmental Prerequisites

There are several preliminary steps which government authorities must take in any nation undertaking the conversion of BW R & D facilities:

1. Most obviously, end all aspects of offensive BW work, both research and development. It is equally clear however that not all states are interested in conversion; the most blatant example being Iraq. There are also the doubts that have dogged Russia since 1992, where the motivation of defense agencies may differ from their research affiliates or at least some of them.
2. The institutes should by and large be removed from control

by military or intelligence ministries. They should be placed under civilian control. The Biological Weapons Conventions permits the retention of research on defensive aspects of BW, and facilities for that purpose. However such facilities should also be open for examination at any time, as the remaining US facilities at Fort Detrick have been since 1972. Civilian control is also crucial to providing full availability of information to the government. The former South African BW programme was an entirely covert, secret programme—as had also been those in the USSR and in Iraq—and was directed by a branch of the military services.

3. All of the institutes in question should be open to international scientific visitors, and secrecy should be removed from any non-military institute.

These are issues of secrecy, transparency, and the sources of government funding and direction. The above three changes are prerequisites for any real conversion: (a) a total end to offensive programmes; (b) a complete turnover of the laboratories to civilian control, and no covert programmes withheld from government knowledge; and (c) an end to secrecy, and complete access to all the facilities by the international scientific community.

APPENDIX 1

The Conversion of US Biological Warfare Facilities

The control of BW provides a very rare case of the recorded conversion of military R & D facilities. This instance is all the more interesting because the nation concerned was producing biological weapons. The USA closed and converted its two main BW R & D facilities, Fort Detrick and Pine Bluff Arsenal, subsequent to the renunciation of the US military biological warfare capability.

Britain also converted its single biological warfare R & D facility, but its size was much smaller and it had not been doing offensive work for many years. After the end of World War II Britain's biological warfare laboratory was renamed the Microbiological Research Department. In 1957 it terminated work on the development of offensive BW programmes and the facility was again renamed as the Microbiological Research Establishment (MRE). In 1979, after the BWC had come into force, the MRE was transferred from the Ministry of Defence to the Department of Health, and became the Centre for Applied Microbiology and Research within the Public Health Laboratory Service. At the same time Britain's Chemical Defence Establishment (CDE) assumed responsibility for work on biological defense as well by setting up a new Defence Microbiology Division, and eventually the entire facility at Porton Down was renamed the Chemical and Biological Defence Establishment (CBDE).[49]

In the United States the process began in March 1969 when the US National Security Council began a major review of US policy concerning biological warfare. The government agencies participating in the review were the Department of State, Department of Defense (DOD), Arms Control and Disarmament Agency (ACDA) and the Office of Science and Technology. Comments were also received from the scientific community and evaluated by the President's Science Advisory Committee. Pending the outcome of the study, the Department of the Army ceased all productions of toxins and biological agents and the filling of dissemination devices with these agents on 15 August 1969. On 25 November 1969 the President issued an announcement of US policy regarding biological warfare which included the following:

1. The USA would renounce the use of lethal biological agents and weapons and other methods of biological warfare;
2. The USA would confine its biological research to defensive measures such as immunisation and safety measures;
3. The DOD would prepare recommendations for the disposal of existing stocks of bacteriological weapons.[50]

On 14 February 1970, a White House announcement extended the policy to military programmes involving toxins, whether produced by biological means or chemical sythesis, and directed the destruction of toxin weapons and stocks which were not required for defensive research programmes.

There were two primary results of these policy decisions: first, "demilitarisation", the name given to the programme to destroy BW stockpiles, and second, the conversion of BW R & D and production facilities. In addition, the US Congress passed legislation which mandated the preparation of a public semi annual and later annual report to the Congress by the DOD on funds obligated in chemical warfare and biological research programmes. Subsequently, when annual Arms Control Impact Statements were initiated by the US Government in the late 1970s, these routinely included in analysis on chemical warfare.

Before turning to an examination of the conversion experience, it is useful to provide a brief description of the "demilitarisation" programme and the subsequent public reporting of the information on R & D activities in the chemical and biological area.

Plans for the destruction of BW stockpiles were approved by December 1970.[51] Stocks were located and would be destroyed at four separate sites. The destruction plans were reviewed by officials from several different federal and state agencies, including the US Department of Health, Education and Welfare (HEW) and the Department of Agriculture. Observers from both of these agencies were also appointed to monitor the entire programme. The agent destruction was carried out with substantial publicity which included press briefings, information releases and public tours.

On 1 May 1972 the former Pine Bluff Arsenal was turned over to the Food and Drug Administration (FDA) of the Department of HEW as a new National Center for Toxicological Research (NCTR).[52] The entire destruction programme was carried out on the schedule outlined in table 8.1.

For some years prior to 1973, the DOD supplied Congress with a semi-annual report on its chemical and biological warfare research, development, training and evaluation (RDT &E) programme. Beginning with the report transmitted in November 1973, classified information was removed from the report so that it could be released publicly.[53] This would presumably enable any interested foreign nation to satisfy itself that the United States was adhering to the restrictions that had been established on its biological warfare R & D programme.

TABLE: DESTRUCTION OF US BIOLOGICAL WARFARE AGENTS

Facility	Stored agents	Destruction completed during
Directorate of biological Operations, Pine Bluff Arsenal	Antipersonnel	10 May 1971 - 1 May 1972
Beale Air Force base	Anticrop	By 10 March 1972
Rocky Mountain Arsenal	Anticrop	By 15 February 1973
Fort Detrick	Anticrop	17 January - 18 May 1972

Subsequent reports were published in the *Congressional Record,* at first biannually, and after a congressional modification in the reporting requirements, annually.[54] A further mechanism which enhanced the transparency of the US chemical and biological R & D programmes were the *Arms Control Impact Statements* submitted to congress beginning in 1978.[55]

The two major facilities affected by the US decision of November 1969 were the Army's Biological Defense Research Center at Fort Detrick, and the Division of Biological Operations, the production facility at Pine Bluff Arsenal. The disposition of the biological complex at Pine Bluff (which was physically distinct from a CW production area that was also in operation at the arsenal) turned out to be the simpler of the two.

It consisted of 33 buildings, in operation since around 1953, at which about 300 people were reportedly involved in as many as 40 R & D programmes.[56] The solution of engineering problems associated with the safety and control of the production processes necessary to culture and store infectious micro-organisms and toxins had resulted in very specialised facilities. Nevertheless on January 12, 1971, very soon after the announcement of the approved plans to destroy the biological stockpiles at Pine Bluff, the Commissioner of the FDA wrote to the Secretary of the Army proposing the FDA take over the installation and establish a National Center for Food and Drug Safety at the site. Pine Bluff Arsenal had by then already been visited by FDA personnel, who found that it could be adapted to the needs of their agency. The new center would "examine the biological effects of a number of chemical substances which are found in man's surroundings, such as pesticides, food additives, and therapeutic drugs. Pine Bluff Arsenal would become a major research center for the purpose of studying the significance to man of toxicological data obtained from animal testing". The main effort would be a very large-scale animal screening programme. FDA officials expressed their intention of employing all of Pine Bluff's professional personnel and most, if not all, of the other personnel. The new center was named the National Center for Toxocological Research and conversion was completed in an opening ceremony in the spring of 1972.

Determining what new research would be undertaken at Pine Bluff Arsenal was free of the decision-making elements that introduced complications at Fort Detrick. Aside from the main effort in animal testing, which seemed clearly established as the primary programme, a study paper proposed a large group of research projects that might be undertaken in the future. These fell into two groupings. One group was clearly based on new priorities and needs, and concerned pollition and pesticide problems, recycling and food production. These included: (a) selective biological destruction of undesirable plants; (b) viral control of algae; (c) disposal of toxic wastes by micro-organisms; (d) destruction of pesticides by micro-organisms; (e) control of oil spills with micro-organisms; (f) utilisation of waste material; (g) production of single-cell protein food supplements, amino acids and growth factors; and (h) production of cellulose and other enzymes.[57]

A second group of research topics, however, was either much more closely related to and/or made more particular use of Pine Bluff's existing specialised facilities: (a) microbial insecticides; (b) viral and rickettsial vaccine production; (c) production of bacterial and fungal vaccines and toxoids; (d) aerosol immunisation; (e) development of decontamination systems; (f) aerosol applications of air pollution control; and (g) development of rapid, automated bioanalysis techniques.

Heden has pointed out that these topics "belong to the areas... where the reciprocity of international cooperation and fellowship programmes might serve a verification purpose", as it is this group that comes closest to classical biological warfare research. He indicated that "it would be unwise to initiate one of the projects: 'production of toxins, specific bacterial pathogens, viruses and rickettsiae', before such international cooperation had started".[58]

It is not clear to what degree, if at all, any of these proposed projects were subsequently carried out at the new NCTR[59] The important context is the intent to convert facilities in which military R & D has been carried out, after weapon development of that type is renounced. If existing biological warfare facilities are converted, it is clearly beneficial from the disarmament and arms control standpoint the more "distance" there is in the projects chosen for the converted facility from those that might have biological warfare relevance as well. Thus *from this standpoint* research in production of cattle feed from paper waste, or of fertiliser, cattle feed or single-cell protein from mixed waste all processes in which there already existed research in 1972—would clearly have been preferable to projects in biological methods of pest control or in aerobiological aspects of disease spread or control, and even more so to viral and rickettsial production, however useful knowledge or products in the latter areas might be. It would simply allay suspicion of intent. This would be so despite the fact that research in biological methods of pest control might already be under way in the laboratories of an agriculture ministry in the same country, and that such research would continue. It would equally be so even if a military agency possessed the ability to monitor and attempted to adapt such research results in the absence of any other analogous research of its own. This example in fact points to the explicit formal relationships that underlie judgements of intent in areas of prospective weapon development.[60]

At the time of the US decision in November 1969 to curtail its biological warfare R & D programme there were roughly 1800 professional, technical and support personnel employed at Fort Detrick.[61] It was a very much larger facility than Pine Bluff Arsenal with 460 individual structures and a value of $190 million. The Army estimated that the number of people required and the basic operating costs of the Fort Detrick facilities—300 people and $6 million annually—were alone equal to the total involved in the transfer of Pine Bluff to the FDA. An additional complication was that the Army had originally hoped to retain all of Fort Detrick's research personnel—which turned out not to be feasible—and that while several different federal agencies expressed interest in the laboratories none of them seemed willing to allot the funds required for the conversion out of their existing budgets.[62] The final resolution was that some portions of the military programme were moved from Fort Detrick, some remained, and several new research institutions arrived with the result that after 1972 some half dozen federal programmes shared the facilities:

1. 240 civilian and 190 military personnel were relocated to the Dugway Proving Grounds in Utah. This represented 19 to 24 per cent of the research personnel in the original installation.[63]

2. The Analytical Science Office and the Biological Defense Material Division studying physical defenses against BW, representing some 40 research personnel, were transferred from Fort Detrick to Edgewood Arsenal, Maryland.

3. Civilian personnel, equipment and facilities of the Plant Sciences Directorate of Fort Detrick were transferred to the US Department of Agriculture in April 1971 to continue the work on defense technology against crop disease, in

accordance with a recommendation made by the President's Science Advisory Committee. This involved 20 researchers.

4. Research in the field of medical defenses against biological agents was transferred to the US Army Medical Research Institute of Infectious Diseases (USAMRID) which remained at Fort Detrick. IN 1977 it had 461 employees.
5. The US Navy also continued to maintain a small medical research detachment at Fort Detrick.
6. The President's decision in October 1971 to establish a branch of the National Institutes of Health (NIH), Department of HEW, at Fort Detrick reflected purely domestic political considerations rather than scientific priority.[64] Once the decision had been made on the major area of new work to be done at Fort Detrick, it was considered desirable for economic reasons to retain as much of the equipment and performance capacities of the former laboratory as possible. Thus the National Cancer Institute concluded that the principal activities for which the facility was suited were: first, the large-scale production and testing of viruses in cancer research, and second, the breeding and location of animals to be used in large-scale screening programmes. The cancer research programme was to be managed by a contractor, and was expected to employ 200 people.

By early 1972, most of the professional staff not absorbed by these various programmes had sought positions elsewhere, and only some 600 support staff remained. As part of the conversion process, it was arranged for two Soviet officials to visit the facilities in August 1972. These were the USSR Minister of Health and the head of a cancer research institute.[65] The visitors portrayed themselves as unimpressed by what they were being shown, and could not reply to a question of whether the Soviet equivalent(s) to Fort Detrick were also being converted. The Soviet Minister of Health replied that he could only state that his ministry had no such facilities, which was at one and the same time a partial truth and a gross evasion. The USSR's Ministry of Health was responsible for the "medical" aspects of the R & D that took place in the USSR's biological warfare R & D laboratories. The laboratories themselves were under the authority of a special section of the General Staff of the USSR, the 15th Directorate, but it was none other than the Deputy Minister of Health of the USSR who was the liaison with them.[66] An offer, first made by President Richard Nixon in October 1971, that several laboratories at Fort Detrick be set aside for use by foreign scientists, including scientists from the USSR, was repeated, but there was no response to this. Official representatives of Soviet ministries—including the then Deputy Minister of Health of the USSR—again visited the laboratories in 1988, at the time of their visit to the US National Academy of Sciences to present the Soviet version of what happened during the epidemic of anthrax in the Soviet city of Sverdlovsk in April to June 1979.[67]

In February 1970, early in the conversion process, a senior US Government official apparently implied that all future defensive biological warfare research would be carried out on an unclassified basis, and was quoted as saying "there will be no need for secret research in this field under this programme"[68] That this was obviously not the case was quickly realised, and either the original reports were incorrect or any such intention was revoked. Classified biological warfare research was still going to be carried out, which rapidly led to criticism that the Army had not reduced or altered the types of research it was doing after the President's renunciation of biological warfare and the announced closing of Fort Detrick.[69] Testimony to Congress, however, indicated the degree of change that had taken place in funding levels.

"Immediately following the President's policy announcement of 25 November 1969, all offensive biological research and development was terminated. The defensive effort has also been reduced as the following figures show (in millions of dollars).

Fiscal year	**1969**	**1970**	**1971**	**1972**
Biological offense	3.158	3.964	0	0
Biological defense	18.378	14.376	15.585	13.200
Total	21.536	18.340	15.585	13.200

The portion of the programme comprising the technological base is unclassified and will remain at Fort Detrick. This will require approximately 160 civilian personnel. The Army considers defensive work in biological alarms and physical defense, and vulnerability analysis to be classified since knowledge of alarm characteristics or US vulnerability to attack would be an invaluable asset to a potential enemy. Consequently, 52 civilian spaces will be transferred to Edgewood Arsenal for the alarm and physical defense work, and 9 civilian spaces will be transferred to Dugway Proving Ground for the Vulnerability analysis work."[70]

From the point of view of arms control, the combination of the conversion of the BW production facilities *as well as* the cessation of the "offensive" R & D programme is almost the best thing that could possibly have happened. Perhaps the very best would have been if it could have taken place even without the stated maintenance of "defensive" research. In the mid 1980s (under the Reagan Administration), US defensive BW research was sharply expanded and substantive questions were raised as to whether particular research projects were of an "offensive" or "defensive" nature. It may also be considered advantageous that the "defensive" biological warfare R&D was not transferred to an existing civil installation, since in that case all the problems of distinguishing between "military" and "civilian" research would descend in turn on the civilian institution. On the other hand there would presumably be much less suspicion of "offensive" research being carried out under a "defensive" guise in such circumstances. In other cases in which limitations on military R & D are conceivable, that is, in regard to weapon systems other than BW, a reduction in production capability might not necessarily be expected. Conversely, in cases in which there were decisions not to produce and deploy a particular system—such as the anti-ballistic missile (ABM) system—continued ABM R & D was frequently supported even by those who strongly opposed the procurement of ABM systems.

APPENDIX II

Soviet and Russian Statements, 1990 to 1994, on Capabilities, Intentions, and Plans for Facility and Research Conversion

What is currently known about relevant conversion capabilities in the former USSR, or what kinds of alternative work have some of these institutes already themselves suggested or embarked on?

1. The USSR had a very large biotechnology programme. Roger Roffey refers to "a national committee and a national programme for biotechnology [which] encompassed 150 institutes in the former Soviet Union".[71] A portion of that programme consisted of extremely large factories for the production of single-cell protein that was packaged for use as livestock food, for cattle, hog and chicken feed.

2. The institute at Obolensk, in addition to its work on BW, was occupied in "the struggle against plant pests and efforts to develop new medicines", and "At present people there are urgently trying to re-quality for the production of medical preparations".[72]

3. For Koltsovo (which admitted to having been "a secret institution" between 1974 and 1990, but claimed not to have been "a BW laboratory"), its product line in vaccines, medical diagnostic aids and molecular reagents is so wide as to scarcely need any change other than to increase its production of these commercial products and to develop its own marketing network within the former Soviet Union. The same holds for its sister institute in Berdsk.

4. In a press conference given in the spring of 1990 by the Ministry of Defense BW facility in Ekaterinburg, now renamed the Sector for Military Epidemiology of the Microbiology Research Institute, the spokesman listed the following capabilities of the laboratory that could be applied to civilian problems:
   - production of nutrients for bacteriological cultures;
   - chemical analysis;
   - monitoring of urban and industrial air quality;
   - assaying and combating damage to metal structures by bacteriological organisms;
   - detection of harmful substances in the environment, and other ecological research.[73]

In June 1992 the Deputy Chief of the Center for Science at the same facility listed a slightly different agenda of projects in the civilian sector that could be performed by their staff.

They are preparing to do the following:

- cleanse maternity hospitals of *Staphylococcus*;
- disinfect poultry farms and railroad cars, removing every kind of fungus or mold;
- destroy chemical weapons and highly toxic waste with their own bacteria, clean up spills of oil and kerosene with micro organisms that will reprocess the oil into

- protein that will go in food for fish, [and]
- set up production of the very antibiotics that hospitals are so catastrophically short of.[74]

Two of these proposals made use of the facility's clean-up capacity, turned to commercial use, two had been proposed earlier, and CW detoxification was newly added. In March 1992 it was claimed that work was under way on two of these technologies. Scientists at the Ekateriburg laboratory had developed a method of eliminating water and soil contamination by fuel oil and other oil products, using micro-organisms that metabolise the fuel oil, and producing a protein product that could be used as a fish food. The process was developed jointly with researchers at the Tyumen Petroleum and Gas Institute.[75] The organisms that degrade petroleum residues have long been known and have been commercialised in the West, and given the degree of oil pollution being reported in the former USSR and in Eastern Europe this should certainly be a viable project. Research was reportedly also under way on the bacterial degradation of the active agents in chemical weapons.

5. Perhaps of greater immediate significance and need, in a country where virtually every single hospital in every city and town claims a near-total absence of antibiotics, there should be little problem in finding immediate useful application for former biological warfare R & D laboratories. There appears to be atleast some attempt to move in this direction, but antibiotics production is clearly needed on a massive scale.[76] Any medium or large-scale fermentation capacity could be immediately put to use: "to avoid unemployment... the scientists in Obolensk established ten small enterprises for the production of medicines that are very short supply."[77]

The same practice has been followed at Koltsovo: the staff has been divided into smaller production units and shareholder companies, and these have funds with which to procure the raw materials needed for production.

The point is that biological conversion does not necessitate major expenditure. Strains, serums, vaccines, and other output developed previously can with comparative ease become the basis of new medical preparations. The money spent on biological weapons research will not have been wasted, it will start to work for people.

Such programmes already exist. The Yekaterinburg [Ekaterinburg] Center's buildings and laboratories are being transferred to medicine production. It is already ready to produce medicine. A special programme is also being formulated for the institute at Sergiyev-Posad, which develops vaccines that are of great interest to foreign countries. Scientists at the center in Kirov are also working successfully on conversion and are providing their city with great assistance in creating its agro-industrial complex.[78]

In a June 1992 interview General A. T. Kharechko, then the head of the Ekaterinburg laboratory, reported that "the cantonment is now geared exclusively for medical needs and is going to produce antibiotics and artificial blood substitutes",[79] and in April 1992 the General Staff claimed that "We have an entirely realistic plan to set up a production facility for antibacteriological agents based on the Yekaterinburg [Ekaterinburg] Center"[80] The problem is that these various statements were at best imprecise and more frequently contradictory.[81] It is not possible to know what is a "plan" and what is now "geared... to produce". As late as October 1993 General Kuntsevich was still using phrasing that appeared to be anticipatory of first "formulating programmes": Now we are formulating programmes to retool the relevant scientific centers, former participants in military-biological programmes, for general scientific activity to supply our population with effective medicines".[82] In an earlier statement in September 1992, he had them already so occupied: "This facility [Ekaterinburg], like similar centers in Kirov and Sergiyev Posad, are now only working on creating means of protection against biological weapons...Apart from the protective means, the facilities are today engaged in developing new highly effective medicines".[83]

In contrast to all of the above, General Yevstigneyev in April 1992) claimed that "the [Ekaterinburg] plant that produced vaccines and diagnostic substances was closed down"[84]. As if to display these contradictions to the maximum, General Kuntsevich stated that "Having invested our scientists and specialists efforts in this, we are preserving a great intellectual and scientific and technical asset", and only a single sentence later, "Now that the equipment has been completely dismantled and the resources that previously were used for military purposes have been destroyed..."[85] Few competent observers have been to any of these facilities and there is no real understanding of which of them is doing what, either of previous work or of "conversion" projects. US and British Government officials that have visited some of these facilities have not seen fit to make any public report which would help understand exactly what the institutes are currently doing. Similarly, Western pharmaceutical companies that have visited at least some of the sites since 1990 investigating commercial possibilities, of course, do no make public reports about their visits either.

Biological warfare R &D laboratories that formerly produced vaccines—the Ekaterinburg laboratory was one of these—could also produce vaccines for domestic livestock; cattle, hogs and poultry are all routinely vaccinated against various diseases. Other projects could also be applied to the production of protein feed and fertiliser utilising bacterial cultures which transform paper waste to cattle feed or to fertiliser. Pilot plant installations for these were developed as long ago as the mid-1960s in various US universities. For each of the sites a list of between five and ten alternative research or production priorities should be drawn up. These would fall into two broad categories: (a) the production of commodities and services which the institutes by their nature can produce and may already have produced in the preceding years, and which are in the most dire need in Russia and in the other new republics of the former USSR, and (b) interdisciplinary problem-oriented research in the fields of public health and ecological and environmental problems.

The entire territory of the former USSR—as well as most of Eastern Europe—suffers from gross pollution in every category: air, water, land and, over large areas, also radiation.[86] Any products these institutions could produce that served environmental protection or recovery would also have immediate application.

Bibliography

1. Barry J.Erlick (Department of the Army) in *Global Spread of Chemical and Biological Weapons*, Hearings, Committee on Governmental Affairs, U.S. Senate, 101st Congress, 1st Session, February 9, 1989, page 33.
2. John Holum, addressing parties to the BWC on November 1996, quoted in OMRI Daily Report, November 27, 1996.
3. Jordan, Lebanon, and Saudi Arabia are parties to the BWC. Egypt and Syria have signed but not ratified. Iraq ratified only after the end of the Gulf War and the UNSCOM process began. Israel has neither signed nor ratified. Iran has ratified, but is widely assumed to be developing biological weapons, and

to be in violation, as Iraq was previously. It was after the failure of that diplomatic effort that the Bush administration inserted a few sentences on the BW capabilities of several of the Middle Eastern states in the non-compliance report that it released in January 1993. Israel, however, was not mentioned. See also, W. Seth Carus, "The Poor Man's Atomic Bomb,' Biological Weapons in the Middle East," Policy Papers No. 23 Washington Institute for Near East Policy, 1991.

4. Adm. C.A. H. Trost, House Armed Services Committee, February 20, 1990, p. 5.

5. Rear Adm. T.A. Brooks, House Armed Services Committee, March 14, 1990, p. 54.

6. The Chemical and Biological Weapons Elimination Act of 1991 (P.L. 102-182) requires an annual report by the president to congress that contains a complete list of known or suspected BW programmes, including those that are classified.

7. "Proliferation Issues: A New Challenge After the Cold War, Proliferation of Weapons of Mass Destruction," *Russian Federation Foreign Intelligence Report,* (translation), JPRS-TND-93-007, March 5, 1993.

8. "Biological Weapons Proliferation," *Technical Report,* U.S. Army Medical Research Institute of Infectious Diseases and the Defense Nuclear Agency, April 1994. See also, "Technical Aspects of Biological Weapons Proliferation," Chapter 3 in *Technologies Underlying Weapons of Mass Destruction,* Office of Technology Assessment, U.S. Congress, December 1993, PP. 71-117, and *Proliferation of Weapons of Mass Destruction: Assessing the Risks,* Office of Technology Assessment, U.S. Congress, 1993.

9. "Adherence to and Compliance with Arms Control Agreements," June 23, 1994, US Arms Control Disarmament Agency, p. 12.

10. In 1994, a Congressional Research Service report (J.M. Collins *et al, Nuclear, Biological and Chemical Weapon Proliferation: Potential Military Countermeasures,* Congressional Research Service, 1994-528S, July 5, 1994, p. 2.) Included a table of nations either possessing or having "programmes" of weapons of mass destruction. For biological weapons it listed Russia as the only nation with "possession confirmed," Iraq as "clear intent" (which by 1994, should also have been in the "confirmed" column), China, India, Pakistan, North Korea, Taiwan, Iran and Syria as "probable possession" and Egypt and Libya as "suspected programmes." The interesting—or anomalous—listings are of India and Pakistan, which have not otherwise been included in any unclassified official U.S. listings.

11. "Israel Vows Action Against Iraqi Germ Research, " *Washington Times,* January 19, 1989. (Israeli sources presumably provided the information that was carried in several ABC-TV news reports at roughly the same time.)

12. In its 1995 submission under the BWC confidence building measures, South Africa claims that its BW programme was solely for defensive purposes.

13. Milton Leitenberg, "Biological Weapons Arms Control," *Contemporary Security Policy* 17:1 (April 1996): 1-79, particularly pages 12-43. In the monograph version, *Biological Weapons Arms Control,* Center for International and Security Studies, University of Maryland, 1996, see pages 16-56.

14. "The One That Nearly Got Away; Iraq and Biological Weapons—An Interview with Rod Burton," *Pacific Research*, 9:2 (March 1966): 31-35.
15. Raymond Zilinskas, "Iraq's Biological Weapons: The Past as Future", *Journal of the American Medical Association*, 278:5 (August 6, 1997) 418-424.
16. Barbara Crossette, "Years After War, Iraq is Probably Still Hiding Arms, UN Says," *New York Times*, June 13, 1996.
17. Barbara Crossette, "UN Chief Approves a New Plan to Allow Iraq to Start Selling Oil," *New York Times*, July 19, 1996. The figure of over $120 billion was used by Rolf Ekeus at a proliferation conference in Washington, DC, April 10, 1997.
18. "UN Arms Ispector Seeks Pressure on Iraq", *New York Times*, June 19, 1997.
19. John M. Goshko, "Russia Accedes, Clearing Way for Security Council Resolution Warning Iraq", *Washington Post*, June 22, 1997.
20. Report by the Secretary-General on the activities of the Special Commission established by the Secretary-General pursuant to paragraph 9(b)(i) of Resolution 687 (1991). S/1997/774: October 6, 1997.
21. R. Jeffrey Smith, "Iraqi's Resistance Grew for Months", *Washinton Post*, November 9, 1997.
22. "The One That Got Away...", *ibid.*, and the sequential series of UNSCOM reports. The figure of over $ 120 billion as of March 1997 was used by Rolf Ekeus at a conference in Washington, D.C. on June 10, 1997, and since UNSCOM reckons Iraq's forgone oil export earnings to be roughly $20 billion per year, by November 1997 the figure would have reached $ 130 billion.
23. More detailed descriptions of the BW programme of the former USSR appeared in Milton Leitenberg, "The Biological Weapons Programme of the Former Soviet Union", *Biologicals*, 21:3 (September 1993), PP. 187-191. Milton Leitenberg, "The Conversion of Biological Warfare Research and Development Facilities to Peaceful Uses," in *Control of Dual Threat Agents: The Vaccines for Peace Programme*, Stockholm International Peace Research Institute and Oxford University Press, 1994, pp. 77-105, and Milton Leitenberg, *Biological Weapons Arms Control*, PRAC Paper no. 16, Center for International and Security Studies, University of Maryland, May 1996. Another useful source is Anthony Rimington, "From Military to Industrial Complex? The Conversion of Biological Weapons Facilities in the Russian Federation," *Contemporary Security Policy*, 17:1 (April 1996): 80-112 Some Additional information appeared in a paper by L.C. Caudle, he Biological Threat," Chapter 21 in F.R. Sidell et al. (edited), *Medical Aspects of Chemical and Biological Warfare*, Washington, DC: Office of the Surgeon General, 1997, particularly pp. 454-455.
24. Vyacheslav Yankulin, "Plague Syndrome, or the Purgatory of One of the Creators of Bacteriological Warfare", *Izvestiya*, October 15, 1997, FBIS Translation.
25. Raymond A. Zilinskas, "The Other Biological Weapons Worry", *New York Times*, November 28, 1997.
26. Richard Stone, "An Antidote to Bioproliferation", *Science*, 278 (November 14, 1997), p. 1222. Stone however adds that"... an estimated 150 to 200 scientists still work on biological defense related projects at four centers", which seems

to remove any credibility to his estimates. If that were the case, 9,800 researchers would already have been "converted" in one way or another.

27. A. Devroy and R.J. Smith," U.S., Russia Pledge New Partnership," *Washington Post*, February 2, 1992.

28. J.T. Dahlburg, "Russia Admits Violating Biological Warfare Pact," *Los Angeles Times*, September 15, 1992, and M.R. Gordon, "Russia and West Reach Accord on Monitoring Germ Weapon Ban," *Washington Post*, September 15, 1992.

29. M. Leitenberg, "The Conversion of Biological Warfare," *op. cit.*, and M. Leitenberg, *Biological Weapons Arms Control, op. cit.*

30. For overall surveys, see:

31. John R. Deni and Anne M. Harrington, "Beyond Brain Drain: The Future of 'Nonproliferation Through Science Cooperation' Programmes," Conference Paper, March 30-31, 1995, Center for International and Security Studies, University of Maryland, quoting FBIS-SOV-93-067, April 9, 1993, p. 39, and *Izvestia*, April 30, 1993, p. 15, in FBIS-SOV-93-085, P. 40.

32. Dan Vergano, "Rough Times in Russia: Post-Soviet Science Faces a New Crisis," *Science News*, May 10, 1997.

33. Maurizio Martellini, "Scientific and Technological Research Complex of Russia: Some Figures", V. Kouzminov et al. (editors), *Science for Peace Series*, Vol. 2 (UNESCO, Venice), 1997), 85-92.

34. Irina Dezhina (Institute for the Study of Economies in Transition), quoted in Michael Heylin, "Russian Science Doing Poorly, But Not Dead," *Chemical and Engineering News* (April 7, 1997), p. 47.

35. Dr. Niklas Ohrner, et al., "Some Views on Russian Biotechnology in Academic Institutions, State Research Centres, and Industry," Report from IVA's biotechnology Delegation to Moscow and Obolensk, 11/16-22/96. It should be noted that there is no external verification of these estimates, which were provided to the visiting Swedish team.

36. Deni and Harrington, 1995, *op. cit.*, p. 7.

37. *Current and Projected National Security Threats to the United States and its Interests Abroad*, hearing before the Select Committee on Intelligence, United States Senate, 104th Congress, 2nd Session, February 22, 1996, Washington, DC: US Govt. Printing Office, 1996, p. 198.

38. There have been suggestions that Iran, libya and Syria may have obtained aid for their BW programmes from the former USSR. Most of the indications concern Iran more recently, in the *post*-1992 years. There are also indications that the USSR. supplied some assistance to Iraq's BW programme before 1990, as did a sizable number of West European countries. See M. Leitenberg, *Biological Weapons Arms Control, op. cit.*

39. *Concepts for future Utilisation of the NCTR Facilities*, 1 June 1972.

40. Heden, C.G., "Global monitoring of the environment, a factor in the control of CBW-research", Seminar on Chemical and Biological Disarmament, Helsinki, Finland, 20-30 September 1972, PP. 13-14. See also Heden, C.G., "The

opportunity for a redeployment of military microbiology", Seminar on Microbiological Control of Insects, Helsinki, Finland, 8-9 June 1977, mimeo.

41. Carl Göran Heden, "The 1991 Persian Gulf War: Implications for Biological Arms Control", in Raymond Zilinskas, editor, *The Microbiologist and Biological Defense Research: Ethics, Politics, and International Security, Annals of the New York Academy of Sciences*, Vol. 666(1992): 1-8. As early as 1963, a conference on the Global Impacts of Applied Microbiology took place in Stockholm under the aegis of Dr. Heden (Impact [UNESCO], 17:3 (1967):187-208), and in the same year, the United Nations Conference on Science and Technology made available an inventory of the benefits that different scientific and technological disciplines could bring to bear on problems of developing nations. These were repeated in several of the ACAST reports—the Advisory Committee on Science and Technology in Development of the UN—produced in the early 1970s. In 1979, SIPRI published *The Fight Against Infectious Diseases: A Role for Applied Microbiology in Military Redeployment*. This book was also authored by Dr. Heden.

42. Daniel Greenberg, " A Pittance to Fight Malaria," *Washington Post*, January 4, 1998; Nigel Williams, "Consensus on African Research Projects," *Science*, vol. 278 (November 21, 1997): 1393.

    Nigel Williams, "Tropical Discases: Drug Companies Decline to Collaborate", Science vol. 278 (December 5, 1997): 1704.

43. Dr. Stephen Hoffman, Briefing on United States Department of Defense Malaria Research Programmes, Naval Medical Research Institute, Malaria Programme, May 14, 1996. The irony is that the total US Government budget for malaria research is on the order of $25 million per year, with malaria a major medical concern for US military forces, while the US Congress mandates that the US Army spend $ 135 million per year on breast cancer research, which should not be considered within the purview of military medical concerns at all.

44. Richard Stone, "Russia: An Antidote to Bioproliferation", *Science*, 278 (November 14, 1997), p. 1222.

45. Pamela C. Ronald, "Making Rice Disease-Resistant", *Scientific American*, 277:5 (November 1997), pp. 100-105.

46. W.W. Gibbs, "Biotechnology: Human Antibodies Produced by Field Crops Enter Clinical trials", *Scientific American*, 277:5 (November 1997), p. 44.

47. Kathy Sawyer, "With DNA, Simple Cells Perform Acrobatic Feats of Computation", *Washington Post*, December 15, 1997.

48. (*See Science*, December 5, 1997.)

    "Ebola Gene Vaccine is Promising in Animals", *New York Times*, December 30, 1997.

49. Carter, G.B., "75th Anniversary: the Chemical and Biological Defence Establishment (CBDE), Porton Down, 1916-1991", *ASA Newletter*, no. 4 (August 1991); Carter, G.B., "The Microbiological Research Establishment and its Precursors at Porton Down, 1940-1979" *ASA Newsletter*, no. 6 (December 1991); Carter, G.B., "The Microbiological Research Development and Establishment, 1946-1979". *ASA Newsletter*, no. 1 (February 1992).

50. *Statements by the President* (While House, Office of the Press Secretary: Washington, DC, 25 November 1969), 2pp., mimeo.

51. *Pentagon Press Briefing on the Biological Demilitarisation Programme*, 21 pp, mimeo; US Department of Defense, "Demilitarisation of biological agents and weapons", *Press Release*, no 1025-70 (18 December 1970), 1p.

52. "Demilitarisation" (annex L), *US Army Activity in the US Biological Warfare Programme*, vol. 2 (US Army: Washington, DC, 24 February 1977),pp. L-1 - L-11.

53. "Department of Defense Semi-Annual Report on Chemical and Biological Warfare Programmes", *Congressional Record*, Senate, 2 November 1973, pp. S-19818—S-19830.

54. *Congressional Record*, Senate, 11 March 1974, pp. S-3334-S-3348; *Congressional Record*, Senate, 12 September 1974, pp. S-16508—S-16521; *Congressional Record*, Senate, 17 March 1975, pp. S-3999—S-16349; *Congressional Record*, Senate, 19 September 1975, pp. S-16348—S-16357; *Congressional Record*, Senate, 11 February 1977, pp. S-2709—S-2723; *Congressional Record*, Senate, 17 March 1978, pp. S-4025—S-4040; *Congressional Record*, Senate, 19 July 1979, pp. S-9921—S-9939.

55. For example US Congress, Joint Committee Print, *Fiscal Year* 1980: *Arms Control Impact Statements*, "Chemical Warfare", 96th Congress, 1st Session (US Government Printing Office, Washington, DC, March 1979), pp. 218-283; US Congress, Joint Committee Print, *Fiscal Year 1981*: *Arms Control Impact Statements*, "Chemical Warfare", 96th Congress, 2nd Session (US government Printing Office, Washington, DC, May 1980). pp. 306-338; *Fiscal Year* 1982: *Arms Control Impact Statements*, "Chemical Warfare",, 97th Congress, 1st Session (US Government Printing Office, Washington, DC, May 1980), pp. 306-338; *Fiscal Year 1982*: *Arms Control Impact Statements*, "Chemical Warfare", 97th Congress, 1st Session (US Government Printing Office: Washington, DC, February 1981), pp. 34717-34752.

56. *Problems Associated with Converting Defense Research Facilities to Meet Different Needs: The Case of Fort Detrick*, Report to the Congress, by the Comptroller General of the United States, B-160140 (US Government Printing Office: Washington, DC, 16 February 1972), pp. 9-10. An unpublished study was prepared for a UN Secretary-General's working group on military R & D in 1983-84; see Longwell, V., *Fort Detrick: From Biological Warfare to Cancer Research*, Mimeo., undated. It was not used in preparing this paper.

57. *Concepts for Future Utilisation of the NCTR Facilities*, 1 June 1972.

58. Heden, C.G., "Global monitoring of the environment, a factor in the control of CBW-research", Seminar on Chemical and Biological Disarmament, Helsinki, Finland, 20-30 September 1972, pp. 13-14. See also Heden, C.G., "The opportunity for a redeployment of military microbiology", Seminar on Microbiological Control of Insects, Helsinki, Finland, 8-9 June 1977, mimeo.

59. Schmeck, Jr., H.M., "Deadly arsenal turned into a health research center", *New York Times*, 23 February 1975.

60. These interrelationships were investigated in a larger study by the author, *Research and Development in Chemical and biological Warfare: An Examination*

of the Possibility of Distinguishing Between Civil and Military, Offensive and Defensive R &D. This study was prepared for the Ministry of Foreign Affairs, Sweden, in 1983, and an earlier examination of some of the same issues was presented by the author in a paper to the International Congress of Microbiology, Mexico City, 1970.

61. *Problems Associated with Converting Defense Research Facilities to meet Different Needs: The Case of Fort Detrick* (note 29), pp. 16, 27. The exact number given in the GAO report to Congress is 1792. Other numbers have appeared in the literature. 1595 civilians and 650 military for a total of 2245 in Goldhaber, S.Z., "CBW: Interagency conflicts stall administration action", *Science*, vol. 169, no. 3944 (31 July 1970), pp. 454-456; and 1685 civilian plus 160 military, for a total of 1845 in Boffey, P.A., "Fort Detrick: a top laboratory is threatened with extinction", *Science*, vol. 171 (22 January 1971), pp. 262-264.

62. See Boffey, (note 13).

63. See Goldhaber, (note 13).

64. A Republican, President Nixon, made the decision to pre-empt the calls by a major Democratic opponent, Senator Edward Kennedy, for a national cancer research programme. See Meyer, L., "US to use Fort Detrick for research in cancer", *Washington Post*, 15 October 1971; Meyer, L., "Cancer research lab planned at Ft. Detrick", *Washington post*, 19 October 1971; Semple, R.B., "Nixon counting on conversion of military facilitie:", *New York Times*, 19 October 1971; Greenberg, D., "Washington view: modern cancers", *New Scientist*, vol. 54, no. 796 (18 May 1972), P. 393; "NCI announces award for Fort Detrick", *Science*, vol. 176, no. 4042 (30 June 1972), p. 1402; Schmeck, Jr., H.M., "Litton will run cancer unit lab", *New York Times*, 25 July 1972.

65. Wade, N., "Russians reserve doubts: Is Fort Detrick really detricked", *Science*, vol. 177, no. 4048 (11 August 1972), p. 500; Auerbach, S., "Russians see once-secret germ lab", (from *Washington Post*), *International Herald Tribune*, 5-6 August 1972; "Fort Detrick converting to major medical research center", *Army Research and Development News Magazine*, vol. 14, no. 4 (July/August 1973), pp. 16-19.

66. Leitenberg, M.,"A return to Sverdlovsk: allegations of Soviet activities related to biological weapons", *Arms Control*, 12:2 (September 1991), pp. 161-190.

67. ibid, p. 187.

68. See Goldhaber (note 13). The official who made the remarks was president Nixon's National Security Advisor, henry Kissinger.

69. US Department of Defense Statement. "US curtailing chemical biological weapons research", 6 December 1971, *Press Release*, no. 350, 8 December 1971; "Fort Detrick's future disputed", *Baltimore Sun*, 20 June 1970; Smith, R.M., "US slow to dismantle germ-war arsenal despite Nixon's stand", *New York Times*, 10 November 1970; Getler, M., "US issues plan to destroy biological warfare stocks", *International Herald Tribune*, 19-20 December 1970; Schmidt, D.A., "Senator asserts Army plans to expand biowar research", *International Herald Tribune*, 25 January 1971; lewis, F., "Full speed ahead on chemical and biological weapons", *Washington Post*, 6 December 1971.

70. Authorisation For Military Procurement, Research and Development, Construction and Real Estate Acquisition for the Safeguard ABM and Reserve Strengths, Fiscal Year 1972, "Research and development", Hearings before the Committee of Armed Services US Senate, 92nd Congress, 1st Session (US Government Printing Office: Washington, DC, February-May 1971), Part 3 of 5 parts, p. 2605. Under the Reagan Administration in the 1980s, US expenditure for all chemical and biological weapons R&D increased by an order of magnitude, including that for defensive BW R & D. See also footnote 44, below.

71. R. Roffey, "Bioteknisk forskning i Ryssland: målinriktad satsning, ökande konkurrens" [Biotechnical research in Russia: target-oriented venture, increasing competition], *FOA Tidningen*, 30:3 (November 1992), pp. 16-17. [add references to Rimington's papers.]

72. V. Nedogonov, "Directed-Energy Plague, or What the Secret Scientific Research Institute near Moscow Did,"*Komsomol' skaya Pravda*, September 24, 1991, in FBIS-SOV-91-191, October 2, 1991, pp. 48-50.

73. S. Bogomolov, "19th cantonment: a reporter is a rare guest here", Ural'sky Rabochiy, March 12, 1990 in JPRS-TAC-91-025, October 24, 1991, pp. 53-56.

74. K. Belyaninov, et al., "Anthrax epidemic linked to biological war", *Komsomol'skaya Pravda*, June 10, 1992, in JPRS-TND-92-022, July 10, 1992.

75. L. Usacheva, "Nine hours behind the barbed wire: reporting from the 19th military installation", *Poisk*, March 7-13, 1992.

76. The actual picture in the early 1990s appeared somewhat confusing. Virually all foreign observers report a massive deficit in all kinds of pharmaceuticals in the former Soviet Union, with bribes required to obtain medication, or seeking it on the black market, etc. None other than Arkady Volskiy, formerly the head of an organisation of the USSR's former defense industrialists, included in the four main "conditions" that he put to President Yeltsin on October 28, 1992 a demand for "the provision if medicines, of which only 60 per cent of the needed amount was available" (J. Lloyd and L. Boultan, "Industrial czar puts Russia's leader on spot", *Financial Times*, October 29, 1992). Most Western observers would consider the figure of 60 per cent a gross overestimate, and most likely a number that was selected at random. On the other hand, the facility at Koltsovo reported in the autumn of 1992 that it was producing antibiotics and that its storage facility was full, but that local communities and hospitals did not have the funds to purchase them. At the same time as the USSR may have simultaneously had both overcapacity and insufficient production of pharmaceutical products and antibiotics for *civilian* use in the past, there were reports for many years that it was buying antibiotics from Cuba, and may have continued to do so in the early 1990s.

77. V. Kaisyn, "Visiting the Caged Beast," *Pravda*, February 4, 1992, in JPRS-TAC-92-008, March 9, 1992, pp. 47-48.

78. V. Litovkin, "Yeltsin Bans Work on Bacteriological Weapons. This means: Work Was Under Way, and We Were Deceived," *Izvestia*, April 27, 1992, in FBIS-SOV-92-084, April 30, 1992, p. 7. Regarding the vaccines which the Sergiyev Posad facility hoped to export, the article added the following: "Our specialists claim that during the Gulf war the Americans were studying the possibility

of buying these vaccines because of their great effectiveness", which suggests that the vaccine in question was for anthrax.

79. V. Chelikov, "A weapon against their own people", *Moscow News*, no. 23, June 7-14, 1992.

80. A. Pashkov, "Military deny involvement in mysterious illness: chief of CIS joint armed forces top secret military laboratories answers journalists questions for the first time", Izvestia, April 17, 1992; see also FBIS-SOV-92-082, April 28, 1992, pp. 4-5.

81. Statements from different Russian sources, military and civilian, as to whether the Vorozhdeniye Island BW test site was not closed, and if so when it had ceased operations, were also widely contradictory during 1991 and 1992. A military official claimed that "all testing on the island...ceased" in June 1991, whereas another report stated that "in November 1991, at a scientific council in Sergiyev Posad, a decision was made to terminate the experimental work on the island definitively", while other reports had the site "still operating... in January 1992", and on April 17, 1992, the CIS General Staff offices responsible for these installations reported that although the Kazak Government had taken over responsibility for the island test site, the General Staff had *proposed* "a more fundamental decision, that this institution be shut down". There had already been numerous statements that the site had been shut down for some time. In September 1992 General Kuntsevich claimed that "it had been destroyed" ("Official examines destruction of CBW weapons", *Rossiyskiye Vesti,* September 22, 1992, in FBIS-SOV-92- 186, September 24, 1992, pp. 2-5. In August and September 1992 US authorities did state that the BW testing site was closed. S. Kozlov, "Secret Vozrozhdeniye Island BW Lab Abandoned," *Nezavisimaya Gazeta*, June 23, 1992, in JPRS-TND-92-022, July 10, 1992, pp. 25-26, and D Safulin, "Status of Biological Lab Remains Unclear," *Nezavisimaya Gazeta*, July 1, 1992, in JPRS-TND-92-026, July 31, 1992, p. 27. Detailed maps showing the island and its relation to Aralsk appeared in *National Geographic*, 177:2 (February 1992): 74, 75, 80.

82. "Official surveys CBW destruction process", *Krasnaya Zvezda*, October 22, 1993, in JPRS-TND-93-035, November 10, 1993, pp. 47-49

83. "Official examines destruction of CBW weapons" (note 21), pp. 2-5.

84. "General quizzed on chemical weapons production", *Izvestia*, April 17, 1992, in FBIS-SOV-92-082, April 28, 1992, pp. 4-5.

85. "Official examines destruction of CBW weapons" (note 21), pp. 2-5.

86. M. Feshbach and A. Friendly, Jr., *Ecocide in the USSR: Health and Nature Under Siege*, Bloomington, Indiana: Basic Books, 1992; see also M. Shapiro, "Ex-Soviet Empire the New 'Sick Man of Europe'", *Washington Post*, October 3, 1992.

10

Prevention of Biological Warfare: What Can We Learn from History?

—*E. Geissler, Germany*

Experts from various countries have recently completed a comparative analysis of the history of biological and toxin warfare (BTW) preparations performed in Canada, France, Germany, Japan, Poland, the Soviet Union, the United Kingdom and the United States before and during World War II (WW II) (Geissler and von Courtland Moon, 1998).The major aim of the study was to evaluate:

- why states performed biological and toxin warfare activities;
- whether such activities *became known* to other countries (that is, whether such activities could be successfully hidden) and, if so, to *what extent,* and
- whether and how such knowledge influenced a given country's national BTW programmeme.

In the following some major findings of our evaluation are summarised with special emphasis given to German BTW activities (Geissler, 1998). If not indicated otherwise I will not differentiate between defensive and offensive activities since it is obvious that there is no explicit borderline between defensive and offensive BTW preparations. As early as in 1924 the German department responsible for chemical and biological warfare matters, the Inspectorate of the Artillery, emphasised that prerequisites for the preparation of defense measures are research and knowledge of the methods an enemy

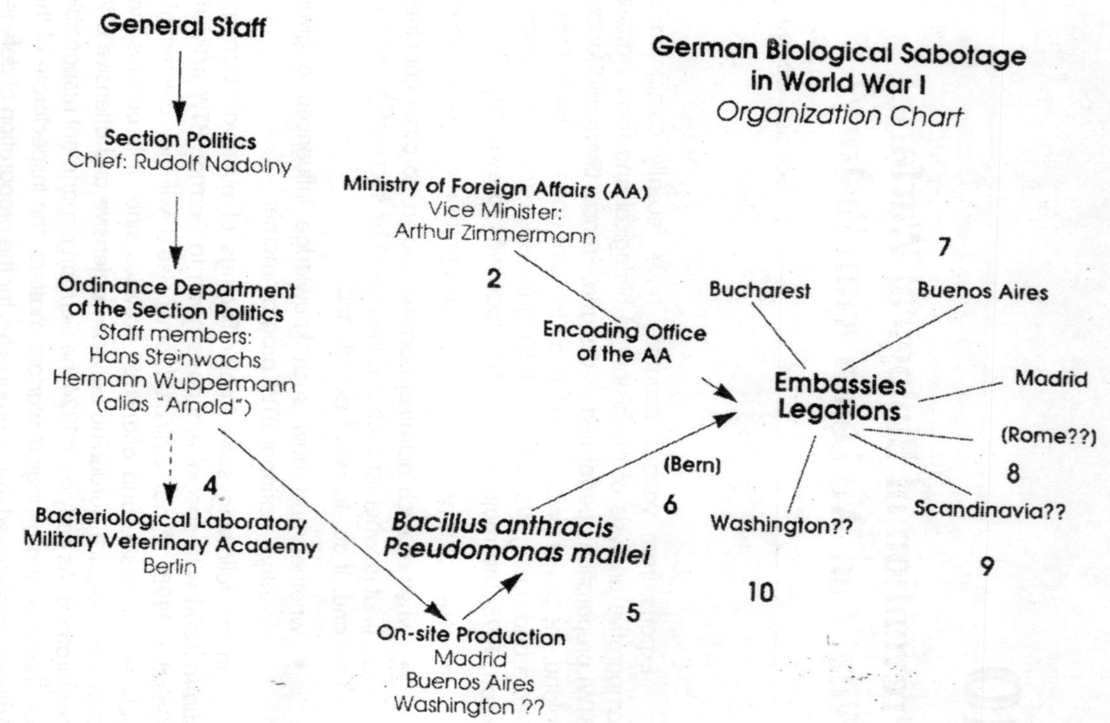
German Biological Sabotage
in World War I
Organization Chart
General Staff
Section Politics
Chief: Rudolf Nadolny
Ministry of Foreign Affairs (AA)
Vice Minister:
Arthur Zimmermann
2
Ordinance Department
of the Section Politics
Staff members:
Hans Steinwachs
Hermann Wuppermann
(alias "Arnold")
Encoding Office
of the AA
7
Bucharest
Buenos Aires
Embassies
Legations
Madrid
(Rome??)
8
Scandinavia??
9
Washington??
10
(Bern)
6
4
Bacteriological Laboratory
Military Veterinary Academy
Berlin
Bacillus anthracis
Pseudomonas mallei
5
On-site Production
Madrid
Buenos Aires
Washington ??

would and could use (Auer, 1924). Likewise, the US Bureau of Consultants Committee concluded two decades later, "preparation for defense necessitates a knowledge of the offense, and if this knowledge is not available from experience, it must come from the results of careful investigation" (WBC Committee, 1942). Consequently the 1969 Us decision to "renounce the use of lethal methods of bacteriological/ biological warfare" did "not preclude research into those of offensive aspects of bateriological/biological agents necessary to determine what defensive measures are required" (Kissinger, 1969).

German BW sabotage activities in World War 1

In modern times, it was Germany, a country at that time leading in bacteriology, which started the hostile use of biological agents (Geissler, 1997, Wheelis, 1998). Between 1915 and 1917 biological sabotage activities were initiated and/or performed in Argentia, Romania, Scandinavia, Spain and the United States (fig.1). The major aim of the actions was to interfere with the supply of lifestock to Germany's enemies by infecting military horses and mules, as well as cattle. Infection of Italian horses was considered too, but obviously not executed (fig. 2) Nadolny, 1915a). At least the activities planned and /or carried out in Argentia, Romania and Spain were initiated and supported by the German military intelligence, the Political Department of the General Staff, with the support of the Foreign Ministry and German diplomates (fig. 1).

From the very beginning the agents causing glanders and anthrax, *Pseudomonas mallei and Bacillus anthracis* were used. These agents, especially anthrax bacilli, are still today top on the priority lists of of biological warfare agents. IN WW II the British tested the efficiency of *Bacillus anthracis* in open air experiments with sheep on Gruinard Island and produced five million cattle cakes contaminated with anthrax spores, to be used, if necessary, for retaliation against suspected German attacks (Carter and Pearson, 1998). The Japanese in 1941 and 1942 annually produced on thousand kilograms of anthrax bacteria (and probably one hundred or more kilograms of glanders bacteria) in "Unit 100" (Harris 1998). They developed bombs munitioned with *Bacillus anthracis* which were tested on humans (Tanaka, 1996, p. 137) but were not operational before Japan's surrender. After the war the Soviets established a BW facility in Stepnogorsk, kazachstan, equipped with several 20,000 liter fermentors made for mass production of *Bacillus anthracis* (Johnson and Weber, 1998; Rimmington, 1998). Likewise 84,250 1 anthrax bacilli have been produced by the Iraqis and, after concentration, weaponised into missile warheads, artillery shells and bombs (United Nations, 1995, pp. 22-28; Kelley, 1997).

While it was claimed "on many occasions since World War I- and even before it " that chemical weapons "are less inhumane than other weapons" (Neild and Robinson, 1997)·[1] anti personal use of infectious agents was out of question during WW I for moral reasons (Nadolny, 1915c). Instructions for the land was issued by the German General Staff in 1902 requested that "certain measures of war which lead to unneccessary suffering are to be excluded [...] To such belong: The use of poison both individually and collectively (such as poisoning of rivers and food supplies), the propagation of infectious diseases" (Großer Generalstab, 1902, p. 10). Therefore, but also for technical reasons, the General Staff rejected proposals to disseminate plague bacilli (*Yersinia pestis*) over London and British harbours (von Boetticher, 1929; Winter, 1942).

Although the taboo against antipersonnel use of BW agents eroded during the mid-1920s it is interesting to note that the British during WW II decided that anthrax spores to be used for possible retaliation purposes should not be disseminated directly but embedded into cattle cakes to ensure, that animals only are the main targets and not humans (Carter and Pearson, 1998).

Unjustified concerns triggered BTW activities

While it is known that a German epidemiologist, Professor Heyer, warned the government already in August 1914 that the enemy might contaminate water supplies (Heyer, 1914), it is totally unknown who in 1915 was the first to consider and to propose the hostile use of bacteria. As the earliest documents available dealt with sabotage acts performed both with explosives and bacteria (Nadolny, 1915a, 1915b) the possibility exists that it was not a scientist but the officer in charge of activities destined to interfere with the enemy's capabilities, namely captain Rudolf Nadolny.

The German sabotage activities acted as a booster initiating BTW activities in several countries. France reacted with BW sabotage activities in 1917 (Unleserlich, 1917; Brose, 1917; Centre d'Études du Bouchet, 1923) and initiated BTW activities in 1922 not least because of the misconception of an assumed German BW threat (Lepick, 1998).

For the same reasons the Soviets started a secret BTW programme in 1926 (Bojtzov and Geissler, 1998).

The Germans, on the other hand, became aware of preparations performed in foreign countries and considered preparations on their own in the mid-1920s. These deliberations did not lead to any German BTW programme, mainly because biological weapons were considered to be unrealiable two-edged swords.

After Hitler came to power German emigrants claimed that Hitler and the Reichswehr were actively preparing BW. These claims were intensified by the well-respected British journalist Wickham Steed (Steed, 1934). In consequence, and on the basis of information collected by intelligence, BTW programmes were initiated at least in Italy (1934) (Morselli, 1944), the United Kingdom (1936) (Carter and Pearson), Hungary (1936) (Faludi, 1988) and Canada (1938) (Avery, 1998). Although Steed's allegations are still taken seriously by some authors, German documents reveal convincingly that the emigrants told lies and that the documents quoted by Steed had been falsified. British intelligence reports obtained during the 1930s on alleged German BW activities were wrong, likewise. Without doubt there were no BTW activities performed in Germany between 1925 and 1940 with the exception of some occasional considerations. Such considerations were trigged by more or less reliable intelligence information on foreign activities. They did not result in BTW activities, however. For example, correct intelligence reports on French anthrax activities obtained in 1938 (Anonymus, 1938) wer thoroughly evaluated by several departments of the army but did not induce German activities in that field. A main reason not to start corresponding activities based, ironically, on faulty expert opinion (von Ostertag, 1938).

On the contrary there were quite extensive programmes both in the USSR and in Japan. The Japanese programme, which included cruel human experimentation involving more than 10,000 human guinea pigs, finally led to the actual use of biological weapons against China (Harris, 1995, 1998). The Soviet programme, in contrast, was severely impaired by Stalin's persecution of the *intelligentsia* including leading BW experts (Bojtzov and Geissler, 1998). The Soviet programme was re-established during WW II. Its extent is still difficult to evaluate because of limited access to the archives.

The situation changed, although only marginally, when the

Germans in 1940 discovered a BTW facility, the "Laboratoire de Prophylaxie" of the "Poudrerie Nationale du Bouchet" and obtained detailed knowledge of the French programme (Kliewe, 1940). This discovery caused the German military for the first time to charge an expert in bacteriology, Professor Heinrich Kliewe, with BW matters and to establish a small BW laboratory, the "Kliewe laboratory". But neither the discoveries made in France nor- usually wrong- intelligence information obtained on alleged enemy preparations did lead to a German BTW programme, mainly because Hitler repeatedly and explicitly prohibilted any BTW preparations including preparations for retaliation in kind (Unleserlich 1942). On a small scale only, bypassing Hitler's order, a few open air experiments with simulants and with foot-and-mouth disease virus and potato beetles were carried out. These experiments did not lead to mass production or weaponisation of BTW agents, however. Germany was not prepared to wage biological or toxic warfare during WW II.

The motives behind Hitler's decision to prohibit BW preparations are unknown. There is reason to believe that Hitler was impressed by the Geneva Protocol. It was only in February 1945, after the devastating air strike against Dresden, that Hitler seriously considered whether Germany should leave the Protocol (Admiral, 1945). But military and legal experts convinced him that such action would endanger Germany much more than her enemies (Anonymus 1945a,b; Cartellieri, 1945). In addition it is conceivable that fear of retaliation might also have influenced Hitler's decision. The Anglo-Americans were prepared, indeed, to retaliate in kind to a German BTW attack, although at quite limited scale. An additional reason behind Hitler's decision might have been his exaggerated anxiety about the constant threat to his health posed by pathogenic bacteria. Taking into account his sophisticated destructive mind I could imagine that Hitler was really afraid that bacteria disseminated by his troops would like boomerangs or hostile carrier pigeons find their way to his headquarters to attack himself.

BTW *defence* measures were repeatedly ordered by Hitler. It is interesting to note, that German defence considerations at that time included also protection of the civilian population. For this reason, a civilian, Professor Kurt Blome, was appointed coordinator of German BW activities, which were so far coordinated by a military committee, the association "Blitzableiter". Nevertheless, and despite increasing concern about enemy preparations and sabotage activities, defence measures developed slowly, because of poor organisation and because the most qualified scientific experts were not involved in such activities. The civilian BW coordinator Professor Kurt Blome, for example, was a dermotologist specialised in cancer epidemiology.

Since this paper was presented at a conference in Como, Italy, it is worth mentioning that cooperation in that field was missing not only between Germany and Japan but also between Germany and her European ally. Similar to their ignorance of the Japanese BTW activities the Germans obviously did not know that the Italians performed BTW activities, too, since 1934—although on a small scale and mainly for defence purposes. Kliewe stated that he "never had any discussion with the Italians about BW. He did hear in 1944 that they had an institute in Rome working on the problem. The Germans did not give the Italians any information" (Alsos Mission, 1945). Colonel Walter hirsch, chairman of the association "Blitzbleiter", told his interrogators that "there had been no collaboration with Italy on BW though they had gone through all the Italian records after the German coup de main in Italy in 1943" (Mills et al., 1945). Hirsch's American interrogators noted that "he did not seem to be aware of the Italian BW experiments with Bacillus prodigiosus spray". Such experiments were described after the war by Lt. Col. Dr. Giuseppe Morselli from the Laboratory of Microbiology of the Italian Ministry of War (Morselli, 1944). According to Morselli ""no plans were ever made by us for the development of BW". According to a confidental report submitted to the Hungarian Government by BW expert Dr. Dezso Bartos in 1955 the Italian facility was directed by Lt. Col. Professor Hugo Reitano. The facility was located on the area of a military hospital and was quite large with working space not only in two floors but also the basement (Faludi, 1997, 1998). Although the Germans took note of an article on bacteriological warfare published by Reitano, 1937), which was reviewed in a German newspaper (W., 1937), they were clearly unaware of the professor's direct involvement in Italian BW activities. Reitano's name was mentioned neither in any document of the German BW committe nor, after the war, in interrogations.

Most intelligence reports were wrong or incomplete

In addition to the ignorance about the BTW activities of their allies, the Germans also obtained mostly wrong information about corresponding activities performed by their enemies. None of the eight British, Canadian and US BTW facilities were known to the Germans. Instead, they erroneously identified two US facilities, Huntsville, Alabama, and Pine Bluff, Arkansas, as places where BW activities were performed. Facilities located at these locations were involved in CW activities only. Only in an additional facility identified by the Germans as a BW facility, Edgewood Arsenal, Maryland, limited BTW activity was carried out at the beginning of the war.

In contrast to the almost total ignorance regarding facilities, some correct information was available on agents studied for warfare purposes. Germany obtained correct information on French and British anthrax activities and knew about Anglo American activities regarding rinderpest virus and botulinus toxin. Germany obtained erroneous information on enemy use of potato beetles and foot-and-mouth disease virus during WW II.

But Germany's enemies were also mislead by insufficient intelligence information. None of the German facilities involved in BTW activities were known to her enemies before the end of the war. The Anglo-Americans, instead, obtained a lot of wrong information regarding an alleged involvement of German Institutes, enterprises and non-scientific civilian places (such as Leuchtenburg castle in thuringia) in BTW activities.

In contrast to the assumption of her enemies, Germany did not work on *Bacillus anthracis* and botulinus toxin for military purposes. On the other hand, the Anglo-Americans were not alarmed regarding German military interest in foot-and mouth disease during WW II which involved one or a few more open air trials with foot-and-mouth disease virus. Likewise, they had no knowledge about the small-scale German activities involving potato beetles.

After WW II US military historians conceded "that the truth about German BW activities was considerably at variance with earlier intelligence reports". Instead of admitting to the inadequacy of the secret services the American experts claimed that "flase reports of German intentions to resort [to] germ warfare had unquestionably been spread as a psychological warfare weapon" (Brophy et al., 1959, p. 114). Although deliberate misinformation played a role in all wars since the invention of the Trojan horse there is not a single trace of a hint that the Germans (or their enemies) tried to delude the other side consciously regarding any BTW preparations. However some actions were performed which can be described as "psychological biological wafare". According to Kliewe "the purpose of bacterial warfare is to render as many soldiers as possible unfit for service, to minimise the productivity of the civilian population, and to starve them" (Kliewe, 1943, p. 77). It occurred to the British Political Intelligence Department and the U.S. Office of Strategic Services that such a goal could not be accomplished only by "mass use of certain types of bacteria" and "spreading bacteria by agents, saboteurs, etc." as suggested by Kliewe, but by tempting people to *simulate* severe illness. In consequence at least 20 different instructions to simulate were disseminated as leaflets or booklets with an unsuspicious cover (e.g. *"Krankheit hilft"* [Illness is helpful] Wohltat, 1944] over the front lines and the German home territory (Kirchener, 1976). The initiator of these actions, psychological warfare expert Sefton Delmer hoped that some Germans would follow the instructions (Delmer, 1962, pp. 538-40). Moreover, he hoped that German physicians might be attempted to assume that some of their patients were simulating in spite authentic illness. It was consequently intended that sick patients would be sent back to work or to the battlefied and to infect others.

The Germans paid her enemies back in her own coin and distributed a translation at least of one of the instructions in form of a match-book entitled *"Better a Few Weeks Ill Than All Your Life Dead"*[2]).

The disadvantageous role of secret services

While it is unknown why Germany started biological sabotage actions in WW I and it is unclear who was the first to propose them it is obvious, that the German initiative induced other states to initiate BTW activities. It was thus the German activities which triggered BTW activities in other countries—perhaps with the exception of Japan. It was not the scientific progress or the initiative of individual bacteriologists. After the programmes had been started scientists became involved in different countries to a different degree. In Germany scientists had been much less involved in BTW activities then in chemical warfare preparations.

An increasing number of countries started to perform BTW activities after Hitler came to power because of wrong information collected by the secret services and of unfounded claims of German emigrants. The disastrous role of the said emigrants should warn those who nowadays hope that violations of the norm against biological and toxin warfare can be cleared up by scientists and technicians acting as whistle-blowers.

Some authors argue that the efficiency of the secret services is nowadays much higher than before and during WW II. As a German, who experienced that the secret services could neither predict the establishment of the Berlin Wall in 1961 nor its breakdown in 1989 and the collapse of the Warsaw Pact, I doubt whether the efficiency of such services has really increased especialy with respect to their ability to uncover secret BW activities.

Moreover, after the introduction of molecular biotechnology, it might be even more easy to carry out prohibited activities in secrecy than fifty years ago. Furthermore there is no explicit borderline between (prohibited) offensive and (permitted) defensive activities. Besides that, BTW agents are dual-threat agents which threaten people, animals and plants also independent to their intended military or terrorist use. BTW agents have to be studied therefore, to develop powerful countermeasures, vaccines, for example. But it is extremely difficult to ascertain whether the development and production of a vaccine against a dual-threat agent is performed for peaceful or for hostile purposes, since such vaccines are necessary also to protect people preparing biological and toxin warafre. It was for this reason that British and US experts examined that immune status of German and Japanese prisoners of war, to evaluate any BW preparations by the enemy.

Inspite the tremendous increase in the BTW threats such indicative measures like immunological tests are not considered appropriate today as measures to verify compliance with the Biological and Toxin Weapons Convention (BTWC). It is totally beyond me that one of the mandates of the Ad Hoc Group developing an additional protocol to the BTWC (see the contribution by Iris Hunger, this volume) is to consider compliance measures which should be, *inter alia*, "as non-instrusive as possible" (United Nations 1994, p. 10).

Immunological tests are rejected not only because they are regarded to be too intrusive but also for "ethical" reasons. However, immunological tests are certainly much less intrusive than DNA fingerprints. Novertheless DNA fingerprints are common today to indict everyday criminals without any ethical reservations. If it is legitimate to investigate daily crimes by a very intrusive method it should be legitimate to investigate severe, international crime namely violations of disarmament treaties by the much less intrusive method concerned in determination the immune status of people possibly involved in prohibited BTW activities. It is not immunological tests that are unethical (if the persons to be tested do not refuse). It is unethical to ignore most indicative measures to demonstrate compliance with the BW Convention.

The Iraq example demonstrates, however, that even most intrusive measures might not be sufficient to prove violations of the norm against biological weapons and to detect hidden stockpiles and munitions. The disclosure of Iraq's BTW programme was only to a limited extent the result of the UNSCOM activities and primarily due to the provision of documents after the escape of Saddam Hussein's son in law. Novertheless, UNSCOM obviously is still searching for weapons of mass destruction including biological and toxin weapons.

But the secret services are not able to perform similar intrusive investigations in other countries.

Throughout the 1990s in addition to Iraq, several countries were regarded at least by the secret services of Russia, the USA and the UK as being potential possessors or developers of BTW agents. According to a survey issued in 1993 by a well-respected independent institution these states included, Russia, China, India, Israel, North Korea, South Africa, Syria, Libya, Iran, Taiwan, Belarus, Pakistan and South Africa (possibly capable of developing BTW agents) see Reed, 1993). These countries denied emphatically that they are involved in such activities or have such intentions. With respect to the activities performed in the Soviet Union and Russia until 1992 (Russian Federation, 1992, p. 82) and the Iraq (United Nations, 1995) the services obtained correct information. South Africa informed to have performed defensive BW activities until 1992 (South Africa, 1995, p. 64). A recent evaluation published by the US Department of Defense (DoD) again mentions some of these countries (Office, 1997), notably iraq, Syria, iran, North Korea, and Pakistan. Even Russia might "be continuing some research related to biological warfare".

But what about other countries? It is obvious that any country with a basic capacity for bacteriology and biotechnology is "potentially or possibly capable of developing such weapons".

Consider the results if similar list had been established before or during WW II by the Anglo Americans. Japan perhaps would not have even been mentioned or mentioned only at the bottom of the list, because the Japanese programme was performed in absolute secrecy. On the other hand Germany certainly would have been placed erroneously on the top of the list. Intrusive verification measures, excerted, for example, by a neutral organisation, could have easily demostrated that the allegations were unfounded and that it was not necessary for other countries to prepare retaliation measures directed against Germany.

Form VFP or ProCEID

Nevertheless most intrusive measures might not be sufficient to uncover violations of the norm against the hostile use of biological agents and toxins, be they exerted by states or by terrorists. The use of BTW agents cannot be completely excluded, unfortunatély. Nobady is able, however, to predict whether the probability of such an event is almost zero or higher. It should also be taken into account in this connection that dual-threat agents do not respect national borders and threaten military personnel and civilians.

We must consider, therefore, not only to establish the most efficient measures to prevent the use of such agents but also how to react in the case of violations of the Geneva Protocol and the BTWC. A global system of biological security is necessary, all the more as we are threatened not only by the hostile use of BTW agents but also by natural outbreaks of emerging and reemerging diseases.

To promote such peaceful international co-operation, Vaccines for Peace (VFP) programme was elaborated between 1989-1992 (Geissler, 1992). It called for establishment of an international programme of development and use of vaccines against pathogens of relevance to the BWC, to be administered by the World Health Organisation (WHO). The proposal, which was favourably acknowledged by the Third Review Conference of the States Parties to the BTWC, was then evaluated by experts in a series of private international meetings. These culminated in a meeting in Biesenthal near Berlin, Germany, in September 1992 involving experts in biotechnology, defense, diplomacy, international development, medicine, molecular biology and vaccinology from Australia, France, Germany, Hungray, India, Peru, Russia, Sweden, the United Kingdom and the United States as well as representatives or observers from the WHO, UNIDO and the United Nations Office for Disarmament Affairs (Geissler and Woodall, 1994). The participants of the workshop agreed that a modified VFP programme should be established, later designated as the "Biesenthal Vaccine Initiative" (BVI) (see Geissler and Woodall, pp. 255-258).

In December 1994 an in-depth analysis of both proposals, VFP and BVI, was carried out at the XVI Kühlungsborn Colloquium on the island of Vilm, near Rügen, Germany, by defense experts, epidemiologists, molecular biologists, and virologists from Germany and the USA. Recognising that changing political, economic, technical and military factors warranted a further evolution of the propsal, participants recommended a modified version, the "Programme for Controlling Emerging Infectious Diseases" (ProCEID) as a set of measures designed to facilitate peaceful international cooperation of States Parties to the BTWC without weakening the international norm against biological and toxin weapons (ProCEID Steering Committee). ProCEID was revised additionally in the first half of 1995, especially during the XVII Kühlungsborn Colloquium 30 June—2 July 1995, by experts from Germany, Russia, the United Kingdom, and the United States. The title was modified to "Programme for Countering Emerging Infectious Disease" (Meeting Report, 1996).

ProCEID and similar programmes might provide means to develop a system of global biological security and to strenghten the norm against hostile use of biological agents and toxins. Iris Hunger will discuss such possibilities in more detail (Hunger, 1998).

References

1. On the other hand it was argued that there is a stigma against using CW which persisted until the early 1940's and contributed to the non-use of CW in WW II (Price, 1995).
2. Facsimiles of *"Krankheit hilft"* and *"Better a Few Weeks Ill Than All Your Life Dead"* are attached to Kirchner, 1976.

Bibliography

Admiral s. B.V [zur besonderen Verwendung] beim Ob.d.M. [Oberbefehlshaber der Marine] (1945), "Teilbahme des Ob.d.M. an der Führerlage am 19.2. 17,00 Uhr". Chefsache! Nur durch Offizier! 20. Februar. In: *Der Prozeß gegen die*

Hauptkriegsverbrecher vor dem Internationalen Militärgerichtshof. Nürnberg 14. November 1945 - 1. Oktober 1946. Amtlicher Text in deutscher Sprache. Nürnberg (1948), 34 pp. 641-642.

Alsos Mission (1945), Report on the Interrogation of Professor H. Kliewe, May 7-11th. Report No. A-B-C-H-H/149, U.S. National Archives, Washington, DC (NAW).

Anonymus (1938), "Avis préliminaire". Bundesarchiv- Militärarchiv, Freiburg/Brsg. (BAMA) RH 12-9/V.62, P. 252.

Anonymus [Wehrmachtführungsstab/ Qu. 2 (I) 1 (1945a), "Betr:Derzeitiger Wert und Unwert völkerrechtlicher Bindungen, wie Genfer Kovention, Haager Landkriegsordnung usw.". Geheime Kommandosache. 20 Februar. BAMA, RW 5/v.335, GP 198/1.

Anonymus (1945b), [Stellungnahme des Oberbefehlshabers del Kriegsmarine zur Frage des Austritts aus der Genfer Kovention]. Vortragsnotiz. "Betr: Genfer Kovention". Geheime Kommandosache. 21. Februar. BAMA RW 5/v.335, GP 198/5-6.

Auer (1924), Schreiben an der Chef des Stabes m. W.b. [mit der Wahnehmung beauftragt] der S In [Sanitäts-Inspektion] Herrn Geeraloberst Prof. Dr. Napp. "Betr: Bekämpfung von Seuchen", Geheim. 24. Januar. BAMA, RH12-9/v.27.

Avery. D (1998), "Canadian biological and toxin warfare research, development and planning, 1925-1945". In: Geissler and van Courtland Moon.

Bojtzov, V. and E. Geissler (1998), "Military biology in the USSR from 1920 to 1945". In: Geissler and van Courtland Moon.

Brophy, L.P., W.D. Miles and R.C. Cochrane (1959), "Biological Warfare Research" In: Conn, S. (ed.), *United States Army in World War II. [X] The Technical Services. [vol. 2] The Chemical Warfare Service: From Laoratory to Field*, Washington, D.C., pp. 101-122.

Brose (1917), Schreiben an das [Preußische] kriegsministerium - U.3, das Bayerische Kriegsministerium, München, das Würtembergische Kriegsministerium, Stuttgrat, das Sächsische Kriegsministerium, Dresden "Betr.: Säbotage durch Rotzbazillen. Geheim". 20. July, Kriegsarchiv (Bayerisches Hauptstaatsarchiv, Abteilung IV), München (KA), 12919.

Carter, G.B. and G.S. Pearson 1998, "British BW and biological defence: 1925-1945". In; Geissler and van Courtland Moon.

Centre d' Ètudes du Bouchet, Vert-le-Petit (1923), Minutes of the bacteriological commission. Plenary Session on 18 May, p. 2 Quoted by Lepick, 1998.

Delmer, S. (1962), *Die Deurschen und ich.* Hamburg, S. 538-40.

Faludi, G. (1997), personal communication, 7 November.

Faludi, G. (1998), "Challenges of BW control and defense during arms reduction". In: Geissler, Gazso and Buder.

Geissler, E. (1992), "Vaccines for Peace: an international programme of development and use of vaccines against dual-threat agents", *Politics and the Life Sciences,* vol. 11. no. 2, pp. 231-43.

Geissler, E. (1997), "'Anwendung von Seuchenmitteln gegen Menschen nicht erwünscht'. Dokumente zum Ensatz biologischer Kampfmittel im Ersten Weltkrieg".

Militärgeschichtliche Metteilungen, 56, pp. 1-49.

Geissler, E. (1998), *Biologische Waffen—nicht in Hitlers Arsenalen. Biologische und Toxin-Kampfmittel in Deutschland* 1915-1945. LIT-Verlag, Münster, Hamburg, London.

Geissler, E., L. Gazso and E. Buder (eds.) (1998), *Conversion of BW Facilities*, Kluwer Academie Published, Dordrecht (in press).

Geissler, E. and J. E. van Courtland Moon (eds) (1994), *The Genesis of Germ Warfare: The Development and Use of Biological and Toxin Weapons from the Middle Ages to the End of World War II*. SIPRI Chemical and Biological Warfare Studies 17, Oxford University Press, in press.

Geissler, E. and J.P. Woodall (eds.) (1994), *Control of Dual-Threat Agents*:

The Vaccines for Peace Programme, SIPRI Chemical and Biological Warfare Studies, no. 15, (Oxford University Press, Oxford).

Großer Generalstab (ed.) (1902), *Kriegsbrauch in Landkriege*. Kriegsgeschichtliche Einzelschriften, no. 31. Ernst Siegfried Mittler, Berlin.

Harris, S. (1995), *Factories of Death: Japanese Biological Walfare: 1932. 1945 and the American Cover-Up*. Routledge, London and New York.

Harris, S. (1998), Japanese biological warfare programme: An overview". In: Geissler and van Courtland Moon.

Heyer 1914, Telegramm an Reichskanzleramt Berlin, 3. August. bundesarchiv Berlin-Lichterfelde (BAL) R 86/2550.

Hunger, I. (1998), "Technical cooperation in the framework of the Biological and Toxin Weapons Convention- conversion aspects", In: Geissler, Gazso and Buder.

Johnson, B. and A. Weber (1998), "Conversion in action: The Stepnogorsk example". In: Geissler, Gazso and Buder.

Kelley, D.C. (1997), personal communication, 27 October.

Kirchner, K. (ed) (1976), *Achtung Feindpropaganda*, Mappe 4. *Krankheit rettet. Psychologische Kriegführung*. Verlag D+C, Erlangen.

Kissinger, H.A. (1969), "National Security Decision Memorandum 35", 25 November. Reprinted in S. Wright, (ed.) 1990, *Preventing a Biological Arms Race*, The MIT Press, Cambridge, MA, London, pp. 403-405.

Kliewe, H. (1940), "Bericht über die Erhebungen in dem 'Laboratoire de Prophylaxie, Poudrerie nationale due Bouchet'" 17. September. U.S. National Archives, College Park, Maryland (NACP), Record Group (RG) 319, Box 1, Folder BW2, pp. 30-50.

Kliewe, H. (1943), "Der Bakterienkrieg". Geheime Kommandosache. NACP, RG 319, Box 3, Folder BW 14, pp. 77-78.

Lepick, O. (1998), "French activities related to biological warfare: 1919-1940". In: Geissler and van Courtland Moon.

Meeting Report (1996), "Programme for countering emerging infectious diseases (ProCEID) by prophylactic, diagnostic and therapeutic measures. Mission Statement, revised". Biologicals, vol. 24, pp. 71-74.

Mills, A.K., H. Columbine, and P. Mrovsky (1945), "Report on the Interrogation of Oberst Hirsch, Abteilungschef Wa Prüf 9 Heeres Waffenamt", May 19-21. NAW, RG 165, E 187, B 137 File "BW."

Morselli, G. (1944), "Summary of work carried out on regarding bacteriological warfare", Rome, 19 July; Answers to questions presented 23 July"; "Answers to questions presented 28 July". NAW, RG 165, E 486, B 67 2345 NW 441.2.

Nadolnÿ, R (1915a), Telegramm in Ziffern an den Militärattaché bei der Kaiserlichen Botschaft in Rom. 13. Mai. Auswärtiges Amt Bonn, Politisches Archiev (PAAA) R 21236, p. 2. Reprinted in Geissler (1997).

Nadolny, R (1915b), Telegramm in Ziffern an den Militärattaché bei der Kaiserlichen Gesandschaft in Bukarest, Geheim. Eilt. 17 Mai. PAAA R 21200, p. 58. Reprinted in Geissler (1997).

Nadolny, R (1915c), Telegramm in Ziffern an den Militärattaché bei der kaiserlichen Gesandschaft in Bukarest, 6 Juni, PAAA R 21200, p. 108. Reprinted in Geissler (1997).

Neild, R. and J.P. Robinson (1991), "Appendix I. The claims that CB weapons are less inhumane than other weapons". In: SIPRI (ed.), *The Problem of Chemical and Biological Warfare, 5: The Prevention of CBW.* Almqvist & Wiksell, Stockholm—Humanities Press, New York, pp. 124-136.

Office of the Secretary of Defense (1997), *Proliferation: Threat and Response*, US Department of Defense, Washington, D.C., November.

Price, R. (1995)), "A genealogy of the chemical weapons taboo", *International Organisation*, vol. 49, pp. 73-103.

ProCEID Steering Committee (1995), "ProCEID-Programme for controllling emerging infectious diseases: Mission Statement", *Politics and the Life Sciences*, vol. 14 no. 1, pp. 89-92.

Reed, J. (1993), *Defence Exports: Current Concerns.* A Jane's Special Brief. Jane's Consultancy Services, Coulsdon, Surrey.

Reitano, H. (1937), *Rivista di fanteria*, January.

Russion Federation (1992), CBM Report, 3 July, United Nations DDA/4-92/BWIII/Add. 3 (22 September 1992), pp. 32-90.

South Africa (1995), CBM Report, 8 August, United Nations CDA/14-95/BW-III/Add.2 (12 September 1995), pp: 51-68.

Steed, W. (1934), "Aerial Warfare: Secret German Plans". *The Nineteenth Century and After*, 116, Nr. 689, pp. 1-16.

Tanaka, Y. (1996), *Hidden Horrors. Japanese War Crimes in World War II.* Westview Press, Boulder, Colorado.

United Nations (1994), *Final Report. special Conference of the States parties to the Convention on the Prohibition of the Development, Production and Stockpiling of bacteriological (Biological) and Toxin Weapons and on their Destruction* (Geneva 19-30 September). BWC/SPCONF/1.

United Nations (1995), *Report of the Secretary-General on the status of the implementation of the Special Commission's plan for the ongoing monitoring and verification of Iraq's compliance with relevant parts of section C of Security Council resolution 687* (1991). S/1995/864, 11 October.

Unleserlich [Inspektion der Kriegsgefangenlager] (1917), Schreiben an das K. stellv. General-Kommando I.B.A.K. "Betreff: Sabotage von Kriegsgefangenen". 23. Mai. KA 12918.

Unleserlich [General der Nebeltruppen] (1942), Schreiben an V In [Veterinär- Inspektion], "Betr.: USA-Versuche mit Bakterien/ Lieferungen nach England". Geheime Kommandosache. 23. mai. NACP RG 319, Box 3 Folder BW 14.

von Boettocher (1929), Schreiben an das Auswärtige Amt, 22. März. PAAA, R 26583.

von Ostertag, R., "Stellungnahme zu den beiden Fragen des Schreibens vom 12-7.38 "Betr.: Milzbrandsporen", BAMA RI I 12-9/v. 62, pp. 241-244.

W., H. (1937), "Betrachtungen über den Krieg mittels Kleinlebewesen "*Deutsche Wehr*, Nr. 35, 26. Juni, pp. 560-561.

War Bureau of Consultants Committee (1942), *Report*, 18 June, quoted by J.E. van Courtland Moon, "United States BW planning and preparedness: The dilemmas of policy". In: Geissler and van Courtland Moon.

Wheelis, M. (1998), "Biological sabotage in the First World War". In: Geissler and van Courtland Moon.

Winter (1942), "Vergangenes und Zukünftiges. Betrachtungen über das Für and Wider des Bakterienkrieges", Dezember. NACP, RG 319, Box 3, Folder BW 14, S. 123-135.

Wohltat (1944), *Krankheit hilft*. Ostensibly 23rd edition, ostensibly published by Reclam-Verlag Leipzig. Faksimile 1975, Supplement to Kirchner (1975).

11

Prevention of Biological and Chemical Warfare; Regional or Zonal Approaches

J. Prawitz, Sweden

Introduction

History of the zone concept

The regional or zonal approach to arms control management has a long history. After the invention of the atomic bomb, the nuclear-weapon-free zone (NWFZ) idea has been the most successful application. Such zones now encompass more than half of all land in the world and 99 per cent of all land south of the equator. The success of the NWFZ idea did inspire several proposals to establish also chemical weapon free zones, although no such proposal has materialised. In 1990, an expansion of the NWFZ concept to include all weapons of mass destruction was proposed.

The concept of nuclear-weapon-free zone (NWFZ), as it has developed from the political deliberation since the mid-1950s, does today cover a spectrum of arrangements, geographically ranging from whole continents like Latin America to smaller areas or even single states, and functionally serving purposes of preventing the spread of nuclear weapons as well as avoiding nuclear war in the respective regions.

The first proposal on regional limitation of nuclear weapons was tabled in 1956[1]. It referred to Central Europe and was proposed by one superpower—the USSR—directed against its adversary superpower. One year later Poland proposed the so called Rapacki-plan on permanent absence of nuclear weapons from the entire territory of several states in Central Europe[2]. The latter proposal was thus made by one of the states within the prospective zone region.

At that time two different approaches to military denuclearisation were pursued in parallel. One was the open ended and global non proliferation approach which finally lead to the adoption, in 1968, of the Treaty on the Non-Proliferation of Nuclear Weapons (NPT)[3]. The purpose of that treaty was to prevent the number of nuclear weapon states to grow beyond the five existing at the time. It was recognised that the risk of outbreak of nuclear war would grow dangerously larger, if the number of "fingers on the trigger" would be permitted to increase.

The first result of the zonal approach was the Antarctic Treaty of 1959 declaring the Antarctic continent a demilitarised zone and by implication also a zone free of nuclear and other weapons of mass destruction as well.[4] Since then several nuclear-weapon-free zones have been established, including four zones in "densely" populated areas[5]. The Tlatelolco Treaty[6] 1967, The Rarotonga Treaty[7] of 1985, the Bangkok Treaty[8] of 1995 and the Pelindaba Treaty[9] of 1996 created such zones in Latin America, the South Pacific, South East Asia, and Africa respectively. Denuclearisation of the Korean Peninsula was declared in 1992 by the two Korean States[10], but the agreement remains to be implemented.

Two multilateral agreements raising barriers to the deployment of weapons of mass destruction in specific new areas and environments were agreed early. One was the 1967 Outer Space Treaty[11] Prohibiting the placement "in orbit around the Earth any objects carrying nuclear weapons or any other kinds of weapons of mass destruction, install such weapons on celestial bodies, or station such weapons on celestial bodies, or station such weapons in outer space in any other manner" (Article IV). The other was the 1971 Sea-Bed Treaty[12] prohibiting the emplanting or emplacing" on the seabed and the ocean floor and in the subsoil thereof, any nuclear weapons or any other types of weapons of mass destruction as well as structures, launching installations or any other facilities specifically designed for storing, testing or using such weapons" (Article I).

Similar proposals which never materialised, however, have been made for the creation of nuclear-weapon-free zones in many other regions of the world, including South Asia, Central Asia, the Middle East, and various parts of Europe[13].

A number of areas have been established as demilitarised zones according to treaties concluded long ago, most of them before the atomic bomb was invented. Among such areas are a number of islands in the Mediterranian as well as in the Baltic and Arctic seas. By implication, such areas should today be considered free of weapons of mass destruction as well.[14]

Reference should finally be made to the possibility envisaged in the humanitarian laws of war to establish by agreement temporarily demilitarised zones[15].

The Mubarak concept

In 1990, President Mubarak of Egypt proposed the establishment of a zone free of all types of weapons of mass destruction in the Middle East[16]. This proposal expanded the scope of earlier NWFZ concepts to include "all weapons of mass destruction without exception", i.e. "nuclear, chemical and biological,...etc".

Although the idea of establishing biological-or chemical weapon-free zones were from time to time mentioned as measures parallel to NWFZs, these ideas did not enter the scene of world politics. In the literature, an analysis of the concept of chemical-weapon-free zones was researched and published by SIPRI in the 1980s, before the end of the Cold War and before the Chemical Weapons Convention had been agreed[17].

President Mubarak's proposal both summarised the emerging understanding that weapons of mass destruction should be considered together in strategic and defence analysis, and gave the concept of a zone free of weapons of mass destruction (WMDFZ) a concrete geographic reference. In this sense, a new arms control concept was introduced in security policy. This concept, specifically referring to the Middle East, was soon to be adopted by the UN Security Council in 1991[18]. The Mubarak formula implies that elimination of biological weapons should be considered and analysed together with other weapons of mass destruction, i.e. nuclear and chemical weapons. This paper will approach regional biological weapon issues in this way.

The Global Treaties

Several arms control treaties with a global application are today sufficiently established to constitute a natural background to any future nuclear-weapon-free zone or zone free of weapons of mass destruction. They are primarily the 1925 Geneva Protocol[19], the 1968 Non-Proliferation Treaty (NPT)[20,] 1972 Biological Convention (BWC)[21], and the 1993 Chemical Weapons Convention (CWC)[22].

The 1925 Geneva Protocol. The use in war of chemical agents on a limited scale has been known since the beginning of history. Attempts were made in the 19th century to prohibit such use by international agreement.[23] But it was during the First World War, that for the first time, such use began on a massive scale. The combination of industrial production of agents and greatly improved transport câpability provided the tools for mass destruction. The general public was horrified and a political basis emerged for a modern and general prohibition.

After some years of negotiation, the 1925 Geneva Protocol was agreed upon. The Protocol refers to the fact that "the use in war of asphyxiating, poisonous or other gases, and all analogous liquids, materials or devices", is prohibited "in Treaties to which the majority of Powers of the world are Parties", and reaffirms these prohibitions and extend them to the "use of bacteriological methods of warfare". The Protocol was intended to become part of customary international law[24].

The Protocol thus prohibits "the use in war" of chemical and biological weapons but not the acquisition and possession of such weapons. No verification system is envisaged, although the UN Secretary General can today investigate reports on possible violations. When ratifying the Protocol, many of the parties filed reservations to the effect that they considered themselves bound by the Protocol's provisions only in relation to other parties observing them. Therefore, the Protocol became a "no-first-use" agreement for these parties.

The implementation of the Protocol did feature some violations but it passed its great test during the Second World War, when chemical weapons were not used. As a consequence of the Vietnam war, the Protocol became widely discussed in the late 1960s and 1970s. Many states, including such states which did not exist in the 1920s, adhered to the Protocol. Some parties which had filed reservations withdrew them making the Protocol a pure "non-use" obligation for them. An attempt was made, not entirely successful, in the UN General Assembly in 1968 to declare the Protocol customary law binding upon all states whether parties or not[25]. Today, however, the provisions of the Protocol could be considered customary law.

The Protocol is thus an important basis for the modern treaties prohibiting the acquisition and possession of biological and chemical weapons agreed upon in the 1970s and 1990s. The Protocol had 132 parties as of 1 January 1998.

The 1968 Non-proliferation Treaty. The development of the non-proliferation regime is very much a success story in the sense that a large majority of the states of the world, including all recognised nuclear weapon states, over the years became parties to the treaty; that after 1968, no additional state established itself as a new nuclear weapon power; and that dramatic nuclear disarmament measures were finally agreed upon. In addition, in May 1995 the parties extended the duration of the NPT indefinitely.[26]

As of 1 January 1998, the NPT had 187 parties, including all nuclear weapon powers. Only seven states of the world were non parties. Four of those are part of nuclear-weapon-free zones and thus covered by a non-proliferation regimes, leaving the well-known "threshold states", India, Israel, and Pakistan, as the only real outsiders.

Therefore, the provisions and implementation machinery of the non-proliferation regime would be a dominating basis for negotiating the establishment of nuclear-weapon-free zones in the future[27]. Whenever a new nuclear-weapon-free zone will be established in the future, most potential states would thus already be parties to the NPT. Indeed, the NPT itself foresees that role in its Article VII also indicating that participation in a nuclear-weapon-free zone would define extended commitments for NPT parties.[28]

At the time the non-proliferation of nuclear weapons was negotiated, in the years before 1968, the attention focused on four principal issues. The major and immediate issue was to what extent the nuclar-weapon powers would be permitted to delegate control of their nuclear weapons to allies. The agreement on the NPT in 1968 solved that problem by prohibiting any control sharing (Art. I). As a result, the world community of states was divided in two spheres: the five recognised nuclear weapon powers[29] and the many non-nuclear weapon states.

The second, not so immediate, but similarly important issue, was that of preventing independent acquisition of nuclear weapons by new states. The NPT agreement started to solve that problem both legally (Art. II) and by means of a continuous effort including several subsidiary arrangements, e.g. the IAEA safeguard system, the Nuclear Suppliers Group (London Club) and the Zangger Committee transfer restrictions, and physical protection regimes.

The third issue was nuclear disarmament to be negotiated {in good faith" (Article VI). At long last, dramatic nuclear disarmament was agreed upon in recent years, i.e. the INF (1987), START (1991) and START II (1993) agreements. Most important for the establishment of nuclear-weapon-free zones is the withdrawal of non-strategic nuclear weapons from theaters of development and from ships, unilaterally declared in the fall of 1991 by the USA and the USSR[30].

The fourth issue discussed in the 1960s was security guarantees to the many non-nuclear-weapon states agreeing to renounce their option to acquire nuclear weapons. This problem was solved by means of separate documents outside the NPT. In June 1968, the UN Security Council adopted a resolution outlining rules for assisting non nuclear-weapon states parties to the NPT subject to attack or threat of attack by nuclear weapons[31]. Later, all nuclear-weapon powers extended unilateral "nagative"[32] assurances that non-nuclear-weapon states will not be subject to attack or threat of attack with nuclear weapons, however, subject to various conditions.[33] In April 1995, the UN Security Council adopted a modernised resolution taking note of both existing negative nuclear assurances and the positive assurances where the five nuclear-weapon states undertake to provide "immediate assistance, in accordance with the UN Charter, to any non-nuclear-weapon state party to the NPT that is a victim of an act of, or an object of a threat of, aggression in which nuclear weapons are used" (Op. 7)[34].

The 1972 Biological Weapons Convention. Since the agreement on the Geneva Protocol, it was understood that biological and chemical weapons would be considered together. However, when the issue of such weapons again entered the agenda of arms control negotiations with the objective of prohibiting not only the use but also the acquisition and possession of them, keeping them together turned out to be impossible. Prohibiting biological weapons was considered relatively straight forward to negotiate, while a matching measure on chemical weapons for obvious industrial reasons was much more complicated to manage.

A split of the two was agreed upon and a separate convention prohibiting biological and toxin weapons was opened for signature in 1972. Toxins are the very poisonous chemicals produced by biological weapon organisms and the ones causing the illnesses being the weapons effects.

The convention prohibits the development, production, stockpiling, or other acquisition of microbial or other biological agents or toxins having no peaceful usage justification. Delivery means for biological weapons are also prohibited (Art. I:1). The parties are obliged to destroy all their biological and toxin agents for military purpose within nine months after the entry into force of the convention (Art I:2). The convention refers to the 1925 Geneva Protocol prohibiting the use of biological and toxin weapons (Art. VIII).

No special verification machinery was established to monitor the implementation of the convention (Arts. V-VII). To compensate for the lack of verification activities and to enhance the confidence in the convention, many parties have in recent years agreed to implement voluntary confidence-building and transparency-providing measures exchanging regular reports on their peaceful activities in the field. A special formal ad hoc conference of the parties was held in 1994 to discuss such matters. The group has continued its work with the aim of drafting a Protocol, to be linked to the BWC, on verification and confidence-building measures[35].

Each party is assumed to provide or support assistance to other parties "exposed to danger as a result of violation of the Convention", while the decision to assist is left to the UN Security Council (Art. VII).

As a 1 January 1998, the Convention had 140 parties.

The 1993 Chemical Weapons Convention. After the entry into force of the biological Convention in 1975, negotiations continued on a similar ban on chemical weapons. But it lasted until 1993, when the much more complicated Chemical Weapons convention could be opened for signature and to April 1997 until it entered into force. A time-consuming problem was of course how to monitor the large chemical industry producing products for peaceful purposes sometimes very similar to the ones useful also for military application.

The resulting convention is very elaborate; about 200 pages with annexes. It prohibits the development, production, other acquisition, stockpiling, and transfer of chemical weapons to anyone (Art I:1a).

The convention also prohibits the use and preparations for use of chemical weapons (Art. I.1b). The parties are obliged to destroy all chemical weapons in their possession as well as those earlier abandoned on the territories of other parties and all its chemical weapons production facilities (Art. I:2-4). Such destruction shall commence no later than two years and be finished no later than ten years after the entry into force of the convention (Arts. IV : 6-V:8). The parties are supposed not to use "riot control agents as a method of warfare" (Art. I:5).

The convention establishes a special "Organisation for the Prohibition of Chemical Weapons" to manage the considerable reporting, monitoring, and inspection activities foreseen to verify the proper implementation of the agreement (Art. VIII) including a considerable machinery for decision making among the parties. Both the work volume of the "Organisations's routine verification activities and its rights to undertake intrusive fact-finding missions significantly exceed those prescribed by the NPT.

States parties attacked or threatened with chemical weapons are entitled to assistance and protection extended by other parties through the "Organisation" (Art. X:8-11). Each party to the Convention is assumed to support the "Organisation" and provide facilities to make possible assistance to victim parties if required (Art. X:7).

A problem for the immediate future is the long time reqired to abolish the chemical weapons in the world. It will last until 2007 before all chemical weapons and all chemical weapons production facilities shall be destroyed.

In the beginning of 1998, the entering into force process was still going on for many states and will probably continue to do so for some time. As of January 1st, 168 states had signed the Convention and the number of parties had grown to 106.

Objectives and principles of WMDFZs

The background setting

The general arms control objective of preventing the proliferation of weapons of mass destruction and of eliminating the risk of mass destruction warfare is well under way with the growing adherence to global treaties. When regional or zonal measures were considered in the early days, the objective was to find solutions to security problems in specific regions while establishment of global arms control regimes seemed postponed into a distant future. When regional measures are considered today, the objective is to address special problems in specific regions and necessarily as a complement to the global arms control panorama. An Objective to improve the Security situation in a region including elimination of all weapons of mass destruction would naturally include a comprehensive regime free from all weapons of mass destruction that must be based upon global regimes already applying in the region. The most prominent such proposal so far, the Mubarak-plan, will be commented on below.

In the background of global treaties, the NPT is the more important and the more complicated treaty, both because nuclear weapons are considered ultimately dangerous, and because the NPT divides its parties in five states with a right to possess nuclear weapons and all the others without such a right. For the establishment of zones free of weapons of mass destruction, the BWC and the CWC would play an equally fundamental role as basic norms as the NPT, a simplifying difference being that these two conventions do not divide their parties in haves and have-nots. When these treaties have achieved global coverage—which today is true for the NPT and might be true for the two others in the future—the whole world would be a biological-and-chemical-weapon-free zone but not a nuclear-weapon-free zone as long as nuclear disarmament has not been implemented.

Zones free of weapons of mass destruction could then be established assuming that there are no biological and chemical weapons around at all, while special provisions would be necessary to manage the existence of nuclear weapons of the nuclear-weapon states in the wider environment of the zone. It should be noted that no multilateral treaty restricts the possession of states of long range missiles although the idea of a "missile non-proliferation instrument" has been discussed.

"Nuclear-weapon-free zone" is a summary concept describing regimes for regional security, independently or as a complement to other arrangements, global as well as regional. The concept has been relatively well researched. Geographical, political and other regional specifics related to nuclear-weapon-free zones would make different zones different. No such zone would be an exact copy of another.

The terms nuclear-weapon-free zone or zone free of weapons of mass destruction would, however, usually imply the fulfillment of certain objectives and the implementation of certain elements of arms control.

The expansion of the traditional NWFZ concept to cover also all weapons of mass destruction is recent and the literature on the widened subject is limited. The new concept may, however easily fit into the legal frames already developed for nuclear weapon-free zones. This analysis, therefore, is primarily and by necessity based on what has been worked out in terms of nuclear weapons.

The general objective for establishing a nuclear-weapon-free zone or a zone free of weapons of mass destruction would be to relieve a zonal area from the threat of being involved in mass destruction war. But there may be a variety of further objectives for the establishment of such zones in specific cases. Regarding proposed zones in Europe, the objective of geographical separation of the nuclear weapons of the major military blocks during the Cold War has been referred to as such a secondary important objective.

The fulfillment of such objectives shall also be considered as a process in time. History has shown that so far, establishment of zones is a process over decades. In addition, the creation of a nuclear weapon-free zone would always be considered a temporary step and a contribution to a process eventually leading to general nuclear disarmament.

Definitions

What is a zone? States participating in a nuclear-weapon-free zone or a zone free of weapons of mass destruction are free to decide what regime they consider appropriate to fulfill the requirements in their specific region. Indeed, each zone established or proposed so far has been intended to serve purposes specific to each case and that will probably be so in the future as well. None the less, a need for general definitions of the zone concept has been met by the General Assembly of the United Nations and may be of assistance in formulating the arrangements for specific future zone projects.

The UN General Assembly in 1975 defined the concept of a nuclear-weapon-free zone as follows:[36]

"I. Definition of the concept of a nuclear-weapon-free zone

1. A nuclear-weapon-free zone shall, as a general rule, be deemed to be any zone, recognised as such by the General Assembly of the United Nations, which any group of States, in the free exercise of their sovereignty, has established by virtue of a treaty or convention where by:

(a) The statute of total absence of nuclear weapons to which the zone shall be subject, including the procedure for the delimitation of the zone, is defined;

(b) An international system of verification and control is established to guarantee compliance with the obligations deriving from that statute.

II. Definition of the principal obligations of the nuclear weapon States towards nuclear-weapon-free zones and towards the States included therein

2. *In every case of a nuclear-weapon-free zone that has been recognised as such by the General Assembly, all nuclear weapon States shall undertake or reaffirm, in a solemn international instrument having full legally binding force, such as a treaty, a convention or a protocol, the following obligations:*

(a) To respect in all its parts the statute of total absence of nuclear weapons defined in the treaty or convention which serves as the constitutive instrument of the zone;

(b) To refrain from contributing in any way to the performance in the territories forming part of the zone of acts which involve a violation of the aforesaid treaty or convention;

(c) To refrain from using or threatening to use nuclear weapons against the States included in the zone."

Three years later, in 1978, this concept of a nuclear-weapon-free zone was again referred to and elaborated by the Tenth Special Session of the UN General Assembly[37].

Which are the weapons of mass destruction? The Various categories of weapons of mass destruction are among the specific terms that may require an explicit definition in a treaty establishing a nuclear-weapon-free zone or a zone free of weapons of mass destruction. The concept of weapons of mass destruction was defined by the UN Commission for Conventional Armaments already 13 August 1948 as "*atomic explosive weapons, radioactive material weapons,* *lethal chemical and biological weapons, and any weapons developed in the future which have characteristics comparable in destructive effect to those of the atomic bomb or other weapons mentioned above"*[38] or expressed in modern terminology as nuclear, biological, chemical, and radiological weapons, or weapons with similar effects.[39]

None of the multilateral treaties of global application concluded so far contains a physical definition of **nuclear weapons**. The regional Treaty of Tlatelolco attempts an elaborate definition in its Article 5. The Rarotonga (Art 1c) and the Pelindaba (Art. 1c) treaties include in their general definitions also unassembled or partly assembled nuclear devices. While there may be a general understanding of what a nuclear weapon is, the countries seeking to establish a nuclear weapon free zone may wish to define the scope of the nuclear weapon concept, in particular, whether the agreed measures would relate to nuclear warheads, to all nuclear explosive devices, as is the case in Rarotonga, Pelindaba, and Bangkok treaties, or whether to include the delivery vehicles carrying nuclear warheads which is not the case in those treaties.

The 1972 Biological Weapons Convention does not include any explicit definition of prohibited **biological and toxin weapons**. The World Health Organisation concluded in 1970, however, that biological agents are those that depend for their effects on multiplication within the target organisms, and that toxins are poisonous products of organisms not able to reproduce themselves.[40].

The concept of **chemical weapons** is defined in great detail in the Chemical Weapons Convention of 1993. Its Article II elaborates definitions of chemical weapons, toxic chemicals, precursers, key components, old chemical weapons, riot control agents, etc., further specified in its Annex on Chemicals.

The idea of **radiological weapons**—to spread out deadly radiating material on enemy forces and territory—is today considered impractical and no convention prohibiting such weapons has been concluded so far. The related measure of prohibiting attacks on nuclear facilities potentially resulting in spreading radioactive material on a mass destruction scale has been subject to negotiation but no global agreement has been reached. The African nuclear-weapon-free zone treaty includes such a prohibition (Art.11), however, as does a bilateral agreement between India and Pakistan of 1988.[41]

In recent years, **long range missiles** have frequently been considered related to weapons of mass destruction both as carriers of warheads of mass destruction and as instruments of long distance conventional surprise attacks. In 1987, seven western industrialised states agreed to establish a "Guidelines for Sensitive Missile-Relevant Transfers" or the "Missile Technology Control Regime" (MTCR) as it is now usually called. Today, some 30 states have adopted these guidelines and become members of the regime. The common guidelines specifies what export of missile equipment and technology should be restricted in order to prohibit the proliferation of delivery vehicles for weapons of mass destruction. The regime focuses on missile systems including both ballistic missiles and cruise missile with a range exceeding 300 kilometres and with a payload capability exceeding 500 kilograms. In 1994, Canada initiated a discussion of a "missile non-proliferation instrument" as a general continuation and codification of the MTCR.[42]

The UN Security Council resolution on Iraq[43], adopted following the 1991 Gulf War, includes definition type specifications on weapons of mass destruction. Besides nuclear (Op. 12), chemical, and biological weapons (Op. 8a), the resolution also addresses "ballistic missiles with a range greater than 150 kilometers" (Op. 8b).

Important objectives

Within the context of *"the ultimate objective of achieving a world entirely free of nuclear weapons"*, as set forth by the UN General Assembly in the Final Document of the Tenth Special Session, several other objectives having regional or, in some cases, also wider significance can be identified and, depending on the circumstances in each case, may be pursued or specified in a zonal agreement. The relevance and relative emphasis of such objectives may vary from one region to another. The subsequent evolution, i.e. development and improvement over time of a zone agreement, would also be possible and, in some cases, feasible. Without prejudice to other objectives, which may be added according to the needs in specific cases, the following general objectives of zones free of weapons of mass destruction could be noted as important:

(a) To spare the zonal states from the use or threat of use of mass destruction weapons;

(b) To contribute to averting potential mass destruction threats and, thereby, to reducing the danger of war, in particular war utilising mass destruction means;

(c) To contribute to the process of disarmament, in particular the elimination of weapons of mass destruction;

(d) To contribute to regional and world stability and security;

(e) To impede proliferation of weapons of mass destruction horizontal, vertical as well as geographical;

(f) To provide increased security at lower levels of armament;

(g) To strengthen confidence and improve relations between zonal states;

(h) To facilitate and promote co-operation in and the development of technologies for peaceful purposes closely related to weapons of mass destruction, within the region and between zonal and extra-zonal states.

Geographical considerations

No precise requirements can be set as regards the suitable size of arms control zones, which could range from whole continents to small areas. Sometimes a zone may be initially established in a limited area and later extended as other countries agree to join in. If large parts of the world are to be kept free from nuclear weapons or other weapons of mass destruction, extending such zones to whole continents would provide the natural way to achieve that aim.

The geographical extent of a zone would depend on the specific characteristics of the region and the precise arms control objectives to be realised.

A single state could declare itself a zone. One example is Mongolia which acted to establish itself a nuclear-weapon-free zone. Normally, however, a zone would comprise the national territories of two or several neighbouring states including their territorial waters and airspace and established by agreement among those states. It would also be possible for states separated from each other by high sea areas or otherwise to form a zone.[44]

Furthermore, a zone might include geographical areas not under the jurisdiction of any state, for instance sea areas beyond territorial waters, or it may in its entirety be established in such areas, Antarctica being a well known example.

Delimitation of a zone would generally depend on two concepts: the zonal outer perimeter and the territory for which the zone agreement is in force. Both concepts were applied in the Tlatelolco and Rarotonga treaties while no specific perimeter was defined for the Pelindaba and Bangkok zones.

The perimeter is a line encircling the entire area of the zone within which all states would be eligible to join the zone agreement. The perimeter may encompass international areas as well. To avoid confusion, the perimeter should, if possible, not cut through the interior of individual states. States outside the perimeter would be considered extra-zonal, but may be invited to assume responsibilities related to the zone and regarding dependencies they may have inside the perimeter.

Where no explicit perimeter is defined, as in the African and the South East Asian cases, the definition of which states would be eligible to be zonal states, would also define an implicit perimeter encircling the combined territories of the cluster of such states. In such cases, the zonal area could, as in the African case, form several sub-areas not in topographic contact.

The Treaty of Tlatelolco defines the outer perimeter of the Latin American zone in its Article 4:2. The Treaty of Rarotonga does generally the same for the South Pacific zone in its Annex 1, as illustrated by a map attached to the Annex. The perimeters of these two zones determine which states are eligible to be parties to the zone treaties and also which extra-zonal states will be invited to assume matching responsibilities regarding their dependencies inside the perimeters. The perimeter of Antarctica is the latitude line 60° S, encompassing also an ocean area and a number of Antarctic islands, all international territory.

The area of application of a zone would be the territory, including land areas, internal, territorial, and archipelagic waters, and air space, of parties for which the nuclear-weapon-free zone treaty and related protocols are in force. Zonal commitments could also be applied to international areas such as Antarctica or parts of the high seas upon entry into force of special legal instruments. The area of application would grow in size as more states adhere to the zone agreement but could by definition not extend beyond the perimeter.

While a zone could thus come into force with a limited number of parties rather early, the addition of further states within the perimeter may (but must not necessarily) go on for a considerable time. The Tlatelolco Treaty was agreed in 1967, and today all but one of the eligible states are parties. The last one, Cuba, is in the process of completing its accession procedure. The Rarotonga Treaty was agreed in 1985 and the entry into force process is still going on. The long time needed for the entry into force process should cause no surprise. Firstly, a fair number of prospective parties would have to adapt to the new commitments. Secondly, nuclear weapon states and other extrazonal states invited to participate by adhering to protocols may also need time to adapt, so more so as those states usually have had an indirect influence only on the drafting of the treaty and the protocols they were invited to subscribe to.

The perimeters of the Tlatelolco, Rarotonga and Antarctic zones encompass large areas of **international waters**[45]. In the Rarotonga Treaty case, the larger part of the area inside the perimeter consists of such waters. As the UN Convention on the law of the Sea (UNCLOS)[46] prescribes, states parties to a nuclear-weapon-free zone treaty (or any group of states) cannot by agreement among themselves institute generally applicable restrictions in such waters which all states have the right to enter and use. Zonal states can agree to restrict only themselves in international sea areas. In addition, they can invite extra-zonal states to subscribe to special provisions and to respect general zonal restrictions, e.g. by signing special protocols to that effect. This is done in the Rarotonga Treaty and the African zone treaty as regards the prohibition of nuclear testing and dumping of radioactive materials.

One element of a zone arrangement could be **"thinning-out"**, i.e. withdrawal or other measures regarding weapons, military forces or military activities in an area adjacent to the zone, the purpose being to enhance the security of the zonal states and the credibility of the assurances extended to the zone by extra-zonal states.

Such security areas adjacent to the zone could be both land and sea areas. They would have to conform to specific conditions in each case and could be based upon agreements reached among the countries directly concerned. Measures of this kind could also be defined in functional terms, that is, in terms of the relations that relevant weapons, forces and military activities could have to the zone. In the latter case the extension of the "adjacency" would implicitly be related to the ranges of these weapons, forces and activities.

With more and more nuclear-weapon-free zones established in the world, it sometimes happens, for geopolitical and other reasons, that proposed zones **overlap** each other. One example is the proposed nuclear-weapon-free zone in the Middle East, two thirds of which will overlap the African zone according to current proposals. Such overlap is certainly possible and could politically be very desirable but would require legal harmony between the treaties establishing the overlapping zones.

An example of the opposite is the provision in the Rarotonga Treaty (Annex 1.B) prescribing that certain Australian islands now part of the South Pacific Nuclear-Free Zone could be withdrawn from the zone, if such areas later become *"subject to another treaty having an object and purpose substantially the same"* as that of the Rarotonga Treaty.

Sometimes, a **territorial dispute** between prospective zonal states or between a prospective zonal state and an extra-zonal state has complicated accession to a zone agreement. It is obvious that a state assuming treaty commitments in relation to its entire territory could not implement such commitments in a part of its claimed territory which is *de facto* controlled by another state. It may also be complicated for a state to assume zone responsibilities for areas under firm national control but claimed by others.

Basic measures and obligations

There would be three measures of central importance for the achievement of the objectives of a zone free of weapons of mass destruction in the general case. These are:

- the **non-possession** of prohibited weapons by zonal States;
- the **non-stationing** of prohibited weapons by any State with in the geographical area of application of the zone, and
- the **non-use or non-threat of use** of prohibited weapons throughout the zone or against targets within the zone.

The meaning of these measures might seem clear enough. However, their legal representation could be complicated, as shown e.g. by the definition of "nuclear weapon" in the Tlatelolco Treaty (Art. 5) and by the definition of chemical weapons in the Chemical Weapons Convention (Art. II).

The non-possession measure would apply to zonal states. Its codification could be much simplified if relying on the concepts of the Non-Proliferation Treaty (Article II), the Biological Weapons Convention, and the Chemical Weapons Convention. If the zone encompasses only territories of states parties to the NPT, the BWC, and the CWC, most of the non-possession requirement would be fulfilled. Only non-possession of long range missiles might require special regulation in detail in the absence of a comprehensive treaty on such missiles. If the zone is to encompass states which are not parties to one or several of these treaties or states which are nuclear weapon states, a special regime must be defined. The same would be true in the special case that only a part of a state will be included in the zone.

The non-stationing measure would primarily apply to the terrirories of zonal states with the exception that zonal states could not by agreement among themselves restrict or prohibit innocent passage (or transit passage) by vessels of nuclear weapon states and other extra-zonal states with prohibited weapons on board in their territorial and archipelagic waters.

Non-stationing measures applying to international land and sea areas would require special legal arrangements.

The founding legal instrument of the zone must also define whether it would be only the warheads that should not be present in the zone or if the prohibition should also include all or some of their delivery vehicles and installations being integral parts of weapon systems.

Related to the non-stationing measure is {transit" of prohibited weapons through zonal territory—an issue primarily related to nuclear weapons. The transit concept would include "innocent" transit over a limited period of time of otherwise prohibited weapons by an extra-zonal state, on land, by air or internal waters including calls at ports by ships or landing of aircraft carrying such weapons. Universal adherence to the BWC and the CWC would limit consideration of transit to nuclear weapons and missiles.

The transit issue was extensively discussed when the Latin American zone was negotiated. The problem was solved by not being solved. Transit was left to the individual zonal states to permit or deny in each case[47]. The South Pacific zone has a similar transit regime as has the African and the South East Asian zones.

A zonal treaty should prescribe if transit would be generally prohibited or arranged in a way similar to the Tlatelolco formula. Transit through zonal high sea areas or through territories which are dependencies of extra-zonal states could not be permitted without making the zonal regime of such areas an illusion[48].

While "innocent transit" has been considered tolerable under all zone regimes so far, "hostile transit" would probably not be accepted, i.e. passage of delivery vehicles with prohibited weapons across zonal territory towards targets beyond the zone. This rule would apply to seaborne and airborne manned or unmanned vehicles and to ballistic missiles in so far as they penetrate zonal air space, while crossing overhead zonal territory in international space could not be prohibited by agreement among the zonal states.[49]

The special transit issue of ships and aircraft which may carry nuclear weapons onboard and call at ports or land at airports in zonal states has been particularly sticky because nuclear weapon powers usually "neither confirm nor deny the presence or absence of nuclear weapons onboard specific ships or aircraft at specific ships or aircraft at specific times"[50]. A political problem of considerable dimension some years ago, the neither confirming nor denying issue has lost most of its former importance following the withdrawal by nuclear-weapon powers of sub-strategic nuclear weapons from navals ships.

The non-use measure would be a commitment by states controlling prohibited weapons. Legally, this provision has been given the form of a separate protocol to existing zone agreements.

Consideration of the non-use measure should be made against the background of ongoing negotiations on general negative security assurances at the Conference on Disarmament in Geneva and the assistance and support to victims of biological and chemical warfare effects prescribed by the BWC and the CWC. All five nuclear-weapon states have made unilateral declarations that they would not attack or threaten at attack with nuclear weapons states that do not possess such weapons themselves or host those of others on their territories. These declarations are not coordinated and include some conditions and reservations[51]. The reservations are linked to the question whether a state can be a member of a nuclear-weapon-free zone and also of a military alliance with a nuclear-weapon state simultaneously. That is certainly possible provided, however, that the two sets of commitments are not contradictory.

Noted should also be the recent UN Security Council resolution taking note of both existing negative nuclear assurances and the positive assurances where the five nuclear-weapon states undertake to provide "*immediate assistance, in accordance with the UN Charter, to any non-nuclear-weapon state party to the NPT that is a victim of an act of, or an object of a threat of, aggression in which nuclear weapons are used*" (Op. 7).

Also significant in this regard are the provisions for assistance to and support of states victims of biological and chemical weapons effects as outlined in the BWC and the CWC.

No general policy commitments related to long range missile attack or surprise attack exist within the international community. For a zonal agreement, such guarantees must thus be drafted from scratch. A zone free of weapons of mass destruction prohibition also on long range missiles from the zone would obviously loose much of its meaning if not matched by a non-use measure covering such missiles.

Linked to the non-use measure has been the idea mentioned above that this measure should be complemented by a "thinning-out" arrangement in areas adjacent to the proposed zone where nuclear weapons are deployed. The "thinning-out" idea implies that such weapons should be withdrawn that are targeted against the zone or that have short ranges and are deployed very close to the zone, thus making them usable primarily against the zone. If such weapons are not withdrawn, non-use commitments would be less credible.[52]

Complaints and Control Procedures

It is traditionally recognised that effective implementation of an agreement on a nuclear-weapon-free zone or on a zone free of weapons of mass destruction would require a system of verification to ensure that all states involved, zonal as well as extrazonal, comply with their obligations.

The precise scope and nature of such a system would vary from zone to zone and depend upon the nature of the obligations prescribed. Generally, a zonal treaty would have to include provisions both for verifying compliance and a complaints procedure for settling issues of suspected non-compliance, should such cases arise. As relevant, the zone verification system could for its implementation be based on verification enforced according to the NPT, the BWC, and the CWC.

In general, subject to verification under a treaty on a zone free of weapons of mass destruction or a nuclear-weapon-free zone should be:

(a) all activities of zonal states related to prohibited weapons to ensure that peaceful activities are not diverted to the manufacture of weapons;

(b) the commitment that no prohibited weapons are present within the zone; special regimes would be required for sea areas;

(c) the removal of prohibited weapons that may be present within the zonal area at the time of entry into force of the zone agreement, possibly also requiring an account of the weapons history of participating zonal states;

(d) the implementation of other measures associated with the zone agreement.

Most verification related to peaceful nuclear activities of zonal states could be entrusted to the safeguards system of the International Atomic Energy Agency (IAEA).

In some regions, the zonal parties may prefer to establish standing organs or special bodies for carrying our verification. In regions where sharp conflicts exist, entrusting the task of verification to an international organisation, perhaps supplemented by bilateral arrangements, might be preferred.

There is also the possibility that an agreement on a zone would provide to any party a right to undertake verification activities in other states parties to the zonal agreement, including on-site inspection. One model for such a system could be the verification system laid down in several arms control agreements adopted within the Conference on Security and Cooperation in Europe (CSCE), i.e. the Stockholm and Vienna Documents on confidence—and security-building measures and the Conventional Forces in Europe Treaty.[53] These treaties give each party the right to undertake inspections in the territory of any other party, subject to annual quotas, and obliges every party to receive and accommodate on short notice such inspections in its own territory.

An Example in Point: The Middle East as a NWFZ or WMDFZ application

The combination of political conflicts and nuclear programmes of size in the Middle East provides both the political incentives and a technological basis for nuclear weapon proliferation in the region. This has been understood for long time. This has also been considered unfortunate for long time. The current conflict pattern in the Middle East, while attracting the involvement of major powers, is regional. The possible ambitions of the countries in the area to acquire nuclear weapons have their roots in this regional context.

Political efforts to change this situation have focused on the possibility to establish a nuclear-weapon-free zone in the area. Back in 1974, Iran supported by Egypt raised the issue in the UN General Assembly[54]. Since that time, the UN General Assembly has every year adopted a resolution recommending the establishment of a nuclear-weapon-free zone in the Middle East (NWFZME). Since 1980, this annual resolution has been adopted by consensus, i.e. with the support of all Arab states, Iran and Israel.[55]

In 1990, President Mubarak of Egypt proposed the establishment of a zone free of weapons of mass destruction in the Middle East (WMDFZME)[56].

Today, the expansion of the Middle East zone concept to include all weapons of mass destruction and also their means of delivery has been politically accepted. In May 1995, the 1995 Review and Extension Conference of the NPT parties adopted a resolution on the Middle East recognising that the current peace process contributed to *"a Middle East zone free of nuclear weapons as well as other weapons of mass destruction"* and claiming upon all states in the Middle East to take practical steps towards *"the establishment of an effectively verifiable Middle East zone free of weapons of mass destruction, nuclear, chemical and biological, and their delivery systems"*[57].

The United Nation's Expert Study

In the fall of 1988, the annual resolution then initiated by Egypt, also requested the Secretary General to *"undertake a study on effective and verifiable measures which would facilitate the establishment of a nuclear-weapon-free zone in the Middle East"*[58]. The report[59] was prepared before Iraq's invasion of Kuwait in August 1990, but submitted to the General Assembly after that invasion. It was, however, welcomed and adopted by consensus that same year.[60]. The UN report includes an account of the history of the issue in the United Nations. The report was later followed up by this author together with James F. Leonard also taking into account the Mubarak plan[61]. The issue of establishing a nuclear-weapon-free zone in the Middle East has also been researched by the Egyption scholar and diplomat Mahmoud Karem.[62] Important analytical contributions were in 1997 made by Shai Feldman and Abdullah Toukan.[63] The effect of a comprehensive test ban on nuclear proliferation risks in the Middle East, including Iran, Iraq, and Israel has been analysed by Eric Arnett.[64]

The Geographical Middle East Concept

The Middle East is a well-known and traditional geographical concept used in everyday political discussion. Defining the geographical extension of the Middle East for arms control application is not trivial, however.

Different definitions have for a long time been used for different purposes. One was introduced in 1989 by the International Atomic Energy Agency (IAEA) when discussing the application of safeguards in relation to the Non-Proliferation Treaty (NPT) or a nuclear-weapon-free zone in the area, i.e. *"the area extending from the Libyan Arab Jamahiria in the West, to the Islamic Republic of Iran in the East, and from Syria in the North to the People's Democratic Republic of Yemen in the South"*[65]. The UN study referred to above found the IAEA concept somewhat limited for its purpose and suggested an area that eventually could encompass *"all states members of the League of Arab States (LAS[66]), the Islamic Republic of Iran and Israel"*.

A definition adequate for legal application of a NWFZME or a WMDFZME regime may or may not coincide with those used earlier for different purposes. Such a definition should encompass all states with a primary security relevance to each other. On the other hand, an ambition to include all states with any security relevance to each other would easily result in a Middle East concept that would include most of the Old World.

The area should thus at least include the actors central to the specific conflicts of the Middle East. The most publicised is the Arab-Israeli conflict. But there are also other conflicts involving many of the same states as demonstrated by the recent examples of the Iran-Iraq war, the Gulf War, and the Polisaria conflict.

As a project definition must be based on the current political geography, it seems relevant, for the purpose of this paper, to define the basic Middle East area according to the UN formula mentioned above. But it should also be understood that the application of a zone regime may begin in a smaller area of a few core states and later expand to finally encompass the entire basic area. It might also be feasible to expand the zone area to include some adjacent sea areas.

Shared views

Although negotiations to overcome the conflicts in the Middle East are difficult, the consultations undertaken when preparing the UN report in the summer of 1990 showed a surprising degree of common view on fundamental matters among many of the states in the area; Arab states as well as Iran and Israel. Among the shared views were:

- The process to establish a NWFZME or a WMDFZME would take several years;
- The geographical concept suggested in the report was generally accepted;
- Positive security assurances beyond those outlined in Security Council Resolution 255 (1968) would be necessary. If a zonal state would be subject of aggression, guarantors should assist the victim, punish the aggressor and provide recovery support as necessary. It is intriguing to notice that such far reaching guarantees did apply just a few months later in order to liberate Kuwait after it had been annexed by Iraq.
- Verification procedures much more far-reaching than those prescribed under the NPT would be necessary. The IAEA operations later undertaken in Iraq under a Security Council mandate show what will be necessary. In addition, Israel indicated a need for bilateral verification rights similar to those prescribed in several arms control agreements adopted within the Conference on Security and Cooperation in Europe (CSCE), in order not to be outvoted within international bodies deciding by majority rule.
- Initial confidence-building measures would be an effective method to support the process of establishing a NWFZME.

- Although Israel was generally considered a nuclear weapon state[67], a view neither encouraged nor denied by Israel itself[68] nuclear weapons were considered political rather than warfighting instruments. Chemical weapons were on the other hand considered as operative instruments of warfighting, as they could easily be manufactured, were known to exist, had been used in recent years in the Middle East.

Because of the above-mentioned common views, a nuclear weapon-free zone or a Mubarak-zone in the Middle East could be considered a realistic project, although the establishment of such a zone would most probably take some time.

Objectives and Measures

A treaty establishing a zone free of weapons of mass destruction in the Middle East, based on the global treaties prohibiting atomic, biological, and chemical weapons (NPT, BWC, and CWC) would share the general objectives of those treaties which is the complete renunciation of those weapons, except that the nuclear weapons of the five nuclear weapon states will remain until nuclear disarmament is completed. In addition, the general objectives referred to above would apply, as would the non-possession, the non-stationing, and the non-use measures.

Furthermore, the Mubarak plan outlined when presented three general components:

(a) All weapons of mass destruction in the Middle East should be prohibited;

(b) All states of the region should make equal and reciprocal commitments in this regard;

(c) Verification measures and modalities should be established to ascertain complete compliance by the states in the region; The proposal also pointed to certain terms to be taken into account:

(d) A qualitative as well as quantitative symmetry of the military capabilities of individual states of the Middle East. Assymetries cannot prevail in a region striving for a just and comprehensive peace;

(e) Increased security at lower levels of armament. Security must be attained through political deliberations and disarmament rather than the force of arms;

(f) Arms limitation and disarmament agreements should consider equal rights and responsibilities, and states should equally issue legally binding commitments in the field of disarmament.

The obvious way of designing a draft treaty on a zone free of weapons of mass destruction is to begin drawing on the application of general arms control treaties in the area. In case there are states in the area which are both non-parties to such treaties and essential for the operation of a zone, such states should, as part of the establishment process of the zone, be requested to subscribe to all those treaties. The situation as regards the global treaties was at the time of writing as follows:

Among the 23 states in the region, as defined above, 17 were parties to the **1925 Geneva Protocol** prohibiting use of chemical and bacteriological warfare, while 6 were not: Comoros, Djibouti, Mauritania, Oman, Somalia, and United Arab Emirates.

All states in the region but Israel were parties to the **Non-Proliferation Treaty**.

12 states in the prospective Middle East zone were parties to the **Biological Weapons Convention**, while 11 were not: Algeria, Comoros, Djibouti, Egypt (signatory), Israel, Mauritania, Morocoo (signatory), Somalia (signatory), the Sudan, Syria (signatory), and United Arab Emirates (signatory).

The entry into force process of the **Chemical Weapons Convention** is currently under way. 16 states within the prospective zone have signed the convention: among them, 10 have become parties. 7 states have not signed: Egypt, Iraq, Lebanon, Libya, Somalia, the Sudan, and Syria.

Among the 23 states in the region, all but Jordan are coastal states. Among the latter, 12 were not parties to the **Sea Bed Treaty**, i.e, Bahrain, Comoros, Djibouti, Egypt, Israel, Kuwait, Mauritainia, Oman, Somalia, the Sudan (signatory), Syria, and the United Arab Emirates. Their accession to the treaty would be a desirable contribution as long as one nuclear-weapon-power, France, is not bound by the treaty.

Also basic for drafting the non-use provisions of a Middle East zone agreement would be the security guarantees provided to NPT parties by the UN Security Council resolutions S/RES/984 (1995) and S/RES/255(1968) as well as the unilateral negative guarantees extended by the five nuclear-weapon powers. The Biological and Chemical Conventions include non use provisions.

The establishment of a WMDFZME building on a general subscription in the region to the global treaties must be complemented by several other provisions.

One important such addition would be special commitments not to possess or deploy ballistic and cruise missiles with ranges exceeding 300 (or 150) kilometers.

Two special guarantee protocols would be desirable. The nuclear weapon powers would be invited to sign one of them to commit themselves not to use or threaten to use weapons of mass destruction against zonal states—the Biological and Chemical Conventions include non-use provisions. Neighbouring states would be invited to sign the other to commit themselves to support the zone regime and to assist in its implementation particularly as regards border policies. In addition, both protocols should include a commitment not to direct prohibited missiles against targets in the zone. The guarantee protocols might also define possible restrictions applying to sea areas adjacent to the zone.

Special provisions prohibiting nuclear weapon testing, dumping of radioactive waste, and attacks on nuclear facilities containing large amounts of radioactive material could be included in the zonal legal instruments.

Special organisations for implementation and supervision of the zone arrangements would have to be instituted.

A verification machinery could be based on those of the general arms restriction treaties applying in the region as complemented by special verification rights similar to those operating according to other zonal treaties and the OSCE regime.

An organisation for cooperation in the field of nuclear energy production for peaceful purposes in the region, similar to Euratom and ABACC, could contribute substantially to confidence-building in the region. This aspect has been analysed by Dr. Mustafa Kibaruglo[69].

References

* The Author is a visiting scholar at the Swedish Institute of International Affairs (P.O. Box 1253, S-111 82 STOCKHOLM, Sweden; tel + 46-8-23 40 60, fax +46-8-20 10 49; email prawitz @ ui.se). He was a Senior Research Fellow at the Swedish National Defence Research Establishment, the Division of Nuclear Weapons Physics, in 1995-96, and a Special Assistant for Disarmament in Sweden's Ministry of Defence 1970-1992. He served as an Advisor in Sweden's Disarmament Delegation 1962-1992.

** This work was supported by **The Swedish Agency for Civil Emergency Planning**. Views and opinions expressed in this paper are those of the author and do not imply the expression of any position on the part of the Swedish Institute of International Affairs. **Final text released 12 January 1998.**

Bibliography

1. UN Document DC/SC.1/41.
2. Un Document A/PV. 697, also called the Rapacki-plan after the Minister for foreign Affairs of Poland at the time. Mr Adam Rapacki (1906-1970) was Poland's Foreign Minister 1956-1968.
3. UN Documents A/RES/2373 (XXII) and S/RES/255. The Treaty on the Non-Proliferation of Nuclear Weapons (*UN Treaty Series*, Vol. 729, No. 10485) was opened for signature on 1 July 1968 and entered into force on 5 March 1970.
4. The Antarctic Treaty (*UN Treaty Series*, Vol. 402, No. 5778).
5. The term "densely populated, area is frequently used to distinguish the Latin American and the South Pacific zones from the Antarctica with some states for political reasons prefer tod designate as a "populated" area rather than the unpopulated" place it is otherwide considered to be.
6. The Treaty for the Prohibition of Nuclear Weapons in Latin America and the Caribbean (*UN Treaty Series*, Vol. 634, No, 9068).
7. The South Pacific Nuclear Free Zone Treaty (*UN Treaty Series* No. 24592).
8. Treaty on the Southeast Asia Nuclear Weapon-Free Zone signed by member states and potential member states of ASEAN at a summit meeting in Bangkok on 15 December 1995. For text, see *SIPRI Yearbook 1996*, Oxford University Press, 1996, pp 601-609.
9. The Pelindaba Text of the African Nuclear-Weapon-Free Zone Treaty signed at an OAU meeting in Cairo on 11 April 1997. For text, see *SIPRI Yearbook 1996*, Oxford University Press, 1996, pp 593-601.
10. Joint Declartion of South and North Korea on the Denuclearisation of the Korean Peninsula, signed on 31 December 1991 and entered into force on 19 February 1992. For text, see UN Document CD/1147, or J. Goldblat, *Arms Control: A Guide to Negotiations and Agreements*, Sage Publications, London, 1994, pp 643-644.
11. Treaty on Principles Governing the Activities of States in the Exploration and

Use of Outer Space, including the Moon and Other Celestial Bodies (Outer Space Treaty; *UN Treaty Series,* Vol. 610) was opened for signature on 27 January 1967 and entered into force on 10 October 1967. As of 1 July 1997, the treaty had 95 parties including all nuclear-weapon states.

12. Treaty on the Prohibition on the Emplacement of Nuclear Weapons and Other Weapons of Mass Destruction on the Sea-Bed and the Ocean Floor and in the Subsoil Thereof (The Sea-Bed Treaty; UN Document A/RES/2660 (XXV), Annex) was opened for signature on 11 February 1971 and entered into force on 18 May 1972. As of 1 July 1997, the treaty had 93 parties including all nuclear-weapon states but France.

13. Compare status report in document NPT/CONF. 1995/PC. III/5 (12 July 1994).

14. Compare S.P. Subedi, *Land and Maritime Zones of Peace in International Law,* Clarendon, Oxford, 1996; H. Coutau-Begarie, *LeDésarmement Naval* (Naval Disarmament, In French), Ecconomica, Paris, 1995, in particular Part I, *Le Désarmement Géographique,* pp 23-164; and L. Hanikainen, F. Horn (Ed.), *Autonomy and Demilitarisation in International Law: The Åland Islands in a Changing Europe,* Kluwer, 1997, in particular C. Ahlström, *Demilitarised and Neutralised Zones in a European Perspective,* pp. 41-56, based on his earlier paper *Demilitariserade och neutraliserade områden i Europa* (Demilitarised and Neutralised Areas in Europe, In Swedish), Meddelanden från Ålands Högskola Nr 7, Mariehamn, 1995.

15. Protocol Additional to the Geneva Conventions of 12 August 1949 and Relating to the Protection of Victims of International Armed Conflict (Protocol I), Art. 60.

16. UN Document CD/989, 20 April 1990.

17. R. Trapp (Ed.), *Chemical Weapon Free Zones*?, SIPRI, Oxford University Press, 1987.

18. UN Document S/RES/687 (1991), Op. 14.

19. The Protocol for the Prohibition of the Use in War of Asphyxiating, Poisonous, or Other Gases, and of Bacteriological Methods of Warfare *(League of Nations Treaty Series,* Vol. XCIV (1929), No. 2138), was signed on 17 June 1925.

20. The Treaty on the Non-Proliferation of Nuclear Weapons (*UN Treaty Series,* Vol. 729, No. 10485) was opened for signature on 1 July 1968 and entered into force on 5 March 1970.

21. The "Convention on the Prohibition of the Development, Production, Stockpiling, and Use of Chemical Weapons and on their Destruction" was opened for signature on 13 January 1993 and entered into force on 29 April 1997.

23. The 1868 St. Petersburg Declaration to the Effect of Prohibiting the Use of Certain Projectiles in Wartime, prohibited the use between the parties of *"any projectile of a weight below 400 grammes, which is either explosive or charged with fulminating or inflammable substances",* The 1874 Brussels Declaration concerning the laws and Customs of War, Prohibited the "employment of poison or poisoned weapons" (Art. XIII); The 1899 Hague Declaration (IV, 2) concerning asphyxiating gases, added a prohibition of the *"use of projectiles the sole object of which is the diffusion of asphyxiating or deleterious gases";* and The 1907 Hague Convention (IV) respecting the laws and customs of war

on land repeated these provisions (Art. 23a). These early agreements have mostly historical interest today, as some of the contracting parties which signed at the time, do not exist any more, and as most states of today did not exist at the time.

24. The Protocols preamble includes the following para: *To the end that this prohibition shall be universally accepted as a part of International Law, binding alike the conscience and practice of nations;...*

25. UN Document A/RES/2603 A (XXIV) was adopted by a vote of 80 in favour, 3 against, and 36 abstentions (7 absent), a result reflecting the ongoing and controversial war in Viet Nam.

26. Document NPT/CONF. 1995/32/DEC. 3 (11 May 1995).

27. Compare J. Simpson, *Inter-Relations Between Regional and Global Approaches to Nuclear Non-Proliferation,* in A. Mack (Ed.), *Nuclear Policies in Northeast Asia,* Research Report UNIDIR/95/16 (UN Sales No. GV.E.95.0.8).

28. The NPT, ARt. VII reads: "*Nothing in this treaty affects the right of any group of states to conclude regional treaties in order to assure the total absence of nuclear weapons in their respective territories*". For an assessment of this role of the NPT, compare *Implementation of Article VII of the Treaty on the Non-Proliferation of Nuclear Weapons* (Document NPT/CONF. 1995/PC. III/5, 12 July 1994).

29. According to the NPT, Art. IX: 3, "*a nuclear weapon state is one which has manufactured and exploded a nuclear weapon or other nuclear explosive device prior to 1 January 1967*". This provision was intended to prevent a race for nuclear-weapon power status by prospective treaty parties before signing. The formula defines China, France, Great Britain, the Soviet Union (after 24 December 1991 succeeded by the Russian Federation) and the United States as nuclear weapon states. India which is not a party to the NPT, did manufacture and explode a nuclear device "for peaceful purposes" in May 1974 but is usually not considered a nuclear weapon power. In March 1993, it was revealed that South Africa had fabricated six nuclear explosive devices. These devices were later dismantled and, in 1991, South Africa became a non-nuclear weapon party to the NPT.

30. Unilateral declarations by Presidents Bush of the USA and Gorbachev of the USSR, on 27 September and 5 October 1991 respectively. On 29 January 1992, after the dissolution of the Soviet Union, president Yeltsin of Russia declared himself committed to the previous Soviet declaration. For the full text of the Bush, Gorbachev, and Yeltsin-statements, see e.g. *SIPRI Yearbook* 1992, Oxford University Press, 1992, pp 85-92.

31. UN Document S/RES/255 (1968) which i.a. "*welcomes the intention expressed by certain states (USSR, USA, UK) that they will provide or support immediate assistance, in accordance with the Charter, to any non nuclear-weapon State Party to the Treaty on the Non-Proliferation of Nuclear Weapons that is the victim of an act or an object of a threat of aggression in which nuclear weapons are used*" (Op. 2). The value of this guarantee was limited, however, as four of the five nuclear-weapon states at the time were permanent members of the Security Council with a right of veto.

32. "Negative" guarantees imply that the guarantor abstains from nuclear aggression as different from "positive" guarantees implying that the guarantor actively supports a victim of aggression.

33. The content of these unilaterally declared guarantees are summarised in *Compilation of Basic Documents relating to the Question of Effective International Arrangements to Assure Non-Nuclear-Weapon States against the Use of Nuclear Weapons* (UN document CD/SA/WP. 15, 16 March 1993) and in *Developments with regard to effective arrangements to assure non-nuclear-weapon states against the use or threat of use of nuclear weapons* (document NPT/CONF. 1995/PC.III/6, 12 July 1994).. Compare also The United Nations DISARMAMENT YEARBOOK VOL. 14:1989 PP 179 - 180. The "Basic Provisions of the Military Doctrine of the Russian Federation" adopted on 2 November 1993 (Decree No. 1833) does not include the USSR no-first-use declaration of 12 June 1982, however.

34. UN Document S/RES/984 (1995), unaminously adopted on 11 April 1995, Op. 7. The basic declarations were made on 5 and 6 April 1995 by the Russian Federation (UN Document S/1995/261), the UK (S/1995/262), the USA (S/1995/263), France (S/1995/264), and China (S/1995/265). Again, the value of this guarantee is limited as the nuclear-weapon states are also permanent members of the Security Council with a right of Veto. The "Threshold states" are not, however.

35. Compare documents BWC/AD HOC GROUP/38 (6 October 1997) and BWC/CONF. IV/9 Part II (6 December 1996).

36. UN Document A/RES/3472 B (XXX).

37. The Final Document of the Tenth Special Session of the General Assembly states i.a. "60. *The establishment of nuclear-weapon-free zones on the basis of arragements freely arrived at among the States of the region concerned constitutes an important disarmament measure.*

61. The process of establishing such zones in different parts of the world should be encouraged with the ultimate objective of achieving a world entirely free of nuclear weapons. In the process of establishing such zones, the characteristics of each region should be taken into account. The State paticipating in such zones should undertake to comply fully with all the objectives, purposes and principles of the agreements or arrangements establishing the zones, thus ensuring that they are genuinely free from nuclear weapons.

62. With respect to such zones, the nuclear weapon States in turn are called upon to give undertakings, the modalities of which are to be negotiated with the competent authority of the zone, in particular:

(a) To respect strictly the status of the nuclear weapon free zone;

(b) To refrain from the use or threat of use of nuclear weapons against the States of the zone."

38. UN Document RES/S/C. 3/30.

39. It could be noted that the 1977 Convention on the Prohibition of Military or

Any other Hostile Use of Environental Modification Techniques (ENMOD) prohibits certain means of warfare *"having widespread, long-lasting or severe effects"* (Art. I:1). An authrotiative interpretation defined 'widespread' as *"encompassing an area of the scale of several hundred square kilometers"*; 'long-lasting' as "lasting for a period of months, or approximately a season"; and 'severe' as *"involving serious or significant disruption or harm to human life, natural and economic resources or other assets"* (Document CCD/520, 3 September 1976). The 1977 Protocol Additional to the Geneva Conventions of 12 August 1949, and Relating to the Protection of Victims of International Armed Conflicts (Protocol I) refers to *"widespread, long term, and severe damage to the natural environment"* (Arts. 35:3 and 55:1). The similar wordings of the ENMOD Convention and Protocol I are not intended to coincide but were drafted to be two separate concepts.

40. Health Aspects of Chemical and Biological Weapons. Report of a WHO group of Consultants. 1970.

41. The "Agreement Between Pakistan and India on the Prohibition of Attack Against Nuclear Installations and Facilities" was agreed in December 1988 and entered into force in January 1991. For text, see e.g. j. Goldblat, *Arms Control: A Guide to negotiations and Agreements*, PRIO, Sage, 1994, pp 537-538.

42. The situation regarding missile proliferation has been analysed by Aaron karp in *Ballistic Missile Proliferation: The Politics and Technics*, SIPRI, Oxford University Press, 1996.

43. UN Document S/RES/687 (1991). The resolution *"takes note"* that destruction of the weapons specified in the resolution *"represent steps towards the goal of establishing in the Middle East a zone free from weapons of mass destruction and* ***all missiles for their delivery*** *(emphasise added) and the objective of a global ban on chemical weapons"* (Op. 14).

44. Sometimes the circumstances are such that only part of a state will be considered for inclusion in a zone. One clear category is when a considerable part of a state is denuclearised while other parts are not, an example being the territory of the former German Democractic Republic now nuclear-weapon-free and part of unified Germany. Another case refers to dependencies of states being part of a zone while the mainlands being to other regions. Protocols of the Tlatelolco, Rarotonga, and Pelindaba treaties apply to this case. A thrid case refers to states belonging to a nuclear-weapon-free zone but a far away dependency does not. In the discussions on a Nordic nuclear-weapon-free zone, norway was considered an obviours part of the zone while its dependency in the South Atlantic, the Bouvet island, was not. A fourth case refers to the case where a separate part of a country is a denuclearised or a demilitarised entity and the mainland is not. An example is the demilitarised Spitsbergen-archipelago, a dependency of Norway not party to a zone. A final combined zonal and non-zonal case is when an extra-zonal state has a military base in a zone, but the host country has no responsibility for the base. An example is the US base of Guant namo in Cuba. Splitting states in zonal and extra-zonal parts would always require special legal reference in zone treaties.

45. For the purpose of an analysis of nuclear weapons carried by warships and aircraft, the term "international waters" covers the UNCLOS concepts of "exclusive economic zones" (EEZ) and "high seas". Compare UNCLOS Art. 55-58 and Art. 86-93.

46. *United Nations Convention on the Law of the Sea* (UN Sales No. E 83.V.5), signed on 10 December 1982 and entered into force on 14 November 1994.

47. The preparatory commission drafting the Tlatelolco Treaty stated that 'transit' in "*the absence of any provision in the Treaty, must be understood to be governed by the principles and rules of international law*" and that "*it is for the territorial State, in the free exercise of its sovereignty, to grant or deny permission for such transit in each individual case*" (Document COPREDAL/76 p. 8, or UN document A/6663). The transit issue was thus considered a bilateral issue between the flag state and the port state in each case and did, in particular, not contradict the traditional US rights to pass the Panama Canal with nuclear weapons.

48. The fact that three nuclear-weapon powers, France, the UK, and the USA, are parties to Protocol I of the Tlatelolco Treaty for their dependencies in Latin America and the Caribbean poses this problem which has not been raised or referred to politically, however.

49. The problem of drawing a line between the territorial airspace, subject to national jurisdiction of the underlaying state, and the international outer space where the underlying state would have no responsibilities, has been on the agenda of the United Nation's Committee on the Peaceful Uses of Outer Space for very many years. Many difficult issues must be taken into account when defining such a line. Without a final solution, a reasonable assumption would be to assume that the line would be drawn at approximately 100.000 meters above the sea level.

50. For an account of the consequences of these policies, see i.e. j Prawitz, The *Neither Confirming nor Denying*" Policy at Sea in J. Goldblat, (Ed.), Maritime Security: *The Building of confidence*. Document UNIDIR/92/89 (Sales No. GV. E. 92.0.31).

51. Compare footnote No. 33 on page 7.

52. The "thinning out" idea was first suggested by A. Thunborg in 1975 in relation to the proposed nuclear-weapon-free zone in the Nordic area, in *Nuclear Weapons and the Nordic Countries Today—A Swedish Commentary*, A special Issue of Ulkopolitiikka 1975, pp 34-38.

53. The Document of the Stockholm Conference on Confidence—and Security-Building measures and Disarmament in Europe (1986), the Vienna Document 1990 of the Negotiations on Confidence—and Security-Building Measures, the Vienna Document 1992 of the Negotiations on Confidence—and Security-Building Measures, the Vienna Document 1994 on Confidence—and Security-Building Measures, the Treaty on Conventional Armed Forces in Europe (CFE), the Concluding Act of the Negotiatioin on Personnel Strength on Conventional Armed Forces in Europe (CFE-1A), and the Treaty on Open Skies. At the CSCE Summit meeting in Budapest 5-6 December 1994, the CSCE was renamed the Organisation for Security and Cooperation in Europe (OSCE).

54. UN Document A/RES/3263 (XXIX)

55. A recent resolution on the matter was adopted by the UN General Assembly on 10 December 1996, UN Document A/RES/51/41.

56. Document CD/989, 20 April 1990. The Mubarak plan has been described by Mohamed Shaker in *Prospects for Establing a Zone Free of Weapons of Mass Destruction in the Middle East,* Director' Series on Proliferation, No.6 Oct. 1994, Lawrence Livermore National Laboratory (UCRL-LR-114070-6). Compare also MM.Zahran, Towards Establishing a Mass-Destruction-Weapon-Free-Zone in the Middle East, Institute for Diplomatic Studies, Ministry of Foreign Affairs of Egypt, October 1992.

57. Document NPT/CONF. 1995/32/RES/1.

58. UN Document A/RES/43/65.

59. UN Document A/45/435; UN Sales No. E 91. 1X3.

60. UN Document A/RES/45/52, op.8

61. J. Prawitz, J.F. Leonard, *A Zone Free of Weapons of Mass Destruction in the Middle East,* Document UNIDIR/96/24 (UN Sales No. G.V.E. 96.0.19).

62. M. Karem, *A Nuclear-Weapon-Free Zone in the Middle East: Problems and Prospects.* Greenwood Press. new York. 1988. The same author has later published *A Nuclear-Weapons-Free Zone in the Middle East: A Historical Overview of the Patterns of Involvement of the United Nations* in T. Rauf (Ed.), *Regional Approaches to Curbing Nuclear Proliferation in the Middle East and South Asia,* Aurora Papers 16. Canadian Centre for Global Security. December 1992.

63. S. Feldman, *Nuclear Weapons and Arms Control in the Middle East,* MIT Press, Cambridge MA, 1997;S. Feldman, A. Toukan, Bridging the Gap: *A Future Security Architecture for the Middle East,* Carnegie, Rowman and Littlefield Publishers, Lanham MD, 1997.

64. E. Arnett, *Nuclear Weapons After the Comprehensive Test Ban: Implications for Modernisation and Proliferation,* SIPRI, Oxford University Press, 1996.

65. *Technical Study on Different Modalities of Application of Safeguards in the Middle East.* Document IAEA-GC (XXXIII)/887, 29 August 1989. (On 22 May 1990, Democratic Yemen and Yemen merged to form a single state with the name "Yemen"). A similar defintion was suggested in the 1975 UN study *Comprehensive Study on the Question of nuclear-Weapon-Free Zones in all its Aspects.* United Nations Document a/10027/Add. 1, (UN Sales No. E. 76.1.7). para 72.

66. The League of Arab States has 22 member states: Algeria, Bahrain, Comoros, Djibouti, Egypt, Iraq, Jordan, Kuwait, Lebanon, Libya, Mauritania, Morocco, Oman, Palestine, Qatar, Saudi Arabia, Somalia, Sudan, Syria, Tunisia, United Arab Emirates, and Yemen.

67. For a recent account of Israel's nuclear industry, see D. Albright, F. Berkhout, W. Walker, *Plutonium and Highly Enriched Uranium: World Inventories, Capabilities, and Policies,* SIPRI, Oxford University Press, 1997, Part IV *Material Inventories and Production Capabilities in Threshold States,* pp 257-264.

68. This policy of deliberate ambiguity has been said to serve Israel's security interests in three ways: firstly, in times, of gloom, it gives hope to the Israelis; secondly, it may caution the enemies of Israel; and thirdly, it relieves other states from the delicate burden of taking an explicit position on the matter. As reported by Dr S. Freier in C. Atterling Wedar, S. Hellman, K. Söder, (Eds.), *Towards a Nuclear Weapon-Free World. Swedish Initiatives.* (ISBN 91-972128-0-6) Stockholm 1993. p. 181.

69. M. Kibaruglo, *EURATOM & ABACC: Safeguard Models for the Middle East?* In J. F. Leonard, *A Zone Free of Weapons of Mass Destruction in the Middle East* (Annex), Document UNIDIR/96/24 (UN Sales No. GV.E. 96.0.19), pp 93-123.

12

The Goal: Prevention of the Misuse of Biological Sciences, Elimination of Biological Weapons and the Science for Peace Oath

—Y. Becker, Israel

The Goal: Prevention of the Misuse of Biological Sciences

The 20th century was marked by major contributions by scientists to basic scientific knowledge and to the developments of new technologies. During the first half of the century these contributions ranged from dynamite by the Nobel brothers through the research on the atom and nuclear physics, from the chemical synthesis of ammonia by F. Haber to the synthesis of chemotherapeutic agents by Ehrlich, and penicillin to treat bacterial infections by Chen and Fleming. Yet, during the same 50 years the world suffered two major World Wars with World War I ending in a pandemic of influensa (Spanish flu) that killed 20 million people and World War II devastated many more millions of civilians. With the developments in atomic physics, the advancements in basic knowledge also brought the atomic bombs.

The second half of the 20th century started with major breakthroughs in biology such as the success of Joans Salk and Albert Sabin in developing polio virus vaccines that stopped the polio virus epidemics world wide and paved the way to the production of scores of vaccines against virus diseases that improved human life around the world by relieving them from many virus diseases.

The deciphering of the structure and functions of genes in DNA and their mode of action, using viruses as molecular models, starting with 'Watson and Crick, and Wilkins and Rosalind Franklin on the structure of DNA, had led to the development of molecular biology and medical biotechnologies that led to the deciphering of the human genome that followed the determination of the nucleotide sequences of all known viruses, bacteria and yeast. In the second half of this century the wars became regional conflicts and the involvement of economical policies in the sustainable growth of nations led to an open world market. Yet, despite advanced scientific and medical knowledge, new viruses like HIV-1 and Ebola emerged and the AIDS disease spread throughout the world in the last 16 years. We were fortunate that a coalition of the United Nations forces prevented a regional conflict in the Middle East from developing into a catastrophe by attempting to destroy biological and chemical weapons of mass destruction.

The advancements in biology and medicine had a marked effect on human life increasing life expectancy from 47 years of age at the beginning of the century to 75-80 years average life span at the end of the century. The development of vaccines against viruses and bacteria eliminated wide spread epidemics except AIDS. Advances in agriculture by producing genetically improved plants and seeds enriched the food resources of the people around the world. Advances in the understanding of the function of human cells and tissue, together with improved diagnostic techniques, moved medicine forward to treat autoimmune diseases and diseases that require organ transplants. Yet, the vulnerability of the human and animal population to emerging diseases like AIDS, Ebola, Monkeypox, and malaria indicates that more research is required to protect against new and reemerging diseases.

The availability of biotechnological methods to change genes in viruses, bacteria and in the mammalian fertilised eggs provide new fears that the important molecular tools for accessing genes in chromosomes, should not be used for harmful purposes. For this reason, ethical aspects of gene manipulations are needed, some of which will be presented in the Panel's discussions. It is therefore an important goal to devise ways to prevent the misuse of biological sciences, since such scientific ativities can not be monitored and require **an international agreement of the scientists** and International scientific organisations to ensure that **scientific research** on genes should continue while the misuse of the scientific activities to modify the genomes in a harmful way should be prohibited. One such possible misuse of science is the research and development of Biological Weapons (BWs) for mass destruction.

The Goal: Elimination of BWs of Mass Destruction

In the 15th SIPRI Chemical and Biological Warfare Studies (1) F.C. Calderon (2) discussed the treaties to prevent BWs since the 1925 Geneva Protocol, the 1972 Convention on the Prohibition of the Development, Production and Stockpiling of Bacteriological (Biological) and Toxin Weapons and on their Destruction (BWC) and the 1980, 1986, and 1991 Review Conferences. Milton Leitenberg (3) provided information on the member states that signed the BWs treaty and yet had continued to develop BWs of mass destruction. One of these countries is still monitored by the UNSCOM for concealed BWs of mass destruction, as traced by calculating the extent of BWs produced from the raw materials that were purchased by that country.

The above information provided evidence that the build up of such weapons can be accomplished under cover of participation in the BWs treaty and demonstrates the state of mind of the BWs developers and the leadership that biological agents, viruses or bacteria, can determine the outcome of a war in favour of the aggressor. Those who plan to use BWs in a war do not think that the pathogenic agents can cause epidemics in nearby populations that have no relation to the conflict and also jeopardise the health of the entire world populations, not to mention the immoral aspects of a biological war. It is hoped that the development of new vaccines may result in making the BWs agents obsolete (1,4).

To ensure the prevention of the use of biological weapons it is needed to add to the international conventions on BWs the involvement and activity by the scientists to speak up against the BWs in a similar way that led the atomic physicists to convince the leaders of the two superpowers to agree to reduce atomic weapons. It is now the time for the scientists to work together to prevent the misuse of sciences.

The Goal: Involvement of Scientists in the Prevention of R & D on Biological Weapons of Mass Destruction; Scientists for Peace

The developments in virology, microbiology and molecular biology that led to developments of biotechnologies that utilise the basic know-how for genetic manipulations during the last 50 years were published in the scientific literature.

Scientists from all around the world from advanced and developing countries, travelled to the centers of research where the biotechnologies were developed and participated in the major scientific endeavours. Yet, some of them were those who participated in the development of BWs and thus misused the scientific tools to create weapons of mass destruction.

The stocks of viruses and bacteria in culture collections were open to scientists seeking to study pathogenic agents, assuming that they will contribute to the knowledge on the properties of these agents to find a cure against the harm they may cause. To eliminate the misuse of the biological sciences a strategy should be developed since "the minds that created the scientific breakthroughs and knowledge are the minds that must prevent the misuse of the biological sciences." It is hoped that the International Panel at the UNESCO International School of Science for Peace will discuss the ways to prevent the misuse of biological sciences and to obtain public support to this activity.

An international network of Scientists for Peace that will involve the international and national societies of biology, microbiology, pharmacology, and chemistry may lead to the education of the people in each country so they will demand that the hazards of man-made diseases will be eliminated and that protection against emerging and reemerging diseases by vaccines will be an international endeavour and that vaccines will be available to all nations.

The Goal: The Science for Peace Oath

The discussions during the Second International Symposium on Science for Peace in Jerusalem 20-23 January 1997 led to recommendations that were presented in the "Jerusalem Statement on Science for Peace" that "efforts be undertaken to develop Science for Peace Oath for young scientists to take when accepting their degrees, similar to the Hippocratic Oath that is taken by medical school graduates".

The education of the young scientists to think positively that scientific accomplishment should be used to benefit people and not to harm them, and that ethical considerations should provide the moral guidelines in research, must begin at the time the students develop their scientific skills. The "Science for Peace Oath" taken by graduates may remind them as they develop in their scientific life to be a moral light house. This will not only make the scientists of the 21st century aware that the prevention of the misuse of biological sciences is necessary, but will remind them to raise their voices loudly against the misuse of science and deter those that do not adhere to the rules.

The success of the movement against land mines (JCBL) led by Ms. Judy Williams which convinced the public to demand that antihuman mines should not be used, demonstrates the influence of public opinion. The same can be done to eliminate biological and chemical weapons of mass destruction.

References

1) E. Geissler and J.P. Woodall (Eds). *Control of Dual Threat Agents: the Vaccines for Peace Programme,* No. 15 SIPRI Chemical and Biological Warfare Studies, Stockholm International Peace Research Institute, Oxford University Press, 1994.

2) F.C. Calderon. *The Missing Link in the Implementation of the BWC: The War against Pathogens.* In: E. Geissler and J.P. Woodall (Editors) Control of Dual Threat Agents: the Vaccines for Peace Programme, No. 15 SIPRI Chemical and Biological Warfare Studies, Stockholm International Peace Research Institute, oxford University Press, 1994, pp 41-47.

3) M. Leitenberg. *The Conversion of Biological Warfare Research and Development Facilities to Peaceful Uses.* In: E. Geissler and J.P. Woodall (Editors). Control of Dual Threat Agents: the Vaccines for Peace Programme, No 15 SIPRI Chemical and Biological Warfare Studies, Stockholm International Peace Research Institute, Oxford University Press, 1994, pp 77-105.

4) Y. Becker. *The Consequences of a Biological War: Can We Protect Humans and Animals by Synthetic Peptides and DNA Vaccines?* 1997.

13

The Role of Microbiologists in the Prevention of the Use of Biological Weapons

R. Atlas, United States of America

Introduction

Microbiologists, individually and collectively through scientific societies, have important roles in preventing the misuse of microorganisms as biological weapons. Microbiologists should adhere to a code of ethics that ensures that research is conducted for the betterment of humankind.

There is a consensus among microbiologists today that steps should be taken to prevent biological warfare and that openness of scientific research and global surveillance of disease outbreaks can significantly increase transparency for detecting development of biological weapons. The American Society for microbiology (ASM) has recommended increased attention to and efforts directed toward global surveillance of disease outbreaks, not only to aid public health organisations in improving human health, but also to establish baseline data against which unusual disease outbreaks can be assessed. The ASM also has recognised the possible dual faces of work with pathogenic microorganisms, holding that medical research and the diagnosis, treatment, and prevention of disease must not be impaired by efforts to control potential biological weapons. Clearly there is no consensus of opinion among microbiologists on their appropriate roles in biological defense research, nor is it likely that there ever will be complete agreement.

The Public and Scientific Affairs Board (PSAB) of the ASM is playing an important and active role in monitoring and responding to legislative and regulatory issues involving biological weapons in the United States. Issues of how best to increase global security and to achieve a scientifically based verification protocol of the Biological Weapons Convention are important and continue to be addressed by the ASM on behalf of its over 42,000 members.

Taking an Ethical Stance Against Biological Weapons

The American Society for Microbiology (ASM) and microbiologists around the world have increasingly become involved in trying to prevent the threat of warfare and terrorism using biological weapons. As a major scientific society with over 42,000 members, the ASM has taken a strong stance against such possible misuse of microorganisms The organisation's involvement with the issue of biological warfare dates back to the 1940's, when microbiologists served as advisors to the U.S. War Department's Biological Defense Research Programme and participated in the Biological Warfare Committee of the U.S. National Academy of Sciences. The ASM governing council in 1970 called for an end to offensive biological weapons programmes by the United States when it resolved that: "the Council affirms support of President Nixon's action on November 25, 1969, and February 14, 1970, to end our involvement in production and use of biological weapons." In 1993 the ASM Council passed another resolution renewing the resolve of the society "to work to prohibit the possession, development and use of biological and toxin weapons by all nations."

The ASM's commitment to take a leadership role in trying to prevent the misuse of microorganisms and microbially produced toxins as weapons of warfare or terrorism was made after heated debates over what military research should be conducted by governments to offer protection against biological weapons and what roles microbiologists should play in these activities.

Because infectious diseases have killed more soldiers in some wars than bullets, biomedical research and the development of vaccines and treatments for many diseases, including exotic diseases that do not normally occur in the United States, has been very important to the U.S. Department of Defense. Many microbiologists have and continue to work for the U.S. military in an effort to prevent and to control infectious diseases, some had previously worked in the U.S. biological weapons programme that ended in 1969. Of the 91 Presidents of ASM, 21 have spent some part of their scientific careers at the U.S. Army Medical Research Institute of Infectious Diseases (USAMRID) at Fort Detrick, Maryland, the main U.S. biological defense research facility and once the site of the U.S. biological weapons programme.

The subject of biodefense and military related research on pathogenic microorganisms has often been contentious among the ASM's diverse membership. It is a thin line that is often seen between such defensive research and the generation of results that could be misused for biological weapons development. While only a few individuals would argue that biological warfare might be a humane form of war, a more problematic question has been whether microbiologists should engage in national defense efforts involving research on biological weapons. A particularly thorny question has been whether research for defense against biological warfare can ever be truly separated from research that effects offensive capability in this area.

The ASM has faced the philosophical dilemma of whether a scientific society can appropriately tell its members what types of otherwise legal research to do or not to do; that is, should or can a scientific society provide real moral direction to its membership and order them not to conduct research that could lead to the misuse of microrganisms. In this regard the Society's code of ethics, published in 1985, contains two relevant sections that seek to discourage ASM members from participating in biological weapons development: (i) "Microbiologists...will discourage any use of microbiology contrary to the welfare of human kind" and (ii) Microbiologists are expected to communicate knowledge obtained through their research through discussions with their peers and through publications in the scientific literature." This code of ethics clearly establishes that it is the individual responsibility of every microbiologist to conduct research that is beneficial to humankind. The ASM provides ethical leadership but does not have the enforcement capability to ensure ultimate compliance with these principles by individual microbiologists.

Beyond endorsing ethical principles against biological warfare and setting codes of ethics for conducting research by microbiologists, the main role of the ASM has been to provide scientific advice on means for preventing the misuse of microrganisms as biological weapons. The ASM has endoresed the concept of transparency as a key element in eliminating offensive biological weapons development. As early as 1970 the Council of the society stated that "...the health of science is enhanced by non-secret research and free movement of scientists." It held that openness of research activities collectively provides the transparency necessary to help prevent activities that could result in the misuse of microorganisms as biological weapons. It recognised the power for role that scientific societies can play in influencing individual scientists to act for the betterment of humanity.

As part of its 1970 statement of principle against biological warfare and the development of biological weapons the ASM Council also stated: "Because of our concern for humanitarian application of microbiological science, we urge that all nations convert existing offensive biological warfare facilities to peaceful uses."ASM members have been working with groups such as the U.S. National Academy of Sciences, the Institute of medicine, and the National Research Council to develop programmes that will fund the diversion of biological weapons activities into peaceful uses for improving human health and welfare. Recently a committee of the National Academy of Sciences developed recommendations for a new U.S.-Russian cooperative programme that engages specialists from the former Soviet Union's biological weapons complex in research on highly infectious diseases. The committee of the U.S. National Academy of Sciences concluded after extensive consultations with American specialists and numerous Russian organisations and experts that this programme could serve important U.S. and Russian national security and public health goals. The committee recognised that some of the same medical and biotechnology expertise for diagnosing and combating highly infectious diseases also could be used to support the development of biological weapons. The recommended programme, called the "Pathogens Initiative," would support research projects at Russian institutes focusing on the prevention, diagnosis, treatment, and epidemiology of highly infectious diseases. Russian specialists would collaborate with scientists from U.S. Government, academic, and private laboratories. The cooperative effort would cost the United States approximately $ 38.5 million over the five-year duration of the programme. By combining their expertise, researchers could improve understanding of the microbiology and epidemiology of dangerous pathogens; strengthen capabilities to prevent, diagnose, and treat these infectious diseases; and improve international surveillance of global infectious disease trends and outbreaks. The ASM actively supports this programme and other such efforts to ensure the elimination of biological weapons programmes.

Verifying Compliance with the Biological Weapons Convention

Given the complexity of issues related to biological weapons, a special task force on biological weapons was formed in 1994 in large part to offer advice to the U.S. Government in developing scientifically sound approaches to biological arms control that would enhance security against biological warfare. The task force is comprised of members from academia, industry, and government who encompass the breadth of expertise on biological weapons found within the ASM. Some of the members of the task force also belong to other groups, such as the Federation of American Scientists and industrial groups representing pharmaceutical manufacturers and biotechnology industry, thus, the task force represents the diverse views of ASM members. A principal aim of the task force is to seek scientifically sound ways of enhancing global security by limiting the likelihood of the misuse of microorganisms as biological weapons.

The task force has struggled with the question of how the Biological Weapons Convention (BWC) of 1972, which prohibits the development and stockpiling of biological weapons, can be strengthened through a verification regime. It has examined a number of possible measures that have been proposed to help ensure compliance with the BWC. The focus has been on the scientific validity of each proposed methodological approach, the likelihood that it could detect biological weapons and related activities, and the impact it would have on scientific research and medicine. A guiding principle has been that global security depends upon preventing biological warfare and on the control of naturally occurring infectious diseases so that it is essential to protect incentives for advancing biomedical research. Throughout its deliberations about biological weapons ASM task force on biological weapons considered it essential that efforts aimed at offering protection against the misuse of microorganisms as biological weapons not inhibit efforts to control and to prevent infectious diseases.

Based upon extensive reviews of the risks and benefits of a mandatory verification regime for the BWC, the ASM has advised the U.S. Government and other negotiating parties in Geneva that detecting offensive biological weapons development activities is very complex and should be based upon scientific principles that would permit differentiation of legitimate biomedical activities from those activities related to biological warfare that are prohibited under the BWC. The ASM has emphasised that microrganisms that can be used as biological weapons may be the same as naturally occurring pathogens or microorganisms used for beneficial purposes, such as vaccine production. The same fermentors that are used for the production of pharmaceuticals could be used to grow large numbers of microorganisms for use as biological warfare agents, as could flasks and other cell culture systems found in academic research laboratories. Natural occurrences of diseases such as anthrax can be mistaken for accidental or deliberate releases of biological warfare agents. Likewise, accidental or deliberate releases of such agents could be misinterpreted as outbreaks of naturally occurring diseases. Finding spores of the anthrax bacillus may very well be an isolated natural occurrence or it may be associated with a contrived outbreak of the disease from biological weapons. A system of verification that does not enable differentiation of legitimate activities and natural occurrences from offensive biological weapons development would be ineffective and would produce a false sense of world security. Clearly, the potential dual nature of research on pathogenic microrganisms, for control of disease on the one hand and as potential agents of biological warfare on the other, makes verification very difficult. Effective verification ultimately rests with determining intent. The experience of inspection teams of the United nations Special Commission (UNSCOM) in Iraq following Gulf War shows that even the most invasive inspections may be unable to detect specific evidence of biological weapons development without the aid of human intelligence information. Yet such inspections provide a critical deterrent against biological weapons programmes and violators of the BWC.

The ASM has offered a series of guidelines based upon scientific principles for use as a framework for consideration of proposals for verifying compliance with the Biological Weapons Convention. The points in these guidelines are aimed at enhancing global security by limiting the potential development of biological weapons while simultaneously allowing the development of pharmaceuticals, vaccines, and diagnostic procedures aimed at reducing the incidence of infectious diseases.

The first principle is that detection of accidental releases or deliberate use of biological weapons requires comparison with the natural occurrences of disease. Therefore, there should be an enhancement of existing epidemiological databases to ensure that causes of disease outbreaks can be determined with a high degree of confidence. Existing databases should be extended to include pathological data that indicate routes of transmission. The ambiguity of the 1979 outbreak of anthrax in the Soviet Union demonstrates that without an adequate database of baseline information, epidemiological investigations are of only limited use in providing evidence of biological weapons. Initially, it was erroneously concluded that this was a natural outbreak of anthrax but a later epidemiological investigation conclusively showed that it had resulted from an accidental release from a Soviet biological weapons development facility. The initial investigation had failed to detect the source. The unusual nature of the 1979 disease outbreak at Sverdlovsk was the high number of fatal cases of anthrax. Recognising such unusual outbreaks depends upon having an adequate epidemiological database so that unusually sudden high rates of mortality can be differentiated from natural variations in disease incidence.

The ASM has called for increased disease surveillance and epidemiological investigations in order to increase global security and human welfare within the context of BWC verification and beyond. The Programme for Monitoring Emerging Diseases (ProMED) initiative started by the Federation of American Scientists is an excellent start in this direction. Besides its role in investigating unusual disease outbreaks that could be associated with biological weapons development facilities, strong global epidemiological databases would also aid public health organisations in improving the health of humankind. The development of an expanded and integrated world-wide epidemiological data base, including signatories to the Biological Weapons Convention, would be a major step to improving global human health. It also is essential if assessments are to be made with confidence about accidental or deliberate releases of biological weapons agents that might result in unusual outbreaks of disease.

Concerning the question of how to recognise an unusual outbreak of disease that could trigger an investigatory action under a BWC protocol, the following definition may be useful in defining a disease outbreak, an unusual disease outbreak is a case of a diverse disease or collection of cases at a time and place when informed experience would indicate that they should not be seen at all or at a much lower frequency or severity. Both mortality and morbidity should be considered. Cooperative scientific programmes between nations and scientists can contribute to a better understanding of infectious disease problems in countries which may not have strong epidemiological databases, on a shared basis with other countries.

The second principle is that incentives for scientific discovery must not be removed. If laboratories become subject to declarations and inspections under the BWC, the included protective mechanisms should be to ensure that incentives for research and developmental activities are not eliminated. The quest for being first with a new discovery is an important motivating factor in research. For academics, some experiments are not publicly revealed until the research is completed so that students can claim originality on their dissertations, and the faculty can receive credit for specific advances in the state of knowledge. For industry there is a need to protect investment in research and development, intellectual property, and commercial assets. Proprietary information must be protected at the same time that measures that increase transparency should be taken to ensure compliance with the Biological Weapons Convention. Achieving these goal simultaneously is seemingly paradoxical. Openness in research is an ethical principle of the ASM code of conduct for microbiologists. However, the ASM recognises that one of the incentives for research and development is the protection of intellectual property rights.

The ASM has been very sensitive to the argument put forth by the Pharmaceutical Research Manufacturers of America (PhRMA) that a mandatory verification regime could lead to the loss of valuable proprietary information and loss of industrial profits. Inspections must not be used for industrial espionage. Disclosure of information during an inspection by an international biological weapons inspection team must not be construed as public disclosure that would preclude obtaining patent protection. Under current international patent law such revelations to an inspection team could well prevent obtaining patent protection. International inspection teams would have to be charged with maintaining confidential information disclosed to them, and nations sending inspections would have to be held accountable and in violation of the Biological Weapons Convention with appropriate sanctions in place, should their inspectors not hold in confidence proprietary information. Greater openness can be expected only if mechanisms are in place to protect proprietary information. Until that can be ensured many industrial groups will oppose many of the proposed measures for mandatory verification of the BWC.

Despite its concern with protecting intellectual property and maintaining incentives for research that will further biomedical science, the ASM taken the position that inspections to confirm compliance with the Biological Weapons Convention should be conducted when there is adequate cause. If there is evidence for noncompliance with the biological weapons Convention, inspections—despite their intrusiveness—are warranted for such cases. There should be a recognised international authority that can authorise challenge inspections under the Biological Weapons Convention. This authoritative body should determine if, where and when, inspections or other intrusive verification activities should be conducted. Conducting inspections for proper cause is important both for the cost of implementing the Biological weapons Convention and because of the potential disruptions to academic and industrial research.

The ASM Biological Weapons Task Force has been evaluating how declarations and site investigations could be conducted that would detect biological weapons without jeopardising the loss of confidential business information and proprietary microorganisms. A variety of options have been debated that have focused on sampling and analysis. Managed access has been proposed by PhRMA as the preferred means of preventing loss of proprietary organisms. Various technological approaches might also lessen the risk of loss of proprietary microorganisms. Using on-site analyses based upon gene probe or serologic detection conducted with killed samples would lower the risks of losing valuable microorganisms compared to off-site analyses conducted using live culture techniques. Such an approach would also lower the risk of inadvertently spreading disease as a result of the transport of pathogens. For the pharmaceutical industry test validation and compatibility with FDA regulations are especially important considerations. For groups of concerned scientists, including those in the pharmaceutical industry, the ability to detect biological weapons and to deter their development is of critical concern. The ASM hopes through its efforts that both national economic security and global security through the elimination of the threat of biological warfare can be achieved.

As a third principle, the ASM has stated that lists of equipment and organisms cannot adequately define the scope of appropriate surveillance for complance with the Biological Weapons Convention. The ASM Biological Weapons Task Force has been searching for a signature that would define a biological weapons facility. It has yet to find such a signature short of actually finding armaments filled with biological agents. Laminar flow hoods, fermentors, and freeze dryers are examples of equipment that are found in many microbiological laboratories as well as in numerous other facilities. BL2 and BL3 facilities are found at most universities. These are not specific to biological weapons research development. Finding such equipment is of limited significance in identifying biological warfare verifications. All clinical and pathology laboratories work in pathogens and exercise appropriate safety containment practices. The ASM opposed the U.S. Department of Commerce's development of a list of equipment used in biological research that would be prohibited from shipment to countries which have not signed the Biological Weapons Convention. Much of the equipment proposed for the list, such as laminar flow hoods, are commonly used in medical laboratories and could be obtained from many sources outside the United States. The advice of the ASM to the Commerce Department was that the list was unlikely to serve any useful purpose in terms of limiting research on biological weapons, but would instead open the U.S. for criticism for inhibiting medical care and research in developing countries.

Possession of a microorganism on a prescribed list of pathogens has been considered as a possible trigger for requiring the filing of a declaration under a mandatory verification regime of the BWC. Consideration for establishing such a list for restricted export arose in the United States shortly after the Gulf War when it was learned that the American Type Culture Collection had shipped cultures of bacillus anthracis to iraq. Even though these shipments had been in conformity with U.S. law and had been approved explicitly by the Department of Commerce, the unfavourable publicity was generated by these shipments. This was because of concern about the potential use of biological warfare agents by Iraq led the Commerce Department to start to generate a list of organisms whose export would be restricted.

The approach to defining organisms or concern as potential biological weapons based on defined lists is relatively simple because it sets easily defined objective criteria. However, this approach is fraught with problems if one considers the full breadth of pathogens that could be used as biological weapons against humans, non-human animals, and plants. Virtually all the pathogens on such a list are naturally occurring disease causing agents that are detected in clinical laboratories and studied at biomedical research facilities around the world where the aim is to control the diseases caused by those microorganisms. Many of those pathogens are grown in pharmaceutical industry factories to produce vaccines which prevent diseases of humans and other animals. In the biological control industry many organisms that might be placed on a restrictive list are used to limit plant pests that threaten world food supplies. Establishing specific lists of controlled microorganisms could serve as a prescription for which organisms to choose for illicit biological weapons programmes—either the ones on the list because they are considered by the world to be the best candidates for use as biological weapons or those not on the list because they are not explicitly prohibited and their detection would not constitute a violation of the declaration process. Concern has also been raised about recombinant microorganisms that might be created to include toxigenic, invasive, or resistance factors that would make them superior biological weapons which would not appear on a list.

Producing large volumes of pathogens, presumably in well contained reactors, is another possible criterion that could help define a facility producing biological weapons. Relying on such a criterion could exempt all clinical laboratories and most academic laboratories but would certainly place a burden for filing declarations on the fermentation industry and vaccine manufacturers. Furthermore, the ASM Task Force on biological weapons has recognised that microorganisms can be stored in one facility and grown in another. Also, while extreme containment would characterise the production of large volumes of a deadly pathogen in a facility following good manufacturing procedures, production could be carried out under conditions of poor containment if the surrounding population were vaccinated or if little care was given to possible disease spread into the populations surrounding the production facility.

Weaponisation is perhaps the most intriguing criterion in terms of being able to distinguish legitimate biomedical facilities from those involved in biological weapons development. Here, though, is where the members of the ASM Task Force have the least expertise. Past major biological weapons development programmes have focused on aerosol dissemination so that the possession of milling equipment and other hardware that could form small enough aerosols for weaponisation has been considered for triggering a declaration and inspection process. The history of biological warfare, however, points to lower technological delivery systems, such as the Tartars catapulting bodies of victims of plague into the walled city of Kaffa in the fourteenth century and the Japanese attempting to float hot air balloons containing plague infected flea vectors over the western United States during World War II. These past experiences suggest that there is no single criterion that can be used for triggering requirements for filing declarations in a BWC verification regime. Like all aspects of the BWC, there is a need for balance that increases the likelihood of dissuading would-be violators from developing offensive biological weapons and the legitimate needs of advancing biomedical research and the control of infectious diseases.

Preventing the use of biological weapons by terrorists

Beyond its concern about preventing biological warfare, the ASM has become very concerned about the possible usage of biological weapons by terrorists. Several recent incidents, such as the attempts of the Japanese Aum Shinrkyo cult to acquire anthrax bacilli, botulinum toxin, and Ebola viruses, have raised concerns about the illegal use of biological weapons by terrorists. It has been illegal since 1990 to engage in biological weapons development within the United States. The ASM played an important role in developing the final version of the Biological Weapons Anti-Terrorism Act of 1989 which was signed into law by President Bush on May 22, 1990. In particular, the ASM helped Senator herb Kohl (D-Wisconsin) develop an amendment that removed specific language that would have shifted the burden of proving innocence to an accused scientist and that could have been used to restrict peaceful, humane research. The amended bill specifically stated that nothing in this Act is intended to restrain peaceful scientific research or development. The congressional report included language stating that Congress is aware that many scientists conduct peaceful research with potentially dangerous agents and toxins and that the bill will not interfere with such activities. While protecting legitimate biomedical research, the Biological Weapons

Anti Terrorism Act of 1989 made it a crime to possess a biological weapon or to help a foreign nation acquire one. It authorises the Attorney General to seek the civil forfeiture of biological weapons before they injure people or harm the environment. It imposes criminal penalties against those who would employ or contribute to the dangerous proliferation of biological weapons.

In 1996, legislators were prompted to amend the U.S. Criminal Code covering biological agents after learning of a 1995 incident involving the acquisition of bubonic plague cultures through the mail by a suspicious purchaser in Ohio. The ASM along with the Centers for Disease Control and Prevention (CDC) and American Type Culture Collection testified before the Senate Judiciary Committee on the interstate transport of human pathogens. On April 24, 1996, President Clinton signed into law anti-terrorism legislation (PL 104-132) which expands federal power to prosecute and punish certain crimes related to terrorism. A bipartisan amendment was included in the new law making it illegal to develop, acquire or attempt to purchase biological agents with the intent to kill or injure or use them as a weapon. The provision on biological agents amends the 1989 Biological Weapons Anti-Terrorism Act. The law directed the Secretary of Health and Human Services, through promulgation of regulations, to provide for (1) the establishment and enforcement of safety procedures for the transfer of listed biological agents, including measures to ensure proper training and appropriate skills for laboratorians who handle and dispose of such agents, and proper laboratory facilities to contain these agents; (2) safeguards to prevent access to such agents for possible use in domestic or international terrorism or for any other criminal purpose; and (3) the establishment of procedures to protect the public safety in the event of a transfer or potential transfer of a biological agent in violation of the safety procedures established under (1) or the safeguard established under (2), while maintaining availability of biological agents for research, education and legitimate purposes.

The task of developing regulations to control the interstate shipment of pathogenic microorganisms that might be used as biological weapons was assigned to the CDC. ASM assisted the CDC in establishing new regulatory requirements for facilities transferring and receiving select agents which have the potential to cause widespread harm. The ASM offered advice in the development of the list of organisms to be subjected to regulation; this advice was based in part on a survey conducted via E-mail of 11,000 ASM members.

The regulations that emerged from these consultations and the mandate to reduce the threat of terrorist use of biological weapons focused on controlling exchanges of those microorganisms and their toxins that have the highest potential for use as biological weapons. Under these regulations, which came into force on April 15, 1997, public and private laboratories, commercial companies, academic and research institutions, and other facilities that wish to transfer or receive select agents listed in a schedule published by the CDC will be required to register their facilities with CDC. Those facilities will be subject to inspection to verify the information provided at the time of registration (see List of Select Agents). Only registered facilities will be able to transfer or receive select agents, and documentation of each transfer will be required.

As with other issues related to the potential misuse of microrganisms as biological weapons, the ASM offered advice in the development of these regulations that was consistent with ensuring continuance of essential biomedical research and diagnostic activities. In the final regulations, the list of restricted agents is limited so as not to unduly restrict legitimate biomedical research. Clinical laboratories certified under the Clinical Laboratory Improvement Amendments of 1988 (CLIA) that intend to use and transfer select agents only for diagnostic, reference, verification, or proficiency testing purposes are exempt from the requirements of the regulations. Thus, essential medical diagnostic practices can proceed unimpaired.

In conclusion, the American Society for Microbiology is actively engaged in efforts to ensure that microorganisms are never misused as biological weapons. The ASM"s membership has unique knowledge and expertise in microbiology and immunology, and this has given the ASM a particularly important role in decisions regarding U.S Policy on biological weapons and defense. The Public and Scientific Affairs Board of the ASM attempts to provide advice that permits the deterrence of biological weapons development while permitting and encouraging research activities that will benefit humankind and in particular human health. Given the potential dual uses of microbiological research, providing scientifically based advice for preventing biological warfare and biological terrorism to increase global security is a challenge for the ASM and its members. Biological weapons issues are an important issue for society and the microbiology community and they continue to be addressed by the ASM.

14

The American Society for Microbiology: Prevention of Biological Warfare

K.I. Berns, United States of America

Foremost among the issues affecting the use of microbiology for peace is the problem of biological warfare, in which microorganisms are the major weapons. Although the membership of the American Society for Microbiology (ASM) has been against the very notion of biological warfare, the subject has long been a contentious one in several ways. While only a very few would argue that biological warfare might be a humane form of war, a more problematic question has been whether microbiologists should engage in national defense efforts involving research in biological weapons. A particularly thorny question has been whether research for defense against biological weapons can ever be truly separated from research that affects offensive capability in this area. A more philosophical issue has been whether the Society can appropriately tell its members what types of research not to do, so long as that research is legal; i.e., can the ASM provide real moral direction to its membership.

Members of the ASM have a considerable history of bringing scientific and technical knowledge to the issue of biological weapons control and being available to serve in advisory roles to the government. ASM's involvement with the biological weapons issue began in the 1940's, when microbiologists served as advisors to the government's Biological Defense Research Programme and participated in the Biological Warfare Committee of the U.S. National Academy of

Sciences. In 1970, a controversy arising from the ASM's involvement with the issue abated when the ASM Council approved a statement concerning non secrecy and free movement in research. Simultaneously, the society affirmed support for President Richard M. Nixon's action to end the U.S.'s offensive biological weapons programme. The Society's Code of Ethics, published in 1985, contains two relevant sections that seek to discourage ASM members from participating in biological weapons development. The first is "Microbiologists will discourage any use of microbiology contrary to the welfare of human kind." Second "Microbiologists are expected to communicate knowledge obtained in their research through discussions with their peers and through publications in the scientific literature." Clearly, however, the second is equally pertinent to all research done in the burgeoning area of biotechnology. The ASM has sponsored a series of round tables and other activities through its Public and Scientific Affairs Board (PSAB). About four years ago the ASM renewed its commitment to a resolution it passed in the 1970's to support the Biological Weapons Convention of 1972 and to work to prohibit the possession, development and use of biological and toxin weapons.

ASM has become increasingly active in offering its expertise and good offices in several areas to protect against the use of biological weapons. An early example of this occurred shortly after the Gulf War. The United States Department of Commerce was concerned about reports of shipments of anthrax bacillus to laboratories in Iraq from the American Type Culture Collection. These shipments had been in conformity with U.S. law and had been approved explicitly by the Department of Commerce. However, the unfavourable publicity generated by the shipments, because of concern about the potential use of biological warfare agents by Iraq, led the Commerce Department to consider establishing a list of equipment used in biological research which would be prohibited from shipment to countries which had not signed the Biological Weapons Convention or were not in conformity with the Convention. The ASM was asked by the Commerce Department for comments concerning the items proposed for inclusion on the list. Consideration by ASM members of the proposed items to be restricted led to several concerns with this approach. Many of the items, such as laminar flow hoods, were equipment that could be used in medical laboratories and were likely to be able to be obtained from many sources outside the U.S. Our advice to the Commerce Department was that the list was unlikely to serve any useful purpose in terms of limiting research on biological weapons, but would open the U.S. to criticism for inhibiting medical care and research in some developing countries.

Although the United States had signed the Biological Weapons Convention in 1972, in 1989 it was not against the law to engage in biological warfare. In 1989 the Public and Scientific Affairs Board of the ASM worked with key members of Congress to perfect language in legislation that would prevent the use of biological weapons as agreed to by the U.S. in the Convention. Although the ASM vigorously endorsed the intent of the legislation, it raised concern about specific language proposed for inclusion in the Senate bill that would have shifted the burden of proving innocence to accused scientists and that could have been used to restrict peaceful research. In other words, anyone could have accused a scientist of working with either a pathogen to be used in biological warfare or with a vector. Almost any microbe could fit this description. The consequence of such an accusation would have been seizure of all the scientist's materials; to get the materials returned, the scientist would have had to prove in court that he was not engaged in biological warfare. The ASM submitted written statements to Congress outlining its concerns. Subsequently an amended bill was introduced to the Senate; this bill did not contain the language which would have placed the burden of proof on the scientists, rather, the accusers had to prove that the scientists, rather, the accusers had to prove that the scientist was engaged in illegal behaviour, i.e., we were back to the notion of innocent until proven guilty rather than the reverse. The amended bill specifically stated that "nothing in this Act is intended to restrain peaceful scientific research or development." The congressional report included language stating that Congress was aware that many scientists conduct peaceful research with potentially dangerous agents and toxins and the bill would not interfere with such activities. On May 22, 1990 President Bush signed into law the "Biological Weapons Anti-Terrorism Act of 1989." The final legislation acknowledged the contribution of the ASM.

In 1991 the Public and Scientific Affairs Board advised Senator Nunn, the Chairman of the Senate Armed Services Committee concerning agents that potentially could be used by terrorist groups. The Board forwarded a latter to Senator Nunn expressing concern that a provision of the FY 1992 Department of Defense Authorisation

Bill restricted the Biological Defense Research Programme of the Department of Defense to doing research directed only at "validated warfare agents.'" The word "validated" was narrowly defined by the Department of Defense as an agent assessed by the intelligence community as being developed or produced as a weapon. The Board considered this to be a naive effort to either save money or to restrict biological weapons research. The letter to Senator Nunn pointed out that the language did not permit the Biological Defense Programme to prepare defensive measures against emerging or potential agents that might be used by unfriendly nations or terrorist groups. The restriction could have inhibited the Medical Command from conducting research on organisms that might be indigenous to areas where our troops could be sent. The Armed Services Committee of the Senate adopted a provision that would permit the Medical Command to continue research and development on potential agents.

In 1993 the ASM Council (the governing body) adopted a set of scientific principles to guide the verification of the Biological Warfare Convention. The ASM stressed that global surveillance of emerging diseases will contribute to early warnings of biological weapons usage.

In late 1994 the Public and Scientific Affairs Board convened a task force of expert scientists to assist the U.S. government in developing scientifically sound approaches to biological arms control. The task force reflects the breadth of expertise on biological weapons found within the ASM. The task force has reviewed the Royal Society's report "Scientific Aspects of Control of Biological Weapons" from the viewpoint of U.S. security, research and industry interests and is considering issues relevant to effective monitoring to detect as early as possible any employment of biological weapons. In regard to possible usage of biological weapons, the most likely scenario is that they will first be employed by terrorists. In view of this threat, a critical area that has not received sufficient attention on the internal level is how to deter terrorist groups from acquiring and using biological weapons.

In early 1996, the ASM testified before Congress on the transport of human pathogens and assisted the centers for Disease Control in establishing new regulatory requirements for transferring and receiving select agents which have the potential to cause widespread harm. The new regulations are intended to minimise the risk of illicit access to infectious agents. On April 24 President Clinton signed into law Anti-Terrorism legislation (PL 104-132) which expands federal power to prosecute and punish certain crimes related to terrorism. A bipartisan amendment was included in the new law making it illegal to develop, acquire or attempt to purchase biological agents with the intent to kill or injure or use them as a weapon. The provision in biological agents amends the 1989 Biological Weapons Anti-Terrorism Act. Legislators were prompted to amend the U.S. Criminal Code covering biological agents after learning of a 1995 incident involving the acquisition of plague bacillus from the American Type Culture Collection through the mail by a suspicious purchaser in the state of Ohio. The new law also directed the Secretary of the Department of Health and Human Services to promulgate new regulations regarding the acquisition and transfer of certain biological agents.

The Secretary of health and Human Services, in turn, directed the Centers for Disease Control and Prevention to develop implementing regulations. The CDC has done so and published a Notice of proposed rule making. The regulations become effective April 15, 1997.

The new regulations contain the following 1) a list of agents (microbes and toxins) considered likely to be of interest to terrorists and establishes a method for changing the list; 2) safeguards to be followed when the agents are transported; 3) a system for tracking the transfer of agents between laboratories; and 4) a protocol for alerting authorities if an unauthorised attempt is made to acquire of the agents.

Public and private laboratories, commercial companies, academic and research institutions, and other facilities that wish to transfer or receive the agents will be required to register their facilities with CDC and the facilities will be subject to inspection.

It is interesting to note that after the Ohio incident which sparked the most recent efforts I have just described, federal regulations were reviewed and found to cover agents of danger to plants and animals, but there were not regulations covering human pathogens, hence the new regulation I have just described.

The regulations were developed with significant input from the scientific community, most notably the ASM and the American Type Culture Collection. On March 5, 1996, representatives of CDC, the ASM, and the American Type Culture Collection presented testimony at a hearing convened by the Senate judiciary Committee to examine concerns arising from the interstate transportation of human pathogens. The subject of "bioterrorism" was further highlighted at the annual meeting of ASM in May in a symposium "Biological Weapons: Challenges to Health, National Security, Economics, and Science."

More directly, the ASM Biological Warfare Task Force was closely involved with the CDC during almost all stages of the development of the regulation. When it was published as a Notice of Proposed Rule Making, ASM assisted by contacting through the mail and internet more than 11,000 members who might be directly impacted by the new regulations. Many of the comments and suggestions that were received by the ASM, as well as those from other members of the scientific community, have been discussed in the Preamble to the Final Rule, and many were incorporated into the Final Rule.

In 1972 the U.S. signed the Biological Weapons Convention and destroyed its stockpiles of offensive agents. Since then the U.S. programme has been defensive and conducted in the open. Although many other countries became signatories to the Convention, there are claims that some states are still pursuing offensive weapons research and have even produced tactical weapons.

The potential for biological warfare existed in Iraq before the Gulf War and led the U.S. to establish elaborate surveillance systems and to equip troops with special defensive measures. Although biological weapons were not found by inspectors sent to Iraq after Operation Desert Storm, many knowledgeable individuals remain suspicious that Iraq had developed biological warfare agents. Continuing concerns about the biological warfare capabilities of Iraq and other countries, such as Libya, North Korea and Syria have led to renewed debate in federal agencies, the Congress and the scientific community about methods to verify weapons research and development. These concerns extend to prohibiting export of certain kinds of technology. The U.S. Commerce Department has imposed export control over certain technology and material including BL 3 and 4 laboratory facilities, complex media, microencapsulation equipment and selected biologicals.

Subsequent to the Gulf War, 136 nations met in Geneva in September 1992 to review the status of the Biological weapons Convention. The delegates discussed the need for verification provisions. At a previous conference in 1986 four "confidence building" measures were approved by the participants. Each state agreed to exchange information on high containment laboratories, to promote international collaboration through meetings, through publication of relevant research and to report incidents of unusual diseases. To date only 27 nations, including the United States, have submitted reports on high containment laboratories.

At the September 1992 meeting in Geneva the U.S. delegates expressed grave reservations about both the feasibility and utility of a verification scheme. They pointed out that there are thousands of government, industry, and clinical laboratories conducting biological research, and given the relatively simple technology involved in research on certain biological agents, any or all laboratories could be seriously compromised if inspectors were to deliberately or even incidentally reveal proprietary information.

Despite their reservations about the efficacy of verification of the Biological Weapons Convention declared during the Bush administration, the U.S. delegation agreed to assess the applicability and drawbacks of various verification measures that could be adopted under the convention. The assessment was conducted last Spring by a team that included representatives from major pharmaceutical companies, the Industrial Biotechnology Association and the American Society for Microbiology. One working group was charged with identifying those aspects of biological arms control verification measures that may adversely affect U.S. industry and academia if a verification treaty is signed. Other groups worked on the positive aspects of implementation so that a pro-com assessment can be prepared. The U.S. delegation returns to Geneva this June with these evaluations in hand for further discussions on biological weapons verification measures.

Measures examined by the groups ranged from satellite and aircraft surveillance to on site inspection and sampling. The Convention already requires declarations of any research and development related to offensive biological weapons. Further verification measures could well lead to the requirement that all government, industrial and academic institutions that have certain types of equipment—such as fermentors and laminar flowhoods, or that work with certain kinds of organisms—such as pathogens or recombinants, file regular declarations of ongoing activities; these facilities would then be subject inspections which would verify the accuracy of the declarations and detect activities barred by the Convention. Even data that is considered proprietary by many non-governmental organisations might have to be submitted for analyses by other nations. The biotechnology industry has expressed serious concerns about the potential impact on research and development that might occur as a result of such intrusive regimes.

The ASM's involvement with biological warfare began during World War II and has continued to the present. The Society's membership has unique knowledge and expertise in microbiology and immunology, and this has given the ASM a particularly important role in decision regarding U.S. policy on biological warfare and defense. The Society's public and Scientific Affairs Board has played an important role in monitoring and responding to legislative and regulatory issues involving biological weapons, particularly in recent years. As this review has made clear, there is constant stress between the desire to effectively prohibit and monitor biological weapons activity, while at the same time maintaining the freedom to carry out research in an unfettered manner and to maintain the confidentiality of proprietary information in industry. It is unlikely that a perfect consensus will ever be achieved, but the issues are of paramount importance to human welfare and will continue to be addressed by the ASM.

15

Majnoon Island Experience: Consequences of Biological and Chemical Warfare

S. Adams, United States of America

Introduction

The use of chemical or biological weapons in warfare is not new. The early writings of the Greeks (431 BC) during the Trojan War document the use of burning tar pitch mixed with Sulphur to produce a suffocating gas. More recently chemical weapons were used by Germany in 1915 (W.W.I) By the end of the war an estimated 124,000 tons of chemicals had been used by all parties and inflicted over 1 million casualties and 100,000 fatalities. Chemical weapons have also been used by Italy in Abyssinia in the 1930's by Japan in China in the 1930's and 1940's, by Egypt in The Yemen in the 1960's. Chemical weapons were not only used intermittently by Iraq during 1980-1988 in the Iran-Iraq war but also against the Iraq Kurdish population. Biological weapons in warfare also have a long history. Contaminating of water supplies to cause dysentery in the enemy camp dates back also to the Trojan Wars. Other examples of biological weapons in war include the Tartars catapulting the bodies of plague victims into the walled city of Kaffa during the 14th century, the contamination of water supplies during the U.S. Civil War, Germany infecting the horses of the Romanian cavalry with glanders in 1914. During W.W. II the

Japanese attempted to float hot air balloons containing plague infected flea vectors over the western United States. The Japanese also conducted large scale BTW human testing in China 1930's-40's. Biological weapons have also been reportedly used in Iran,[1,2] Cambodia[3,4] and Afghanistan.[2,4] The Majnoon Island experience[1] is unique in several ways. It is the first documented and confirmed modern use of chemical weapons by U.N. experts who visited the site. Also it was the first confirmed attack in which mycotoxins were used in combination with chemical weapons, and it is the only such attack where there is now a description of the chronic complications experienced by long term (over 10 years) survivors of a chemical weapons attack.

Majnoon Island and its surrounding area was a strtegic military location during the 8 years of the Iraqi-Iran war. Its location in the southeast corner of Iraqi was important for both defensive as well as offensives control of the borders between the two nations. In addition to providing an anchor for controlling the border war zone, from Majnoon Island it was able to control shipping access to the Persian Gulf for movement of war material into the war zone as well as for the movement of goods to global markets.

The Majnoon Island Experience

The munitions used by Iraqi military forces during the attack on Majnoon island were of two types. Some of the bombs containing chemicals were contact explosives which detonated on impact, while others exploded at preset altitudes dispersing their contents into the atmosphere. The bombs which exploded on impact produced an overwhelming odor of garlic and acid. The climate was windy and hot and the gases dispersed over an ever increasing area. When the combatants realised what had occurred, the soldiers attempted to leave the field. Because evacuation was not well organised, those injured were exposed to the gases for prolonged periods. Some victims tried to hide in bunkers or other shelters, not realising that the toxic gases would collect and linger in closed environments. The downpour of "toxic rain" continued for several hours causing the distinctive triad of injuries consisting of skin burns, severe eye irritation, and respiratory distress.

Burns typically began with severe itching and erythema, which then progressed to form papules, vesicles, and blisters of the skin.

Regions of severe secondary involvement included the genital area and armpits, where the gases accumulated in perspiration. Third-to-first degree burns covered 20 per cent to 70 per cent of the total skin surface, particularly exposed areas of the face, trunk, and limbs. Ocular exposure resulted in lacrimation, severe conjunctivitis, and temporary loss of vision. Corneal abrasion was nearly always present, and photophobia and blurred vision developed in some cases.

Upper airway involvement secondary to burning of the throat was heralded by pharyngitis and tracheobronchitis. Other symptoms included paroxysmal coughing, expectoration, occasional hemoptysis, and shortness of breath. Generalised wheezing and rhonchi were audible upon physical examination. Approximately 10 per cent of the victims developed adult (or acute) respiratory distress syndrome (ARDS) with pulmonary edema, which almost always proved fatal.

Those who survived the initial symptoms later experienced gastrointestinal complaints of dysphagia, nausea, vomiting and diarrhoea. Transient liver and kidney failure were also common. After five to seven days, hematological problems posed the greatest threat to survival. Numerous patients had been marrow depression manifesting as pancytopenia, when their absolute white blood cell count fell below 500 cu.mm, they became febrile and succumbed to infection, usually an over whelming pneumonia.

Some victims died of massive internal bleeding in the absence of external wounds. In these cases, bleeding from the ears and nose was the only sign of trauma. Still other victims died of heat stroke.

As soon as possible after injury, all battlefield victims were transferred via ambulance to an underground field hospital approximately 50 minutes away, where they were examined, subjected to triage, and decontaminated. Those with severe respiratory problems or burns were then airlifted to Ahvaz University Hospital, 100 km from the field hospital. Less severely ill victims were transferred from the field hospital. Less severely ill victims were transferred from the field hospital via ground ambulance to Takhti Auditorium, in Ahvaz, for observation and supportive treatment. Those whose conditions subsequently deteriorated were airlifted to Tehran. Several of the most severely injured patients were later treated in various European countries.

Verification

March 13-19, 1984 a four member team from the United Nations visited the war zone in Iran including Majnoon island, Shatte-ali and Hoor-ul-huwaizeh. A total of 10 aerial bombs were examined. These bombs were identified as containing nitrogen mustard and tabun. Some patients also presented with signs and symptoms that were consistent with exposure to Lewisite. Dr. A. Hendrick of the Toxicological Institute in Ghent, Belgium, also reported finding mycotoxins in blood, stool and urine samples of five Iranian soldiers. However, there was no other evidence of biological weapons use found, and the presence of mycotoxins could be either from overwhelming sepsis or contaminates.

Toxic Agents and Their Antidotes

Education of the causative agent is fundamental to the effective treatment of chemical-weapon victims. Because the various toxic agents tend to produce similar signs and symptoms, it can be extremely difficult to make an accurate clinical diagnosis and to determine the necessary specific antidote. Although highly specialised detection devices may be available to military personnel, such devices are not always accessible to civilians.

The various chemical weapons are roughly grouped into three classes; nonpersistant agents, semipersistent agents, and persistent agents.[5-13] The nonpersistane agents are all gases that disperse quickly, having a lethal duration of About 20 minutes. The semipersistent agents are organophosphates thickened with polymers; these weapons can be delivered via long range missless or bombs, which can discharge their contents (at present altitudes to produce "toxic rain". The persistent agents (which include some of the original war gases) are all oily liquid substances that can readily penetrate exposed skin, mucous membranes and eye tissue.[14]

An alternative classification of chemical weapons is according to their effects. Choking gases (chlorine, phosgene, chloropicrin) irritate the lining of the respiratory tract by the local production of HCl, causing secondary pulmonary edema and death by suffocation. Blood agents such as cyanogen and cyanide impair the blood's oxygen-carrying capacity and may cause lacrimation, choking, and respiratory distress. Blistering agents (sulphur mustard, nitrogen mustard, and lewisite (Lewisite may also be classified as a blood agent) cause skin and eye irritation including topical blistering and temporary blindness. Nausea and vomiting are common along with respiratory distress. Nerve gases (tabun, sarin, soman, VX) are the deadliest of the chemical poisons; when inhaled or absorbed through the skin, these agents disrupt the nervous system, causing death with 15 minutes after contact.

Some dissipate by means of evaporation and dilution. Others "weather" over a period of hours to days. For instance, mustard gas has a long half-life, especially in rubber, oil, grease, and dry materials; it may last for only 1 to 2 days, however in normal soil in the summer.

Because an extensive review of chemical and biological weapons is beyond the scope of this article, we will discuss only the agents used in the Majnoon Island attack; nitrogen mustard, tabun, lewisite, and trichothecene mycotoxins.

Nitrogen Mustard

Like the other mustard gases, nitrogen mustard is an organic sulfide compound that acts as a blistering agent or choking gas. Contact with these components, in either vapour or liquid from, can cause blisters that rupture and disable the victim for weeks to months. The mustard gases are all bifunctional alkylating agents capable of forming covalent bonds, thus cross linking double strands of deoxyribonucleic acid (DNA). Because the functional group tends to react with thiol groups, the purine bases are most often affected. Although the chemical reactions occur quickly, symptoms are often delayed for several hours. Corrosion and necrosis of the skin, eyes, and respiratory tract is common.[15] Death usually results for damage to the mucous membranes of the lungs. Delayed pulmonary edema and secondary anoxia are frequent. There is no specific antidote against mustard gas exposure. The best approach is to prevent such exposure by wearing protective clothing and gas masks. The most effective therapy consists of rapid decontamination and supportive care. Decontamination is achieved by washing the affected areas with water.[15-17] Skin blisters should be punctured and drained, and a moist, open or occlusive dressing should be applied. The eyes should be irrigated with copious amount of water for at least 5 minutes; an antibiotic containing hydrocortisone ointment should then be applied, along with a mydriatic agent. Inhalation injury necessitates the used of 100 per cent oxygen.[18] In the presence of ARDS, prednisone (60 to 120 mg daily) administered intravenously for several days may be of benefit. Fluid replacement, the management of pulmonary edema, and the use o antibiotics are essential supportive therapy.

Tabun

Tabun is one of several nerve gases (organophosphates) which are odorless and extremely toxic and may be either semipersistent or persistent. All are irreversible cholinesterase inhibitors. Their mechanism of lethality is to poison nerve endings with excess acetylcholine by blocking the metabolism of this naturally occurring neurotransmitter. The buildup of acetylcholine results in a cholinergic crisis or excessive stimulation of end organs, followed by paralysis.

Tabun gas or droplets are rapidly absorbed through the skin, across pulmonary membranes, and from the gut. The onset of toxicity is so rapid that emetic agents, such as activated charcoal, or cathartic agents are of no practical use.[19] The specific antidote is pralidoxime (Protopam, 2-PAM) and symptomatic relief can be achieved with atropine (2 to 5 mg, intravenously, every 10-15 minutes).[20] The usual pralidoxime dose is 1 to 2 g, intravenous, at a rate of 0.5g/ minute bolus injection. Continuous infusions (500mg/hour mixed in 0.9% normal saline and infused for several days) may be more beneficial than bolus doses. Other antidotes include the iodide (PAM) and methanesulphonate (P2S) analogs of pralidoxime.

Underdosing is the major cause of atropine failure during therapy. Ins severely toxic patients, doses may be repeated every 10 to 30 minutes, and atropinisation must be maintained until all the absorbed organophosphate has been metabolised; this may necessitate doses ranging from 2 to 2,000 mg over a period of hours to days. High doses of atropine are not recommended: during periods of extreme heat, as was the case of Majnoon Island. High-dose atropine administration resulted in excessive drying of mucous membranes and skin, as well as elevated body core temperature, thus exacerbating dehydration and heat stroke. Therefore, caregivers should maintain low-dose atropine therapy until the victim can be moved to cooler surroundings, such as in air-conditioned hospital. Prostigmine has been proposed as a useful agent, both for the pretreatment of combatants and for treatment after exposure, in combination with other agents. Benazctizine is a component of the TAB (toxogobin, atropine, benactizine) Autoinjector were issued to both United States and Soviet troops. Seizures are a major complication of organaphosphate poisoning. Standard treatment using diazepam, or other benzodiazepines and phenytoin are effective. Respiratory failure, pulmonary edema, and hypotension are usually responsible for death after nerve gas exposure. These conditions should receive immediate treatment, with continual monitoring and supportive care.

Lewisite

Lewisite (chlorovinyldichloroarsine), both a blistering agent and a blood agent, and is often combined with mustard or nerve gases, as in the Majnoon Island attack. Unlike the other gases, lewisite is not an irritant but depletes erythrocyte glutathlone. The resulting instability of the red blood cell membranes produces rapid, massive intravascular hemolysis. Symptoms include abdominal pain, hematuria and jaundice leading to acute renal failure.[21] Because this single agent contains both a hemotoxin and a pulmonary toxin, its synergistic effect is overwhelming.

The chelating agent BAL (British anti-lewisite, dimercaprol), administered intramuscularly in does of 2 to 4 mg/kg every 4 hours for 48 hours, is a specific treatment for any heavy-metal toxicity and lewisite exposure. Chelation does not protect against red cell hemolysis resulting from acute exposure, however, the treatment of choice is exchange transfusion.[21] Supportive care includes fluid replacement and the administration of vasopressors.

Trichothecene Mycotoxins

Biologic weapons are potentially much more destructive than any other non-nuclear munitions. Biologic agents may include bacterial, fungal, or viral microorganisms, or toxins derived from these organisms. The catch all phrase "yellow rain" can refer to any of 40 types of mycotoxins that have been used in combination with chemical weapons.

Trichothecenes are naturally occurring products of several fungal geniuses including Fusarium, Myrothecium, Trichoderma, Cephalosporium, Verticimonosporium, and Stachybotrys. Episodic outbreaks have occurred from naturally occurring exposures from the ingestion of contaminated of corn and wheat[23]. Chemically, these agents are all sesquiterpenoids that contain the trechoteccane ring and are subdivided on the basis of this ring structure.[24]

The trichothecene toxins all affect multiple organ systems simultaneously. The initial injury resembles that produced by nitrogen mustard. Other sites of injury include the skin, the gastrointestinal tract, and the hemopoietic system. Patients generally present with burning and irritation of the mucous membrane, including conjunctivitis, rhinitis (with or without bleeding), pharyngitis and laryngitis. Coughing and shortness of breath is accompanied by burning of the lungs and secondary pulmonary edema. Vomiting (unresponsive to current antiemetics) is common, as is diarrhoea (or bloody diarrhoea) and dehydration. Dermatological necrosis is often extensive; signs range from erythema and inflammation to blistering and bullous formation with desquamation and necrosis. Hematologic derangements such as bone marrow depression are severe. In victims who survive the initial symptoms, death results from leukopenia, sepsis and multiple hemorrhages.

The profound toxicity of the trichthecenes is due to their inhibition of protein and DNA synthesis.[24] The exact mechanisms and cell-specific sites are not yet entirely known. When these agents react with the sulhydryl residue on susceptible enzymes, a "whole-cell" toxicity is produced. Another known toxic mechanism is irreversible blockage of protein production owing to the inhibition of polypeptide chain initiation through the degradation of the polyribosomes.

Because there are no known antidotes to the trichothecenes, therapy is limited to supportive care. Decontamination may be accomplished if the victim washes extensively with copious amounts of water, but this must be done immediately after exposure.. Emesis is usually contraindicated, but the administration of super-activated charcoal (1 g/kg/dose) may be of some benefit. Fluid and electrolyte therapy, as well as cardiovascular and respiratory support, are fundamental.

Anthrax

A well known biological agent, bacillus antracis is a highly virulent gram positive spore-forming rod. The presentation of illness is very dependent upon the route of exposure. The disease can present in one of three manifestations, cutaneous, inhalation or intestinal.

Cutaneous anthrax can present in two forms, one being mild with little to no sequelae and the other being quite malignant. Itching at the skin site of exposure is the initial presentation usually occurring 2-3 hours after inoculation. The mildest from of cutaneous anthrax a necrotic eschar will then develop in 2 to 4 days that will fall off leaving no scar.

The more serious form will develop edema around the eschar with necrosis and blistering. Here the bacilli, endotoxin or both may spread systemically through blood or lymphatic system leading to a papillary necrosis, renal failure, septic shock and death. Inhalation anthrax initially presents as a mild upper respiratory infection. Three to five days after the first symptoms there is progression to pulmonary edema and respiratory failure, accompanied by high fevers and shock. Death usually occurs in 5 to 7 days.

Intestinal anthrax is unusual and presents with abdominal pain with ascetics. High fever, sepsis syndrome and choler-like diarrhoea will also be present. Death usually occurs as a result of massive fluid and electrolyte losses. Treatment consists of symptomatic care with replacement of fluid and electrolytes and antibiotics. Penicillin, tetracycline or erythromycin are all effective agents.[25]

Treatment of Specific Chemical injuries

A successful outcome depends on immediate diagnosis and treatment. In addition to being influenced by the type of chemical agent involved, the prognosis is affected by the type of chemical agent involved, the prognosis is affected by the degree and duration of exposure; the severity of wounds and their combined effect; the presence of associated injuries such as heatstroke, wounds, etc; the availability of a specific antidote; and efficient triage and evacuation.

Burns

A major difference between thermal and chemical burns is the length of time the tissue is exposed to injury. Thermal injury ceases shortly after the victim is removed from the heat source, chemical agents continue to cause progressive damage until they are deactivated. Tissue destruction by chemicals may be limited, however, if large volumes of water are applied to dilute the offending agent and reduce the contact time. Compared with thermal burns, severe, full-thickness chemical burns may initially appear superficial with only mild brownish discoloration of intact skin during the first few post-exposure hours. The immediate treatment of thermal burns is opposite that of chemical burns. In thermal burns, systemic therapy (Particularly the restoration of normal physiological parameters) precedes treatment of the wound itself. In chemical burns, however, immediate attention is directed toward treating the wound by removing the causative agent and irrigating the wound with water.

Long-Term Experience of Majnoon Island Survivors

In an attempt to define the long term consequences of the use of chemical and biological weapons Dr. H. Kadivar, has reviewed as many long-term survivors has possible and has made the following observations:

- Inhalation of aerosolised or chemical weapons vapours can lead to respiratory failure and death.
- Injury occurs to the respiratory epithelium from the nasopharynx to the bronchiole. The damage is concentration dependent.
- At low concentrations only the upper respiratory tract/large airways are affected leading to obstructive lung disease.
- At low concentrations only the upper respiratory tract/large airways are affected leading to obstructive lung disease.
- At higher inhaled concentrations the lower respiratory tract is affected.
- Incapacitating airway injury occurs at vapour exposures significantly lower than those that cause severe skin blistering.

In patients with mild exposures symptoms consisting of cough, or cough with minimal sputum production, burning sensation in the chest, hoarseness and throat irritation. Over time the hoarseness and throat irritation which usually improves, however, in the majority of survivors these symptoms of chemical tracheobronchitis may persist for years.

The pathological findings from inhalation injuries revealed non-specific bronchitis with metaplastic changes, bronchiolitis or pneumonitis, plus alveolar damage, diffuse or focal with or without hyaline membrane. Desquamative interstitial pneumonitis, interstitial fibrosis and bronchiectasis can be noted after the initial injury. All three types of bronchiectais were seen. Cylindrical with dilated bronchi of regular outline, varicose with great dilation and irregularity and saccular and cystic with bronchial dilatation that increases towards the periphery and results in a balloon like outline of the bronchi.

The presentation of illness after chemical exposure follows a predictable pattern. After a delay of several hours after an exposure, victims experience tracheobronchitis with complaints of sore throat, hacking cough, hoarseness, shortness of breath, and chest pressure. Sinus pain and rhinitis develop as symptoms progress over the next 12 hours. Severe exposure there is bronchospasm, sloughing and ulceration of large areas of the airway mucosa can lead to partial obstruction at all levels resulting in suffocation. Emergency cricothryroidotomy or tracheostomy may be required when a large laryngotracheal pseudomembrane becomes dislodged into the airway. Severe exposure leads to non-cardiogenic pulmonary edema, pneumonia and respiratory failure after 24-48 hours. The clinical problems and potential complications associated with Adult Respiratory Distress Syndrome are present as well. Chronic pulmonary effects which would then develop include chronic purulent bronchitis, generalised bronchiectasis and nonspecific bronchial hyperactivity (reactive airway dysfunction syndrome). Asthma and bullous emphysema may occur. Extensive stenosis of the entire tracheobronchial tree may also occur with life threatening sequelae. After a delay of up to a year, scars, ulcers, granulomata and strictures may develop in the airway. Additionally patients having had a severe exposure often has sufficient systemic mustard absorption to suppress leukopoiesis. An initial leukocytosis may be followed by a marked leukopenia with high morality.

Patients having a low dose exposure will present with edema and erythema of the mucous membranes of the pharynx and tracheobronchial tree. Moderate exposure may progress to fibrinous purulent pseudomembrane formation within the airway similar to those seen in diphtheria associated with sloughing of epithelium and lower airway obstruction. Extensive epithelial necrosis occurs after high dose inhalation.

In other patients with moderate exposure who initially presented with more severe symptoms of pulmonary injury later develop a distressing cough with excessive expectoration, recurrent episodes of hemolysis, repeated pneumonitis and dyspnea of varying severity, who required repeated hospitalisation. These patients, with chronic pulmonary complications such as chronic purulent bronchitis with or without bronchiectases, stenosis of tracheobronchial tree, and pulmonary emphysema, should be treated accordingly. Finally patients with severe exposures died acutely with hemorrhagic pulmonary edema, secondary pneumonia, acute respiratory syndrome, with respiratory failure. Chest x-rays may be unremarkable despite severe bronchographic findings even in the most severe cases with severe abnormalities of pulmonary function tests. Almost all patients who complained of dyspnea had abnormal spirometry with varying degrees of restriction and obstruction pulmonary disease. We found that spirometric findings rather than the chest x-ray correlated more closely with the clinical condition of the patient.

Summary

The immediate effects of exposure to chemical and biologic agents used in warfare resulted in a triad of injuries to the eyes, respiratory tract and skin burns. The similarity of clinical signs following exposure to different chemical agents made an accurate bedside diagnosis of the specific agent nearly impossible. The degree of pulmonary injury and neutropenia were the major determinates of patient outcome. Prompt treatment and specific antidotes, the use of protective measures when possible, and effective decontamination of both casualties and equipment are of great importance and are the key for prevention of acute and chronic complications. Chronic complications of survivors tended to be exposure dependent and primarily respiratory in nature. These long term injuries ranged from non-specific bronchitis, bronchiolitis, desquamative interstitial pneumonitis, interstitial fibrosis and bronchiectasis. It appears that once the pathologic process is established, they become irreversible.

Bibliography

1. Kadivar, H. Adams, S.C: Treatment of Chemical and Biological Warfare Injuries: Insights Derived from the 1984 Iraqi Attack on Majnoon Island. Military Medicine Vol. 156 April 1991.

2. Heyndrickx, A. Heyndrickx, B: Comparison of the Toxicological Investigation in Man in Southeast Asia, Afghanistan and Iran, Concerning Gas Warfare. Proceedings of the First World Congress on Biological and Chemical Warfare, March 1984, pp 426-434

3. Sammard B: Chemical and Biological Warfare in Cambodia. Proceedings of the First World Congress on Biological and Chemical Warfare, March 1984, pp 381-383.
4. Report From Secretary of State George P. Shultz: Chemical Warfare in Southeast Asia and Afganistan: An Update, November 1982 Report No. 104.
5. Gilchrist H.I.: A Comparative Study of World War Causalities from Gas and Other Weapons. Washington D.C. U.S. Government Printing Office, 1928, pp 1-28
6. Mershon M.M, Tennyson AV: Chemical Hazards and Chemical Warfare. J. Am. Vet. Med. Assoc. 190: 734-745, 1987.
7. Treatment of Chemical Agent Casualties: Technical Manual 8-285 Washington D.C. Department of the Army, 1975, pp 3-4
8. Stroykov Y.N.: Medical Aid for Toxic Agent Victims in Moscow, Moscow, Meditsina Publishing House, 1970 in U.S. Army Foreign Science and Technology Center translation FSTC-HT-23-1047-73
9. Military Chemistry and Chemical Compounds, Field Manual, Washington D.C.., Department of the Army, 1975, pp 3-7.
10. Sterlin R.N, Yemelyanov V.I, Zimin V.I: Chemical Weapons and Defenses Against Them. Moscow, Soviet Military Press, 1971, In U.S. Army Foreign Science and Technology Center translation, FSTC-HC-23-1000-73.
11. Stoessel W.J. Bakes P.J. Bresesinski Z, et al: Report of the Chemical Warfare Commission, Washington D.C, U.S. Government Printing Office, June 1985, pp iii-104.
12. Stringer H.L., Welch T.J.: Deterring Chemical Warfare: The Next Step, New York, NBC Def Tech Int. May 1986, pp 18-25.
13. Burner B: Protection against NBC weapons: A Swiss View, New York, NBC Def Tech Int, May 1986, pp 32-39.
14. Volvodic V, Milosavljevic Z, Boskovit B, et al: The Protective Effect of Different Drugs in Rats Poisoned by Sulphur and Nitrogen Mustard. Fundam Appl Toxicol 5:S160-S168, 1985.
15. Dabney B.J: Mustard gas, in TOMES Medical Management, Micro Medex, 1990 vol 64
16. Aasted A. Darre E. Wulf H.C: Mustard gas: Clinical, Toxicological and Mutagenic Aspects Based on Modern Experience. Ann Plast Surg 19: 330-333 1987.
17. Requena L., Requena C., Sanchez M. et al: Chemical Warfare: Cutaneous Lesions from Mustard Gas. J. Am Acad Dermatol 19:529-536, 1988
18. Joy R.M: Organophosphates in Pesticides and Neurologic Disease. Edited by Ecobichon DJ. Joy RM. Boca Raton Fl. CRC Press, Inc. 1982 p 132.
19. Morgan D.P. Krenzelok E.P., Kulig K. et al: Organophosphates. In TOMES Medical Management, MicroMedex 1990 vol 64.
20. Haddad L.M., Winchester J.F: Clinical Management of Poisoning and Drug Overdose. Philadelphia. W.B. Saunders Co. 1983, pp 656-664.

21. Fowler B.A., Weisberg J.F: Arsine Poisoning. N Engl J Med 291: 1711, 1974.

22. Bamburg J.R., Strong FM: Trichothecene Mycotoxins, in Microbial Toxins, Edited by Kadis S. Ciegler, A. Ajl SJ. New York, Academic Press, 1971, vol 7 p 207

23. Rodericks J.V., Hasseltine C.W., Meclman M.A: Mycotoxins in Human and Animal Health. Park Forest South, IL. Pathotox Publishers, 1977.

24. Spoerke D.G: Trichothecene Mycotoxins. In Poisindex Toxicology Management. MicroMedex, 1990 vol. 64.

25. Poisindex Editorial Staff: Anthrax: In Poisindex Toxicology Management. MicroMedex, 1990 vol. 64.

Brief Historical Overview of Chemical/Biological Weapons

431 B.C. Trojan War—burning of tar pitch mixed with sulphur to produce a suffocating gas

1899 - Hugue convention outlaws the use of chemical weapons

1914 - Germany uses biological weapons to infect horses in the Romanian cavalry with glansers

1915 - April near Ypres Belgium; Germany during WWI explode canisters of liquid chlorine. 5,000 French troops killed, 10,000 injured in that single battle

End of WWI—an estimated 124,000 tons of chemicals were used in warfare by all parties.

A total of 91,000 killed, 1.2 million injured due to chemical weapons

1925—Geneva Protocol for the prohibition of the use in war of asphyxiating poisonous or other gases and of bacteriological methods of warfare

1930's—Possible use of mustard gas by Italy against Ethiopian Troops

1930's-40's—Possible use of chemical agents and biological weapons by Japan in China

1961-73—Viet Nam - U.S. widely uses defoliants and lachrymatory agents. U.S. position held that herbicides are not anti-personnel weapons and not covered under the articles of CWC.

1966-67—International Red Cross reports that a lethal gas probably mustard gas is used in the Yemeni Civil War

1970—Luska Summit Conference banned the development, production and accumulation of chemical weapons

1972—Biological and Toxins Weapons Convention prohibits the possession and use of biological weapons

1982-November—U.S. Secretary of State George P. Shultz, reports on Chemical Warfare in Southeast Asia and Afghanistan. Report identifies chemical and biological weapons being used in Afghanistan, Loas, Kampuchea/Thailand.

1984—A. Hendrick reports in the proceedings of the First World Congress on the use of chemical and biological warfare in Southeast Asia, Afghanistan and Iran.

1984—B. Sannard reports in the proceedings of the First World Congress on the use of chemical and biological warfare in Cambodia.

1991—U.S. troops vaccinated against anthrax prior to deployment in the Persian Gulf War.

1995—First major terrorist use of chemical weapons when the nerve gas serin is released in the Tokyo underground.

16

Improving Biological Security : The process of Strengthening the BTW Convention

—I. Hunger, Germany

Introduction

Ensuring safety from biological and toxin weapons (BTW) is a more complex issue than controlling other weapons of mass destruction. This is due to the character of the relevant technologies. More than chemical technologies and much more than nuclear technologies, biotechnologies are of dual-use, i.e. the same technology can be used for civilian and permitted military defensive purposes as well as for prohibited offensive or terrorist purposes. In the biotechnological field it is sometimes only the intention that distinguishes permitted from prohibited activities. Moreover, BTW, at least unsophisticated BTW, are relatively easy to produce. The necessary technologies do not need to be either very advanced or expensive and they are already existent in a lot of countries. These technologies are available on the open market and will spread rapidly because they have numerous important applications in the civilian area.

Because of the complexity of the problem the system for preventing and countering the misuse of biotechnologies for offensive military or terrorist purposes must be equally complex in order to be effective. It will be impossible to establish a single regime that will successfully solve all problems of controlling and countering BTW.

Instead a system of regimes is necessary that integrates mechanisms at the personal, national and international level, and at the level of prevention as well as reactions after violations of the Biological and Toxin Weapons Convention (BTWC) have occurred (Table 1).

TABLE 1: DIFFERENT LEVELS OF MECHANISMS FOR ENSURING SAFETY FROM BTW

	International	National	Personal
Prevention	• BTWC and its future Protocol (e.g. declarations and visits) • other international arrangements (e.g. by the Australia Group)	• national legislation	• individual scientific oaths • networks of individual scientists
Reactions in case of violations of the BTWC	• BTWC and its future Protocol (e.g. investigations of non-compliance or alleged use, assistance and protection against BTW) • international sanctions • development and production of measures against relevant diseases on an international scale	• penalties and sentences for offenders • development and production of measures against relevant diseases on a national scale	

Papers by other authors of this book describe several aspects of such a comprehensive system for biological security, like the international criminalisation of BTW or the Science for Peace Oath. This paper deals with efforts at the international level, especially with the BTWC, which is the most important regime for regulating BTW issues in international relations. This regime has explicit norms and rules. Specific procedures, however, are badly missing. In this paper I will first outline the history of the BTWC. I will then shortly describe the different approaches towards strengtheing this treaty and the current state of the rolling test of a compliance protocol to the BTWC, that is being developed for establishing the missing procedures. Finally, I will introduce you to a programme that has been developed in order to strengthen both the aspect of prevention as well as the aspect of reactions in case of violations of the BTWC. This is a programme that also supports the technical cooperation in relevant areas of biotechnology under international control, and helps to improve the international health situation.

History

International treaties prohibiting or condemning the use of poison and disease as means of warfare are relatively old. In the last century or so, one finds the Conventions of the First and Second International Peace Conference in The Hague in 1899 and 1907 by which it is forbidden to "employ poison or poisoned weapons".[1] More important, because it mentions BTW explicitly for the first time, is the Geneva Protocol of 1925, that prohibits, among other things, the use in war of bacteriological methods of warfare. One may look at this treaty as a successful one, because with the exception of Japanese use during World War II, BTW have not been used on a large scale since that time. However, several countries became States Parties only with the reservation that they may use BTW as retaliatory weapon. BTW programmes were developed in several states in the 30s and 40s, most of the programmes continued in the 50s and 60s.

In the late 60s the discussion about chemical and biological weapons that had been closely linked until then became more and more separated. This was an important precondition for allowing negotiations on the BTWC. Finally, the BTWC, the Convention for the prohibition of the Development, Production and Stockpiling of Bacteriological (Biological) and Toxin Weapons and on their Destruction, opened for signature in 1972 and entered into force in 1975.[2]

This Convention is a document of a few pages. Comparing this document with the Chemical weapons Convention of 1993 that entered into force in April 1997 and comprises more than 200 pages raises curiosity to what the difference may be. The difference is a variety of loopholes that are visible much more clearly today than at the time the Convention was concluded. One of the most important loopholes is the lack of a verification regime. The BTWC has only unspecific mechanisms for dealing with suspected, assumed or actual violations of its provisions.

That was recently illustrated when the Cuban Government requested a consultative meeting aimed at clarifying events in spring

1996, when a US aircraft had supposedly been observed sprinkling a white or greyish mist over Cuban territory. According to Cuba that incident was the reason for the current Thrips palmi plague in their country.[3] During this consultative meeting delegations moved forward very cautiously stressing all the time that activities during that meeting would serve as examples for future procedures and may therefore be of great influence on the future efficiency of the Convention. It was repeatedly stressed that one had to be careful not to destroy or weaken the credibility and the spirit of the BTWC.

The problem of missing mechanisms to clarify specific doubts has in the past prevented mistrust to be resolved and violations of the Convention to be detected. E.g., especially after a major unusual outbreak of anthrax in the Soviet city of Sverdlovsk in 1979 there were several claims that the former Soviet Union pursued an offensive BTW programme in violation of the BTWC. These claims, however, could never be officially in the framework of the BTWC substantiated because of the lack of respective mechanisms, although Russian President Yeltsin admitted in 1992 that the former Soviet Union had pursued such an offensive BTW programme.[4]

This weakness of the BTWC existed for more than 10 years unchallenged. With the emergence of new technologies in the biosciences and growing suspicions concerning treaty compliance of some States Parties, the Second Review Conference was compelled to take first steps towards providing more, and more specific, procedures for the BTW control regime, setting in train a process of strengthening the Convention that continues until today and became considerably more intensive in the past few years (table 2).

TABLE 2: STEPS TOWARDS STRENGTHENING THE BTWC

Year	Activity
1972	Conclusion of the BTWC
1975	Entering into force of the BTWC
1980	First Review Conference
1986	Second Review Conference (decision to establish CBMs)
1987	Ad Hoc Meeting of Scientific and Technical Experts (development of CBM modalities)
1991	Third Review Conference (improvement of CBMs, establishment of VEREX Group)
1992-93	VEREX Group
1994	Special Conference (evaluation of VEREX Report, establishment of Ad Hoc Group)
1995- today	Ad Hoc Group
1996	Fourth Review Conference

Confidence-Building Measures

During the Second Review Conference in 1986, States Parties decided to establish so called confidence building measures (CBMs). These were seen as modest first steps towards a continuous cooperation among States Parties and towards improvement of transparency, and even as precursors of future legally binding declarations.

"The Conference, mindful of the provisions of article V and Article X, and determined to strengthen the authority of the Convention and to enhance confidence in the implementation of its provisions, agrees that the States Parties are to implement, on the basis of mutual cooperation, the following measures, in order to prevent or reduce the occurrence of ambiguities, doubts and suspicions, and in order to improve international cooperation in the field of peaceful bacteriological (biological) activities".[5]

The modalities for these CBMs were developed during an Ad Hoc Meeting of Scientific and Technical Experts in Spring 1987. They were further developed during the Third Review Conference 1991 "with a view to promoting increased participation and strengthening further the exchange of information".[6] The CBMs ask for the annual provision of relevant data in accordance with agreed formats. Eight such CBM forms have been valid since 1991 (Table 3).

TABLE 3: CBM DECLARATION FORMS (CBMS THAT HAVE BEEN IN EXISTENCE SINCE 1987 ARE MARKED WITH A STAR)

Name	Content
0	Declaration form on "Nothing to declare" or "Nothing new to declare"
CBM A 1*	Exchange of data on research centres and laboratories
CBM A 2	Exchange of information on national biological defence research and development programmes
CBM B*	Exchange of information on outbreaks of infectious diseases and similar occurrences caused by toxins
CBM C*	Encouragement of publication of results and promotion of use of knowledge
CBM D*	Active promotion of contacts
CBM E	Declaration of legislation, regulations and other measures
CBM F	Declaration of past activities in offensive and/or defensive biological research and development programmes
CBM G	Declaration of vaccine production facilities

The existing CBMs are until now the only permanent measure that has been taken by the States Parties to strengthen the BTWC.

When this measure was concluded in the middle of the 80s it was expected that it would positively influence the efficiency of the BTWC, that it would be the first step on the way towards a compliance or even verification regime and that openness, transparency and cooperation among member states would be enhanced. These expectations were mostly disappointing. Neither did all States Parties take part in the information exchange, nor was the information provided of unambigous character, nor was there any feeling of a strengthened biological and toxin weapons control regime. It took up to 1995 before half of the states Parties had at least once taken part in the information exchange (Chart 1).

CHART 1

Number of states that have participated at least once (* 1997: only submissions until 30 May 1997 are registered)

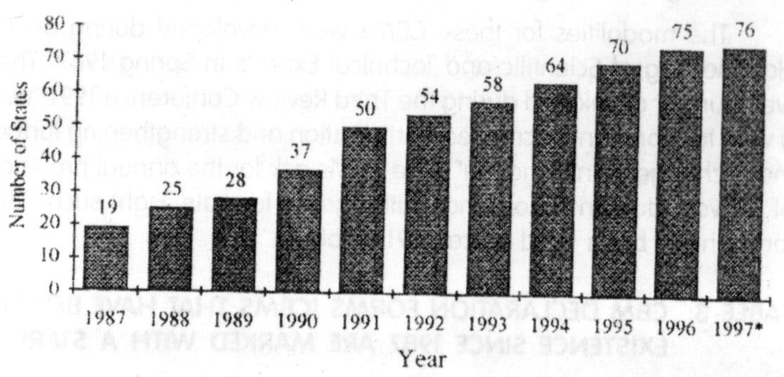

Only 11 states took part without interruption between 1987 and 1996. These are Canada, Denmark, Finland, Germany, The Netherlands, Norway, The Russian Federation, Spain, Sweden, UK and USA. This limited participation is a bit surprising when compared to the number of States Parties that took part in the Second and Third Review Conference where these CBMs were agreed (63 and 78 states respectively) and to the number of States Parties that took part in the Ad Hoc Meeting of Scientific and Technical Experts in Spring 1987,[7] where the modalities for the CBMs were finalised (39 states). It is also interesting to note that the Russian Federation is included in this list, although it, as the former Soviet Union, was in non-compliance with the BTWC. Participation in the CBMs must not be mistaken, therefore, for being in compliance with the Convention. Moreover, this example illustrates that the CBMs are of limited ability to strengthen the Convention because they are only politically binding and not subject to verification.

Encouragingly, the number of states that participated at least once over the years as well as the number of states that provided returns annually as agreed grew steadily over the years (Charts 1 and 2). This is probably due to increased awareness of the problem of BTW as well as knowledge about the existing mechanisms for strengthening respective control.

States Parties that provide complete and properly prepared CBM submissions regularly show that they regard their politically binding obligation under the BTWC as important. They demonstrate that they consider the BTWC to be an important international treaty and do not want to weaken it through dis interest. In fact it seems to be the case that this function of demonstrating interest in the BTWC is by far the most important function of the CBMs. Obviously there is little use of the detailed data provided. States Parties rather focus on analysing pure participation in the information exchange.[8] This illustrates that the detailed data are not as important as the attitude towards cooperation that is shown by participation. As the first disappointment about the inability of the CBMs to strengthen the Convention makes room for a more general view, one tends to see these measures as a "normal" step on the way towards a control regime for the BTWC. That is exactly according to the function that was attributed to CBMs for a long time—they are "paving the way for arms control".

CHART 2

Number of states participating per year (* 1997: only submissions until 30 May 1997 are registered)

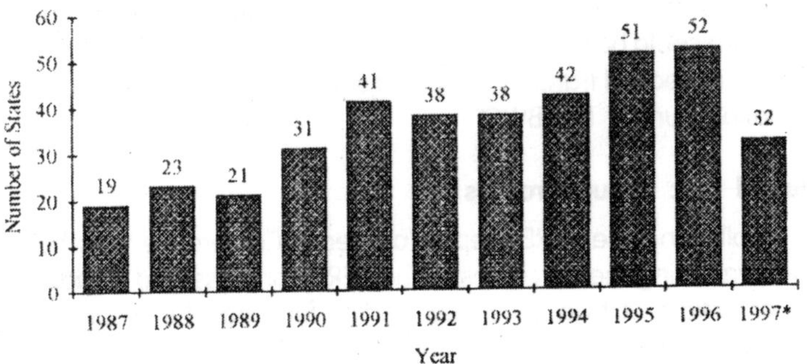

The VEREX process

During the Third Review Conference in 1991 not only were the CBMs extended and improved, but also, with a growing feeling of unease because of the more or less toothless BTWC, the states agreed to establish an Ad Hoc Group of Governmental Experts (VEREX Group) to "identify and examine potential verification measures from a scientific and technical stand point".[9] The mandate of the VEREX Group asks for the identification of measures which are able to determine whether a State Party is in compliance with or violates the Convention. Potential verification measures were evaluated singly or in combination according to a set of criteria including "their ability to differentiate between prohibited and permitted activities", "their financial, legal, safety and organisational implications" and "their impact on scientific research, scientific cooperation, industrial development and other permitted activities, and their implications for the confidentiality of commercial proprietary information".[10]

Four meetings that became known as VEREX meetings were held in 1992 and 1993. The final VEREX report was completed in September 1993 and circulated to the States Parties for discussion. The range of identified potential verification measures stretched from surveillance of publications, multilateral information sharing and declarations to surveillance by aircraft or satellite, sampling and identification, medical examination and continuous monitoring by personnel. As expected by many experts already beforehand, the report did not conclude that a system of verification measures could be developed which would provide 100 per cent security that violations of the Convention could be detected. However, the group was convinced that a system is feasible that would deter cheating by raising the probability of detection in case of massive violations. The report concluded that {potential verification measures as identified and evaluated could be useful to varying degrees in enhancing confidence, through increased transparency, that states Parties were fulfilling their obligations under the BWC".[11]

The Ad Hoc Group Process

Following the VEREX report a Special Conference was held in 1994 evaluating the outcomes of the VEREX process and deciding on further actions. Obviously States Parties felt, that the results of the VEREX process justified further study of a control regime for the BTWC.

The most important result of this Special Conference was, accordingly the establishment of a new Ad Hoc Group mandated to *"consider appropriate measures, including possible verification measures, and draft proposals to strengthen the Convention, to be included, as appropriate, in a legally binding instrument"* and to consider inter alia

"definitions of terms and objective criteria, [...]"; the incorporation of existing and further enhanced confidence and transparency building measures [...]; a system of measures to promote compliance with the Convention, including, as appropriate, measures identified examined and evaluated in the VEREX Report [...]; specific measures designed to ensure effective and full implementation of Article X [...]".[12]

This Ad Hoc Group held an organisational session in early 1995 and seven substantial sessions that became known as the Second to Eighth Ad Hoc Group Session until October 1997. After initial general discussions on the work to be done a Rolling Text of a Protocol to the Convention was introduced during the Seventh Session in July 1997. This Rolling Text was discussed and further developed during that and the Eighth Session. It reflects in extent and contents the emphasis that is placed on compliance measures like declarations, visits and investigations but also the imbalance that was typical for the negotiations so far (Table 4).

TABLE 4: NUMBER OF MEETINGS ON DIFFERENT TOPICS DURING DIFFERENT SESSIONS OF THE AD HOC GROUP

	Number of meetings concerning compliance measures	Number of meetings concerning definitions	Number of meetings concerning Article X measures	Number of meetings concerning confidence building and transparency measures
2nd Session (July 1995)	9	7	2	2
3rd Session (Nov/Dec 1995)	6	5	4	2
4th Session (July 1996)	7	6	3	2
5th Session (Sept 1996)	7	4	4	2
6th Session (March 1997)	8	6	5	0

(Contd...)

7th Session (July /Aug 1997)	11	5	3	0
8th Session (Sept/Oct 1997)	16	3	3	0
Sum (% of all 163 meetings)[13]	64 (39%)	36(22%)	24(15%)	8 (5%)

The Rolling Text of a Protocol

The Rolling Text of a Protocol comprises 23 articles, 8 annexes and 5 appendices so far and is 243 pages long.[14] This massive volume is due to the fact that this text is full of square brackets. Square brackets are used in cases where States Parties are not able yet to agree on a common formulation or disagree with the content. The process of integrating all the different concepts and ideas that are gathered in the text now, into an acceptable final version that can then be submitted for the consideration of States Parties, has still to come. It is already visible, however, that the future protocol to the BTWC will consist basically of the following items: declarations, on-site visits and investigations, and international cooperation and non-transfer arrangements.

Declarations

Declarations will form an important part of the provisions under the future Protocol. Disagreement, however, exists on what kinds of facilities and activities should be declared. The envisaged range of declarations stretches from past offensive and defensive, and current defensive programmes, facilities with P4 containment areas or engaged in aerobiology to declarations on transfers of relevant agents and implementation of Article X. The functions of declarations will include providing background information on relevant activities in a state, identifying targets for non-challenge visits, and providing transparency.

On-site visits and investigations

On-site visits and investigations will form a central part of the provisions under the future Protocol. Agreement exists that there needs to be investigations in cases of non-compliance with the provisions of the BTWC and for alleged use of BTW. Problems in regard to these investigations persist on how such an investigation may be initiated and prepared, how abusive requests may be prevented and how access should be regulated to prevent the loss of commercial proprietary and national security information.

Disagreement still exists on whether any kind of non-challenge visits is necessary at all. If agreement on these types of visits is reachable, it will still be necessary to identify the right filters in order to assure that the optimum between costs and results is reached. Non-challenge visits would fulfill such important functions as provision of training for investigation teams, verifying declarations, providing transparency and providing opportunities for technical cooperation. It is worth remembering that the experience with the CBMs illustrates the uselessness of declaring relevant information without providing mechanisms for verifying them at least on a random basis.

International cooperation and non-transfer arrangements

Provisions for international technical cooperation and non transfer of relevant equipment and agents is one of the most controversial issues. While a lot of states from the non-aligned movement want specific provisions for cooperation and assistance built into the future Protocol, Western states do not consider the Protocol to be the right place for further cooperation provisions in the biotechnological area. However, they are not opposed to cooperative activities that ensure the correct implementation of the Protocol like setting up relevant electronical data bases, assisting in the establishment of national authorities and providing assistance for controlling disease outbreaks.

Concerning non-transfer arrangements, Western countries believe in the usefullness of existing export control regimes like the one pursued by the Australia Group, while some states from the non-aligned movement see these regimes as discriminatory and incompatible with a future Protocol. They insist on a new multilateral regime with equal mechanisms for all States Parties to the BTWC.

These controversies are closely linked and their solution will be one of the most difficult. In this regard it is difficult to understand why meetings on this topic have been rather limited. Work on the aspect of international technical cooperation and non-transfer arrangements in the framework of the future Protocol has to be intensified, especially if states take their repeatedly stated commitment serious to conclude work as soon as possible, favourably before the next century.

The comprehensive countering of the BTW threat is possible and necessary

Despite the prime importance of a strengthened BTWC, it remains only one part of an integrated system designed to counter the threat arising from biological and toxin agents. Threats from biological and toxin warfare, from terrorist activities with biological and toxin agents as well as from the natural occurrence of newly emerging and re-emerging diseases can be effectively countered only through international cooperative efforts. Such efforts should, for economic reasons, try to solve as much of the relevant problems as possible with a single procedure. An example for such an approach is the "Programme for Countering Emerging Infectious Diseases (ProCEID) by Prophylactic, Diagnostic, and Therapeutic Measures".

ProCEID

ProCEID is a proposal specifically designed for the purpose of strengthening the BTWC. It focuses on implementing Article X, one of the most important problem areas as outlined above. However, by implementing this programme a lot of other goals can be reached as will be described in the following.

ProCEID is the result of numerous discussions of an earlier proposal, the Vaccines for Peace Programme.[15] Key objectives of ProCEID include, *inter alia,* 1) an increase of world capacity to produce and make available prophylactic, diagnostic, and therapeutic measures for pathogens of relevance to the BTWC, thereby preventing and combating the respective diseases, 2) the enhancement of peaceful international cooperation in molecular medicine and biotechnology and 3) the strengthening of the BTWC by implementing Article X. In contrast to previous programmes, ProCEID would involve the development and production of all kinds of biologicals necessary to counter and production of all kinds of biologicals necessary to counter diseases of relevance to the BTWC. Nevertheless research on and development and production of vaccines would be an outstanding topic of ProCEID not only because of global health demands but also because of the outstanding role vaccines play in defensive and offensive military biological activities.

How would this programme work? ProCEID would be executed by international teams in institutions located in different parts of the world under conditions of complete transparency. The facilities involved would be provided with the equipment, know-how, and agents necessary to develop and produce pro phylactic, diagnostic, and therapeutic measures necessary to protect from relevant infectious diseases. This measures would be provided on request to States

Parties for countering military or terrorist use of BTW or natural outbreaks of diseases.

Most of the activities involved in such a programme are of a dual-use nature. To minimise the risks of misuse of the programme, ProCEID would only be open to States Parties in good standing with the BTWC, that is, to states that have ratified the future Protocol to the Convention and that are also States parties to the Chemical Weapons Convention. Moreover, one would have to be extremely careful, that participation in ProCEID does not interfere with the BTWC obligation to avoid proliferation of weapons-related agents, technologies and knowledge. This can be ensured by a rigorous transparency regime. In the long run, ProCEID could even contribute to the international control of the spread of relevant agents and technologies, a spread that will definitely take place because of the ongoing biotechnological development worldwide.

One of the outstanding global advantages of ProCEID could be solving the problems of suspicion created by research on and development and production of biologicals protecting from BTWC relevant diseases, and at the same time global provisions of these biologicals so far not or only marginally covered by industry and international health care programmes, without weakening (national) military defence activities. This programme would promote the establishment of an international system of biological security fulfilling at the same time the obligations of Article X of the BTWC.

Additional benefits of ProCEID may include contributing to conversion of former BTW facilities to peaceful purposes under international control. Using former BTW facilities could under certain circumstances not only contribute to reducing costs, time, personnel, and numerous technical prerequisites necessary to establish ProCEID, but would for the first time contribute to establishing an efficient conversion programme while minimising the dangers of misuse and clandestine reversion involved in such undertakings.

A programme that envisages similar procedures as ProCEID and is aimed at converting BTW facilities of the former Soviet Union has recently been developed by the National Academy of Sciences of the USA. Under this programme individual projects in the biological field will be financed and undertaken in tight cooperation with US American laboratories. Efforts are focussed on the core experts of the former Soviet BTW programme and on topics that pose a security threat.

Conclusion

Although it is only one part of a comprehensive system to counter the BTW threat, the strengthening of the BTWC is one of the most important tasks in the next few years. It is urgently necessary to become widely aware that there is no alternative to the protocol and, therefore, to actively move forward the process of concluding it. Without a strengthened arms control regime it will be difficult to maintain the efficiency of other approaches towards controlling BTW.[16] If we are able to develop a cost-effective system for the prevention of BTW development, production and transfer that includes as many countries as possible we would make a big step forward towards limiting the threat deriving from BTW. The cost of relying only on counter measures after non-compliance with the BTWC has occurred is nicely illustrated by the efforts and expenses of the United Nations Special Commission for Iraq (UNSCOM). Cooperative efforts for establishing a system of biological security are clearly the cheaper and more promising approach.

Bibliography

1. CBW and the Law of War. The problem of Chemical and Biological Warfare Vol. III, Stockholm International Peace Research Institute, Stockholm: Almqvist & Wiksell, 1973, p. 152-153.
2. For some further remarks on the history of negotiations see: Aida Luisa Levin: "Historical outline", in: Erhard Geissler (ed.): Strengthening the Biological Weapons Convention by confidence- Building Measures. SIPRI Chemical and Biological Warfare Studies No. 10, Oxford University Press, 1990, p. 5 - 14.
3. Susa Wright: "Cuba case tests treaty", in: Bulletin of Atomic Scientists Vol. 53, No. 6 (November/ December 1997), p. 18-19.
4. Milton Leitenberg: Biological Weapons Arms Control, Project on Rethinking Arms Control Paper No. 16, Center for International and Security Studies at Maryland, College Park, May 1996.
5. United Nations: The Second Review Conference of the States Parties to the Convention on the Prohibition of the Development, Production and Stockpiling of Bacteriological (Biological) and Toxin Weapons and on their Destruction, Geneva, 8-26 September 1986, BWC/CONF.II/13/II, Geneva, 1986, p. 6.
6. United Nations: The Third Review Conference of the States Parties to the Convention on the Prohibition of the Development, Production and Stockpiling of Bacteriological (Biological) and Toxin Weapons and on their Destruction, Geneva, 9 - 27 September 1991, BWC/CONF.III/23, Geneva, 1992, p. 14.
7. United Nations: Ad Hoc Meeting of Scientific and Technical Experts from States Parties to the Convention on the Prohibition of the Development, Production

and Stockpiling of bacteriological (Biological) and Toxin Weapons and on their Destruction, Geneva, 31 March - 15 April 1987, BWC/CONF.II/EX/2, Geneva, 21 April 1987.

8. A hint supporting this assumption may be the following. The preparatory Committee of the Fourth Review Conference requested the United Nations Secretary-General to prepare a document providing data on the participation of States Parties in the CBMs. As requested, the resulting document BWC/CONF.IV/2 does not analyse anything else than participation.

9. United Nations: The Third Review Conference of the States Parties to the Convention on the Prohibition of the Development, Production and Stockpiling of Bacteriological (Biological) and Toxin Weapons and on their Destruction, Geneva, 9-27 September 1991, BWC/CONF. III/23, Geneva, 1992, p. 16.

10. United Nations: The Third Review Conference of the States Parties to the Convention on the Prohibition of the Development, Production and Stockpiling of Bacteriological (Biological) and Toxin Weapons and on their Destruction, Geneva, 9-27 September 1991, BWC/CONF.III/23, Geneva, 1992, p. 17.

11. United Nations: Ad Hoc Group of Governmental Experts to Identify and Examine Potential Verification Measures from a Scientific and Technical Standpoint, BWC/CONF.III/VEREX/9, Geneva, 1993, p. 9.

12. United Nations: Special Conference of the States Parties to the Convention on the Prohibition of the Development, Production and Stockpiling of Bacteriological (Biological) and Toxin Weapons and on their Destruction, Geneva, 19-30 September 1994, BWC/SP-Conf/1, Geneva, 1994, p. 10. Article X of the BTWC calls for technical cooperation for peaceful purposes in the biological field.

13. Meetings that do not appear in the table were allocated to other topics, technical issues or general discussions.

14. United Nations: Procedural Report of the AD Hoc Group of the States Parties to the Convention on the Prohibition of the Development, Production and Stockpiling of Bacteriological (Biological) and Toxin Weapons and on their Destruction, Geneva, 15 September - 3 October 1997, BWC/Ad HOC GROUP/ 38/, Geneva, 6 October 1997.

15. For a short description of the evolution of ProCEID see: Meeting Report. Programme for Countering Emerging Infectious Diseases (ProCEID) by Prophylactics, Diagnostic and therapeutic Measures. Mission Statement. Revised, in: Biologicals Vol. 24, 1996, p. 71-74. For more information on the earlier programme Vaccines for Peace see; Erhard Geissler and John P. Woodall (eds.): Control of Dual Threat Agents: The Vaccines for Peace Programme, SIPRI Chemical and Biological Warfare Studies No. 15, Oxford etc.: Oxford University Press, 1994.

16. Tibor Tóth: "A window of opportunity for the BWC Ad Hoc Group", in; The CBW Conventions Bulletin No. 37 (September 1997), p. 1-5.

17

Reaching for Biosecurity by Countering a New Danger to Democratic Order

R. Zilinskas, United States of America

Introduction

In June 1997, NATO member countries met in Madrid to consider the membership applications of the Czech Republic and Poland. Before the meeting, the Czech President, Václav Havel, made an eloquent speech in which he discussed the need for an expanded NATO to meet the new dangers to democratic order that transcend the boundaries between war, terrorism, and organised crime (Havel, 1997). Although he did not mention biological warfare (BW) or biological weapons, his words provided a focus for the many thoughts I have had about the nature of the threat posed to international security by the BW programme of the former Soviet Union (FSU). I share these thoughts in this paper. Thus, I begin by describing of how these three baneful activities, which throughout history have transpired in isolation from one another, are converging. Unity of the three appears to be occurring in the FSU and, in the process, is creating a new type of security problem. The paper ends with a discussion of how the new security problem may be overcome through the development and application of novel, innovative approaches.

1. *War, Terrorism, and Criminality Converging*

Historically, humanity's three most baneful activities, war, terrorism, and criminality, have transpired in isolation from one another. Even in situations where nations known to support terrorists have been involved in conflict, the dependent terrorists have chosen not to take to the field on their mentor's side, even as saboteurs. For example, there have been no known concerted efforts by any of the many terrorist groups active in the Middle East to affect outcomes of the several wars that have been fought in the region since the establishment of the state of Israel.

In the last ten years or so, borders between these malevolent activities have begun to crumble. Thus, as denoted by the term "narco-terrorist," a synthesis of terrorism and criminality occurred in Latin America and the Golden Triangle when drug producers hired terrorist groups to provide protection. Although not yet reality, as far as is known, it is not farfetched to believe that drug producers some day will put their well equipped, well staffed chemical laboratories to use to produce chemical and, even, biological weapons, which then can be deployed to attack competitors, weaken adversary governments, and arm the terrorists who protect them.

An example of a linkage between war and gangsterism was seen in Panama during late 1989 and early 1990. Beginning on December 20, U.S. Military forces commenced operation "Just Cause," which had as its aim the overthrow of President Manual Noreiga's government and, thereby, stopping its durg-related activities. President George Bush's administration, which was responsible for this drastic action, justified the armed intervention on the principle of self protection, i.e., it claimed that military force was necessary to roust the gangsters responsible for smuggling large quantities of drugs into the U.S. Although the U.S. action was the first of its kind, it is probable that as the problem of drug addiction increases throughout the world, other governments will seek to protect their citizens by mounting similar military operations to interdict drug trade routes and apprehend drug traffickers.

A linkage between terrorism and war might have forged in Iraq. Although we still do not know its intent, Iraq's BW programme seems to have been a "traditional"; i.e., its main function was to produce strategic and/or tactical weapons for the nation's military forces (Zilinskas, 1997a). However, there are certain signs that it might have had an additional dimension. A so-called "anti-terrorism training camp", which actually was a terrorist training camp, was located on the Salman Pak Peninsula, a few hundred meters from Iraq's foremost

BW research laboratory. Further, Iraq's BW programme—as has been officially acknowledged by the Iraqi Government itself-produced several toxins that appear to have scant value for the military, such as aflatoxin and ricin. Might terrorists have been equipped with such toxins? We do not know, but the possibility is not farfetched. The Iraqis claimed to have destroyed all toxins, but the United Nations Special Commission (UNSCOM) has not been able to verify that assertion (Zilinskas, 1997a).

The unsavory two-way relationships described above certainly threaten regional stability and the welfare of individual nations. However, the synthesis of all three components, which threatens international security, appears most likely to eventuate in countries of the FSU.

2. *Implications of the Dissolving BW Programme of the Former Soviet Union*

It is now well recognised that the Soviet Union established and operated a powerful, well-funded secret programme to acquire biological weapons (Leitenberg, 1993, 1994, 1996ab; Rimmington, 1986). Although President B. Yeltsin acknowledged the BW programme's existence in 1992, and decreed that it was to be discontinued and dismantled (Smith, 1992a), it continues to raise concerns. In particular, there are three good reasons why the fate of FSU'd BW facilities in Russia ought to be taken seriously by Western security experts and policy-makers.

First, President B. Yeltsin's 1992 decree could have been disobeyed and secret, BW-related activities contravening the 1972 Biological and Toxin Weapons Convention (BWC) continue in the Russian Federation. Certainly the U.S. Government, among others, fears that this is occurring (United States Arms Control and Disarmament Agency, 1994; United States Department of State, 1995. A Russian BW programme, although smaller that of the FSU, can be expected to build on past accomplishments. As such, it would be far more threatening to Western security than the smaller, more primitive programmes of proliferant nations such as Iraq.

Second, the parts of the FSU's BW programme that has been, or will be, discontinued present a substantial proliferation concern. In particular, the effect of President Yeltsin's 1992 decree as well as the general decrease in support by the Russian Government for the military sector, can be assumed to have led to drastic cut in funding for facilities that once were part of the FSU's BW programme. Although we do not know the full consequences of these acts, presumably some BW facilities were closed down and many others downsized. It is likely that thousands of scientists, engineers, and technicians either were fired or had their wages cut. Those out of work joined a growing army of Russian scientific workers who are experiencing unprecedented hardships, as can be demonstrated by a startling demographic statistic: Of the 3 million persons employed as scientific workers in 1991, just 1.5 million remain so employed in 1996. For those who remain in the scientific field, the situation is inauspicious; the Duma decreased support for basic research by 50 per cent in the summer of 1997 (Freemantle, 1997). At present, the wage of a scientist is only 80 per cent of the wage for the average Russian (Lepkowski, 1997). It is not surprising that a large number of Russian scientists have departed for foreign lands and many more are prepared to follow.

Due to the difficult conditions under which science operates in Russia, and in consideration of the dissolving or diminishing weapons research institutions, the country is likely have a substantial number of disgruntled and frustrated scientists and engineers with expertise in the biological weapons area. Many of these persons are likely to be amendable to recruitment by foreigners. However, only a few of the most qualified scientists and engineers will be hired by legitimate foreign research units and enterprises. The remainder may be approached by representatives from governments and subnational groups intent on acquiring biological weaponry. For this reason the dissolving BW programme poses additional threats to international security (Cooperman & Belianinov, 1995).

Third, little is known about the security of biological research institutions in countries that comprised the FSU, some of which store the most virulent bacteria and viruses known to science. If security were poor or lacking, they would be vulnerable to theft of pathogens, toxins, and other material. Most important, after theft, it would be easy for the perpetrator to hide and transport seed cultures of organisms that could be directly used in biological weapons or to produce toxins. The problems pertaining to the smuggling of biological material abroad certainly would be less difficult than the smuggling of nuclear material, something which has occurred several times since 1991. Criminal groups operating internationally would be particularly well placed and equipped to steal biological material and smuggle it aborad.

It bears nothing that two sets of actions in the recent past by the Russian Government bearing on international biological arms control efforts raises concerns about a continuing Russian BW programme. First, to demonstrate that all BW work has indeed ceased after yeltsin's 1991 decree, Russia entered into a trilateral agreement with the United Kingdom (U.K.) and the U.S. The agreement specified that two of the countries can inspect biological facilities and sites in the third and enjoins the three governments to provide data annually regarding biological defense programmes and other activities (Smith, 1992b; Lacey, 1994; Meridian Corporation, 1994). The implementation of the trilateral agreement commenced in a promising fashion; inspectors from the U.K. and U.S. visited five civilian-operated Bio preparat facilities known to have been part of FSU's BW programme, while Russian inspectors visited one English and two American facilities. However, in 1994 the Russians breached the agreement by refusing U.S. and U.K. inspectors access to military BW facilities and, further, disallowed return visits to the Biopreparat facilities. The trilateral process came to a standstill and has remained inactive since then.

Second, Russia, U.K., and U.S. are working with other State Parties of the BWC to strengthen the Convention (Chevrier, 1995). Most notably, the BWC State Parties have established an Ad Hoc Group to negotiate the establishment of a BWC compliance regime with powers to conduct inspections of suspect facilities and alleged uses of biological weapons. The approach being taken is to develop a protocol to the BWC that will specify the mission of the compliance regime, its organisation and administration, and the powers it will be able to wield. However, these negotiations are hardly in better shape than the trilateral process. In particular, Russia insists on the adoption of three measures that are unacceptable to the Western governments. First, the Russians insist on developing a list of agents that would be included under the BWC's Article prohibition, something that Western governments believe would limit the prohibitions spelled out in the BWC's Article 1. Second, the Russians wish to define threshold limits for allowable quantities of listed BW agents, which Western governments insist is meaningless since large quantities of a BW agent, which Western governments insist is meaningless since large quantities of a BW agent can be quickly grown from a very small seed culture. Third, the Russians refuse to endow the future compliance regime with the authority. to investigate suspicious outbreaks of disease. The intent of these tactics might be to prevent investigations of future epidemics like the one that struck Sverdlovsk in 1979 (Meselson, et al., 1994) and to enable Russian scientists to develop new agents not on the list and, of course, outside the purview of future BWC-related inspections (Zilinskas, 1997b).

From the foregoing, it can be seen that a synthesis of the three baneful activities is possible in the FSU. If part of the FSU's BW remains operational, its inventions and developments can be utilised either for military purposes, perhaps to counter the technological superiority of NATO. Alternatively, its accomplishments may be employed for terrorist purposes, perhaps to subdue breakaway national groups or weaken adversary nations. The dissolving part of the FSU's BW programme has created, and is creating, a unique pool of human resources comprised of scientists and engineers possessing specialised knowledge pertaining to biological weapons. It is not unreasonable to believe that some of these people would be willing to work for nations and/or rogue groups that offer substantial financial enumeration. If that were to occur, more nations would be able to produce and deploy effective biological weapons, in the process adding to the already substantial present threats to our biosecurity. Finally, of all groups, well-organised criminal groups operating internationally are best placed to steal biological material from poorly protected BW facilities in the FSU and sell it abroad to proliferant nations and rogue groups.

3. *Meeting the Threat to Our Biosecurity*

Both governments and non-governmental organisations can take actions to meet the biological threat posed by the FSU's BW programme. In regards to a possible active BW programme, two approaches are immediately available to governments that belong to the BWC. First, they can invoke Article VI of the BWC, which specified that evidence of misconduct shall be presented to the UN Security Council, which shall than investigate and resolve the matter. Or, according to the decision of BWC State Parties at the second and third review conferences, one or more of them can call for a consultative meeting of BWC State Parties and present the evidence of misconduct to that meeting.

Since the UN Security Council may be stymied by a Russian veto, the consultative meeting of BWC State Parties would be the preferred approach. Secret BW programme's are, of course, exceedingly difficult to discover (Zilinsks, 1986, 1996). However, by directing the searchlight of public attention on suspect activities, the Russian Government might be motivated to stop them. Conversely, if Russia has no active BW programme, a consultative meeting would be an appropriate venue for its government to convincingly demonstrate that fact.

As to the dissolving part of FSU's BW programme, governments can seek to ease the difficulties facing Russian weapons scientists and engineers, thus forestalling them from emigrating and offering their service to those who might misuse it. Some programmes are already in place that attempt to do so. The International Science and Technology Center (ISTC), which was established in march 1994, is an example of a programme that tries to ensure that scientists with expertise in weapons of mass destruction are provided with opportunities to use their skills of civilian endeavours (Harrington, 1995; Seltzer, 1996). The major supporters of the ISTC are the European Union, Finland, Japan Sweden, and the U.S; the main beneficiaries are scientists in Armenia, Belarus, Georgia, Kazakhstan, the Kyrgyz Republic, and the Russian Federation (a parallel agreement exists for Ukraine only). However, the ISTC has not been particularly well supported in the past and its present funding is in sad shape (Lepkowski, 1997; Stone, 1997). The U.S. in particular is not doing what it could to give the ISTC full support. Discussions have been underway in Congress since late 1996 about the necessity of funding a more general assistance programme (the so-called Freedom Support Act) for Russia and the newly independent countries at the level of approximately $ 600 million. However, Congress has taken no action so far in 1997 either to fund the Freedom Support Act or replenish U.S. funds for the ISTC. There is little indication at present that Congress will address the issue of special assistance to East European countries in the near future.

Nongovernmental organisations (NGOs) can play an important role in international attempts to achieve biosecurity. NGOs have already taking on significant responsibilities in international arms control activities in non-biological fields. For example, when an unexplained seismic event took place in the Arctic Ocean on August 16, 1997, analysis by its intelligence agencies convinced the U.S. Government that Russia had set off an underground nuclear explosion. However, seismologists constituting an international private network, the Incorporated Research Institutes for Seismology (IRIS), analysed data collected by stations they manned and found that the event, while not wholly explained, most probably had not been a nuclear explosion. In October 1997, the U.S. Government acknowledged that the conclusion of the seismology network was correct. Not only did IRIS prove its worth in this episode, it also has an important role to play in the verification of the Comprehensive Test Ban Treaty.

A few NGOs are beginning to become active in the biological arms control field. The Chemical and Biological Arms Control Institute, headed by Michael Moodie, and the Harvard Sussex Programme on CBW Armament and Arms Limitation, co-directed Dr. Matt Meselson and Dr. Julian Perry Robinson, both provide information and analysis on chemical and biological warfare-related subjects that are independent of official sources. These endeavours are helpful to private analysts especially because it provides them with information that otherwise would be available to those who work for governments and, therefore, are served by substantial information-gathering agencies.

The Federation of American Scientists is being recognised as playing an ever-increasing role in international biological arms control. Two of its activities merit special notice. The first, called "Outbreak," is an Internet-based information service that publishes information about outbreaks that might be caused by biological, chemical, or toxin warfare agents (http;//www.outbreak.org/). It is related to a second private network, called "ProMED", which information about outbreaks of emerging infectious diseases. In either case, the providers of information to Outbreak and ProMED are private persons located in the field throughout the world who subscribe to these services. Should an health event take place anywhere in the world that is caused by a biological or chemical weapon, likely it will be first reported on one of these services, rather than by monitoring systems run by governments or the World Health Organisation. Since it is almost a truism that the quick detection and identification of an incidence of BW has a deterrence effect, private networks such as ProMED and Outbreak have an important role in achieving international biological and chemical arms control.

I end this paper by paraphrasing the person who said that war is too important for generals; in other words, biological arms control is too important to be left to governments alone. In an earlier SIPRI publication, Professor Carl Göran Hedén and I argued that bioscientists, acting by themselves and through their professional organisations, especially the International Council of Scientific Unions, must become directly involved in biological arms control (Zilinskas and Hedén, 1991). Since that was written, a tremendous new development has taken place in information collection and dissemination, namely the rise of Internet. By using this mechanism, we can significantly increase private inputs into official efforts to control biological weapons and their proliferation. The networks I have already mentioned are a beginning. But I can envision a next step, which is to try to form a network of scientists and engineers who were involved in FSU's BW programme, or who knew something about it, and exchange information about this programme which remains mysterious and fear-inspiring to most of the world. Many of these persons have emigrated to Israel, the U.S., the U.K., and other countries but because they are dispersed, the information they possess is difficult to access. This barrier might be able to find out about its intent, accomplishments, methods for transferring results from research to applications, responsible authorities, and names and work programmes of all components. If this information was on hand, it might be possible to discover its present status in Russia, whether active or a fantasy. The importance of such an endeavour is that it would supplement or better similar endeavours that until now have been in the realm of intelligence services only. By doing so, we would improve the policy-making process bearing on international biological arms control, which will serve to ensure a higher level of biosecurity for us all.

Bibliography

Cooperman, A. & Belianinov, K. (1995). Moonlighting by modem in Russia: Hard-up scientists sell their skills abroad. *US News & World Report* April 17:45-48.

Freemantle, M. (1997). Russian science on the rack. *Chemical & Engineering News* 75(51): 25-34.

Harrington, A.H. (1995). Redirecting biological weapons expertise: Realities and opportunities in the former Soviet Union. *Chemical Weapons Convention Bulletin* (29): 2-5.

Havel, V. (1997). NATO and the Czech Republic: A Common Destiny. *NATO Review* 45 (5):8.

Lacey, E.J. (1994). Tackling the biological weapons threat: The next proliferation challenge. *Washington Quarterly* 17(4): 53-64.

Leitenberg, M. (1993). The biological weapons programme of the former Soviet Union. *Biologicals* (21): 187-191.

Leitenberg, M. (1994) The conversion of biological warfare research and development facilities to peaceful uses. In: E. Geissler & J.P. Woodall (eds.), *Control of Dual-Threat Agents: The Vaccines for Peace Programme.* (New York: SIPRI and Oxford University Press), pp. 77-105.

Leitenberg, M. (1996a). *Project on Rethinking Arms Control. Biological Weapons Arms Control.* (College Park, MD: Center for International and Security Studies at Maryland).

Leitenberg, M. (1996b). Biological weapons Arms Control. *Contemporary Security Policy* 17 (1): 1-79.

Lepkowski, W. (1997). U.S. Aid to Soviet scientists falters. *Chemical & Engineering News* 75 (14): 45-47.

Meridian Corporation. (1994). *Report of a Simulated Visit to a US Commercial Facility Under the US/UK/Russian Trilateral Agreement.* (Washington, D.C.: United States Arms Control and Disarmament Agency).

Meselson, M., Guillemin, J., Hugh-Jones, M., Langmuir, A., Popova, I., Shelokov, A. & Yampolskaya, O. (1994). The Sverdlovsk anthrax outbreak of 1979. *Science* 266: 1202-1208.

Rimmington, A. (1986). Warfare and biotechnology in the USSR. *International Industrial Biotechnology (Wales)* 6: 104-108.

Seltzer, R.J. (1996). Moscow science center lauded. *Chemical & Engineering News* 74(51): 14-15.

Smith, R.J. (1992a). Yeltsin blames '79 anthrax on germ warfare efforts. *Washington Post* June 16:A15,

Smith, R.J. (1992b). Russia agrees to inspection of its biological research facilities. *Washington Post* September 15:A 14.

Stone, R. (1997). The perils of partnership. *Science* 275:468-471.

United States Arms Control and Disarmament Agency. (1994). *Threat Control Through Arms Control: Report to Congress,* 1994. (Washington, D.C.: U.S. Arms Control and Disarmament Agency).

United States Department of State. (1995). *Report on the extent of compliance of the independent states of the former Soviet Union with the Biological and Toxin Weapons Convention and other international agreements relating to the control of biological weapons.* (Washington, D.C.: United States Department of State).

Zilinskas, R.A. (1986). Verification of the Biological Weapons Convention. In: E. Geissler (ed.), *Biological and Toxin Weapons Today.* (New York: Oxford University Press), pp. 82-107.

Zilinskas, R.A. (1996). Detecting and Deterring Biological Weapons: Lessons from United Nations special Commission (UNSCOM) Operations in Iraq. In James Brown (ed.) *Arms Control in a Multipolar World..* (Amsterdam: VU University Press), pp. 193-212.

Zilinskas, R.A. (1997a). Iraq's Biological Weapons Programme: The Past as Future? *Journal of the American Medical Association* 278 (5):418-424.

Zilinskas, R.A. (1997b). The other Biological-Weapons Worry. *New York Times,* November 28:A39.

Zilinskas, R.A. & Heden, C.G. (1991). The Biological Weapons Convention: A Vehicle for International Cooperation. In S.J. Lundin (ed.), *Views on Possible Verification Measures for the Biological Weapons Convention,* (New York: Oxford University Press), pp. 71-97.

18

Possible Consequences of the Misuse of Biological Sciences

G. Harigel, Switzerland

1. Introduction: What is a Misuse?

An example from chemistry: *"It's in your Own Home and It's a Real Killer"*[1]: The chemical compound dihydrogen monoxide, or DHMO, has been implicated in the deaths of thousands of Americans every year, mainly through accidental ingestion. In gaseous form, it can cause severe burns. And according to a report, "the dangers of this chemical do not end there".

The chemical is so caustic that it "accelerates the corrosion and rusting of may metals...is a major component in acid rain and... has been found in excised tumours of terminal cancer patients". Symptoms of ingestion include "excessive sweating and urination", and for those who have developed a dependency on DHMO, complete withdrawal means certain death". (Even when completely purified, drunk in short time at larger quantities, it may lead to severe pain and death[2])

Yet the presence of the chemical has been confirmed in every river, lake and reservoir in America.

Judging from these facts, do you think dihydrogen monoxide should be banned? 86 per cent of fourteen year old students voted to ban dihydrogen monoxide because it has caused too many deaths.

Do you think you would get during a poll a different answer

from members of Congress? (Or, if you replace hydrogen by its heavy isotopes deuteron or, even worse trition?[2]).

Let's take an example from physics: Should we ban Radioactivity?

Radioactivity is produced by scientists, and *only* since the beginning of the twentieth century, by builders of nuclear power plants, and by elementary particle physicists. Once there, radioactivity is omnipresent, settles in plants, is eaten by animals, enters the food chain on both ways, is eaten by us and settles in our body, and in the end produces cancer, terrible physical suffering, and leads to an almost certain death.

A vote on this subject would probably give an almost hundred per cent condemnation of this stuff.

Asking the question if it is of any use, we may learn—if we are lucky—that it is sometimes used in medicine, and many people would confirm it is a *good, healthy radiation*, however applied with ambivalent success.

Probably nobody will come up with any of the sensitive and correct answer like:

"The very substance of the Universe, the particles and atoms, are unstable, in a state of continual change and decay. When their lifetime is short—shorter than the age of the Earth—the decay is clear; we call it radioactivity. That decay is vital. Like the phoenix, it gives birth to the future, driving new complexities, creating a new world... and us.

Without radioactivity the stars would have cooled and the Universe would be dark. But with radioactivity the impact heated stars ignited. Then, at the end of their lives, the largest stars, their hydrogen exhausted, collapse to supernovae, and in those fantastic explosions an incredible sequence of neutron capture, radioactive decay and nuclear fission created in seconds all the elements that we known, and beyond. We, too, are made of the stars' radioactive waste.

But radioactivity plays another part. The emitted radiation is the ultimate source of the heat energy driving the Universe and creating light and life on Earth. All our other sources of energy—all the coal and oil, the wind and the weather—are fruits of the Sun's radioactivity. Consider the Earth: the residual heat from its accretion would have dissipated long ago and the cool interior would now see no convection. No mountain creation. All the Earth's land would be at sea level.

Germ mutations take place all the time. Those that are viable and produce offspring (perhaps one in several thousand DNA damage events) are the material of evolution. From the potassium in our bodies, slightly radioactive, several hundred thousand beta particles and gamma photons rear through our cells each minute"[3].

What about biology? Should we condemm all cross-breeding, etc.?

Conclusions from these three example:

1. Scientific education of the public at large is important;

2. Anything can be used for the good and the bad, often depending only on its quantity;

3. Difficulty to separate the branches of science and establish treaties (physics, chemistry, biology, BUT physical chemistry, biochemistry, biophysics, microbiology, etc.);

4. Civilians can have handguns, but no tanks, how about AK47?

5. Curb production of chemical weapons, require identification (e.g. by trace elements in ammunition).

2. Chemical warfare, its history

By their nature, chemical arms have a relatively limited range: they create regional rather than global problems. In this, they are militarily more akin to conventional arms than to nuclear or biological weapons, even though nuclear, biological, and chemical weapons are generally classed together as "weapons of mass destruction".

The Greeks first used sulphur mixtures with pitch resin for producing suffocating fumes in 431 B.C.[4] Attempts to control date back to a 1675 Franco-German accord signed in Strasbourg. The 1870 Brussels Convention. The 1899 Hague Convention, reinforced in 1907. But prohibitions were largely ignored during World War I: at the battle of Ypres, chlorine gas was used by Germany and caused 15,000 casualties in April 1915 (Fritz Haber, nobel price winner 1919 for invention of ammonium fixation, convinced the German Kaiser to use it to end the war quickly). Mustard gas on both sides in 1917, hundreds of thousands of injuries. 1922 Washington Treaty, 1925 Geneva Protocol, which included a ban on bacteriological weapons.

During 1935-36 Italians used chemical weapons in Ethiopia and during World War II the Japanese in China.

Taburn, the first nerve gas, discovered in 1936, followed by Sarin, Soman, and VX.

During the Vietnam War, the United States was accused of using heavy doses of herbicides in much the same manner as chemical weapons.

In 1990, US and URSS agreement to destroy at least 50 per cent of their stockpiles by 1999 and to retain no more than 5000 tons of agent by 2002.

Saddam Hussein's use of chemical weapons against Iraqi civilians as well as against Irani soldiers.

Convention on the Prohibition of the Development, Production, Stockpiling, and Use of Chemical Weapons And on Their Destruction (CWC)[5], signed in 1995 by over one hundred nations, entry into force 1997 after deposit of 65 ratification documents.

The Waste Problem[6]

Should scientists be held responsible for the **Invention, production, use and elimination** of chemical weapons?

Dumping of chemical weapons into the oceans[7]

The captured German chemical weapons after 1945 totalled 296,103 tons.

The US dumped chemical weapons in the Scandinavian region, totalling between 30,000 and 40,000 tons, nine ships in the Skagerrak Strait and two more in the North Sea at depth of 650 to 1,180 meters.

As soon as World War II ended, 46,000 tones were dumped in the Baltic areas known as the Gotland Deep, Bornholm Deep, and the Little Belt (mean depth of Baltic Sea is 51 meters).

The Russians alone have dumped 30,000 tons in an area 2,000 square kilometres in size near the Gotland and Bornholm islands.

Between 1945 and 1949, the British dumped 34 shiploads carrying 127,000 tons of chemical and conventional weapons in the Norwegian Trench at 700 meters depth.

During the 1950s, the US conducted an ambitious nerve gas programme, manufacturing what would eventually total 400,000 M-

55 rockets, each of which was capable of delivering a 5 kg payload of Sarin. In 1967 and 1968 51, 180 nerve gas rockets were dropped 240 km off the coast of New York State in depths 1950 to 2190 meters.

The US is responsible for 60 sea dumping totalling about 100,000 tons of chemical weapons filled with toxic materials in the Gulf of Mexico, off the coast of New Jersey, California, Florida and South Carolina, and near India, Italy, Norway, Denmark, Japan and Australia.

The cost of the current US disposal programme estimated at between $ 8.5 and 9.5 billion.

3. New Chemicals (Non lethal weapons)

Are the following agents chemical weapons?[8]

1. Sticky Foam, Super lubricants ("slickums and stickums");
2. Military smoke agents;
3. Pulsed chemical laser beams;
4. Depleted uranium munitions;
5. Incendiaries, old as war. Producing fires and burns of skin.

The Chemical Weapons Convention (CWC) does not cover sea dumped chemical weapons; in fact it makes a clear exception for them.

Abuse of herbicides is dealt within the **Environmental Modification Convention**, reasonable distinction?

Napalm (thickening gasoline with the aluminum soap of naphtemic and palmatic acids) a herbicide and chemical weapon[9,10]? Napalm burns at higher temperatures than gasoline alone.

White phosphor burns more intensive than napalm.

Masking chemical weapons as like fertilisers, dyes, and pesticides.

A Chemical Weapons Atlas[11]

(Countries in normal print have or actively develop chemical and biological weapons) Defence Intelligence Agency (DIA) in the U.S. reports:

Middle East

Egypt

First country in the Middle East to obtain chemical weapons

training, indoctrination, and material. It employed phosgene and mustard agent against Yemeni Royalist forces in the mid 1960s, and some reports claim that it also used an organophosphate nerve agent.

Israel

Developed its own offensive weapons programme. The 1990 DIA study reports that Israel maintains a chemical warfare testing facility. Newspaper reports suggest the facility is in the Negev desert.

Syria

It began developing chemical weapons in the 1970s. it received chemical weapons from Egypt in the 1970s, and indigenous production began in the 1980s. It allegedly has two means of delivery: a 500-kilogram aerial bomb, and chemical warheads for Scud-B missiles. Two chemical munitions storage depots, a Khna Abu Shamat and Furqlus. Centre d'Etude et Recherche Scientifique, near Damascus, was the primary research facility. It is building a new chemical-weapons factory near the city of Aleppo.

Iran

Initiated a chemical and warfare programme in response to Iraq's use of mustard gas against Iranian troops. At end of war military had been able to field mustard and phosgene. Had artillery shells and bombs filled with chemical agents. Was developing ballistic missiles. Has chemical-agent warheads for their surface-to-surface missiles.

Iraq

Used chemical weapons repeatedly during the Iraq-Iran war. Later attacked Kurdish villages in northern Iraq with mustard and nerve gas. Since end of Guld War, U.N. destroyed more than 48,000 litres of Iraq's chemical agents and 1.8 million litres of precursor chemicals.

Libya

Obtained its first chemical agents from Iran, using them against Chad in 1987. Opened its own production facility in Rabta in 1988. May have produced as much as 100 tons of blister and nerve agent before a fire broke out in 1990. Is building a second facility in an underground location at Tarhunah.

Saudi Arabia

May have limited chemical warfare capability in part because it acquired 50 CSS-2 ballistic missiles from China. These highly inaccurate missiles are thought to be suitable only for delivering chemical agents.

Asia

North Korea

Programme since 1960s, probably largest in the region. Can produce "large quantities" of blister, blood, and nerve agents.

South Korea

Has the chemical infrastructure and technical capability to produce chemical agents, had a chemical weapons programme.

India

Has the technical capability and industrial base needed to produce precursors and chemical agents.

Pakistan

Has artillery projectiles and rockets that can be made chemical capable.

China

China has a mature chemical warfare capability, including ballistic missiles.

Taiwan

Had an "aggressive high-priority programme to develop both offensive and defensive capabilities", was developing chemical weapons capability, and in 1989, it may be operational.

Burma

Its programme, under development in 1983, may or may not be active today. It has chemical weapons and artillery for delivering chemical agents.

Vietnam

In 1988 was in the process of deploying, or already had,

chemical weapons. Also it captured large stocks of U.S. riot control agents during and at the end of the Vietnam war.

Europe

Yugoslavia

The former Yugoslavia has a CW production capability. Produced and weaponised Sarin, sulphur mustard, BZ (a psycho chemical incapacitant), and irritants CS and CN. The Bosnians produced crude chemical weapons during the 1992-1995 war.

Romania

Has research and production facilities and chemical weapons stockpiles and storage facilities. Has large chemical warfare programme, adding that it had developed a cheaper method for synthesising Sarin.

Czech republic

Pilot-Plant chemical capabilities that probably included Sarin, Soman, and possibly VX.

France

Has stockpile of chemical weapons, including aerosol bombs.

Bulgaria

Has stockpile of chemical munitions of Soviet origin.

USA

31,000 tones, 3.6 million grenades[12]

Russia

40,000 tons total, composed of:[13]

- 9,800 ton Sarin, Soman, VX ammunition
- 7,000 tone Lewisite
- 1,500 tone mustard + Lewisite
- 1,500 ton mustard gas
- 30,000 ton phosphorus organic agents

New chemical weapons agents which are 5 to 10 times more dangerous than VX, the most dangerous toxic gas known today[14].

4. Old and New Biological Weapons

History

- Poisoning of water wells by cadavers or chemicals is as old as human history.
- Throwing corpses, infected by small pox, behind city walls started the epidemic desease ("black death") in the middle ages. This techniques continued with cholera or thyphus infected corpses.
- German infected the horses of Romanian cavalry with glanders in 1914.

New developments[16]

Violations[17]

The Biological Weapons Convention (BWC) prohibits bacteria such as salmonella being used against soldiers, it would permit bacteria that are petroleum or rubber for the destruction of equipment.

References

1. *"It's in Your Own Home And It's a Real Killer"*, James K. Glassman, IHT, Oct, 23, 1997.
2. Added by the author (G.H.).
3. Physics World, August 1992, pg. 64.
4. *"No more poison Bullets"*, Bulletin of Atomic Scientists, October 1992, pg. 37.
5. Convention on the Prohibition of the Development, Production, Stockpiling, and Use of Chemical Weapons And on Their Destruction, Arms Control Today, April 1997, pg. 15-28.
6. *"The challenge of Old Chemical Munitions and Toxic Armament Wastes"*, in... Edited by Thomas Stock and Karlheinz Lohs, SIPRI, Oxford University Press, 1997.
7. The Bulletin of Atomic Scientists, September/October 1997, pg. 40-44, Ron Chepesiuk, *"A Sea of Trouble?"*.
8. New Scientists, 18 October 1997, Laura Spinney, *"A fate worse than death"*.
9. Arms Control Today, Oct. 1992, 8-24, Ambass. St. J. Ledogar, *"The End of Negotiations"*.
10. The Bulletin of Atomic Scientists, September/October 1997, pg. 35-39, E.J., Hogendoorn, *"A Chemical Weapons Atlas"*.
11. Encyclopaedia Britannica, 1988 edition, Vol. 29, pg. 580.

12. Arms Control Today, February 1996, U.S. Unitary and Binary Chemical Stockpiles, pg. 34.

13. SIPRI Yearbook, 1993.

14. International Network of Engineers and Scientists for Global Responsibility, INES Newsletter No.6, May 1993.

15. Iris Hunger, *"Successful Biological Weapons Control is a Comprehensive Task and the Responsibility of All"*, Statement of the INES Working Group on Biological and Toxin Weapons Control, prepared for the Fourth Review Conference of the Biological Weapons Convention 25 November -6 December 1996, Geneva.

16. *"Make This Evidence of Saddam's Poison Public"*, Jim Hoagland, IHT, Nov. 20, 1997.

17. The Bulletin of Atomic Scientists, Nov./Dev. 1997, pg. 18-19, *"Cuba case tests treaty"*.

19

Prevention of Misuse of Biological Sciences

V. Volkov, Russian Federation

State Research Center for Applied Microbiology (SRCAM) was established specially for the research of pathogenic bacteria and development of means for prevention (vaccines), diagnosis and treatment of dangerous infectious diseases. The ensurance of the safety of staff, population and environment was the paramount problem from the beginning of works with pathogens. In Russia there is a severe system of sanctioning such works, and state official services strictly control the observance of the fixed rules. However the efficiency of such control depends in many respects on social, economic and political conditions in the country and in the world.

In the present report the potential danger of unauthorised use of the pathogenic substances and information received during research in modern conditions is analysed. Information on the participation of the scientists of SRCAM in Russian and international projects, change of the character of research and frame of financing is submitted. Some generalisations about the possibility to use bacteria in terroristic acts and measures to prevent this are made.

Research Centre and satellite town Obolensk are located in a picturesque region 100 kms to the south from Moscow and approximately 20 kms from Serpukhov.

Distinctive feature of Obolensk Research Center is that

infrastructure of the Center and modern laboratory base (building N 1) enable realisation of research works with pathogenic micro-organisms, including causative agents of especially dangerous infections in humans and animals.

The constructional characteristic of the main laboratory building N 1 allow to conduct a full cycle of research, beginning from strain-producer production and finishing with the development of technologies for preparation of ready commodity forms of the preparations. At all stages there are essential levels of physical protection of staff and environment (BL2, BL3) according to the international WHO requirements.

In SRCAM a team of skilled scientists and technicians was set up. The work with dangerous materials is performed by the experts who finished special additional courses under the programme prescribed by federal management of medical-biological and extreme problems of Ministry of Health of Russian Federation. Of the Center's research the works prevail, which objective is the creation of new preventive, medical and diagnostic immunobiological preparations and advanced technologies by using the methods of molecular biology, genetic engineering and microbiological synthesis. Of them, priority is given to the research directed on the development of new generation of vaccines and preparations for diagnosis of bacterial infections.

During the beginning of the 90s, the level of experimental research of many pathogens was as high as that of research conducted in the laboratories of western Europe and USA, and in some directions even advanced them. This was evidently shown at international conferences and workshops held in 1995 in Obolensk, Pokrov, Porton-Dawn (Great Britain), Uhmea (Sweden), Vienna (Austria), Yamagata (japan, 194.) etc.

Scientists of SRCAM have conducted fundamental research on identification of pathogenicity and immunogenicity factors of causative agents of such diseases as plague, anthrax, diphtheria, tularemia, whooping-cough, legionellosis, pasteurellosis, melioidosis, coliderhea, erysipeloid, etc.

Cloning of major pathogenicity determinants of causative agents of these severe infections was performed, nucleotide sequences of these determinants were investigated. Expression of cloned gene synthesis products, their chemical nature and role in the pathogenicity of infections was investigated. Principally new knowledge of structure-functional organisation of causative agents and toxins at molecular level was received. Unique knowledge gained by our scientists on the properties of natural and genetic-engineered bio-agents is applied with a humanistic objective for prevention, diagnosis and treatment of a lot of dangerous diseases. However we fully realise that the given knowledge as well as achievements in this field and related technologies can be used for diversion or terroristic purposes. This causes certain concerns and stipulates the necessity of solving his problem cooperatively with the international community.

Taking into account obviously insufficient financing of R & D as well as a new business strategy of the Center directed on the diversification of main activities, a decision was made to establish self-supporting branches, which mainly attend to industrial activity on principles of self-repayment. The branches provide the employees with the salary, provide payments for the maintenance of buildings and structures, including accounts for all kinds of energy supply.

Since 1991, in connection with the aggravation of the financial situation in the Center, "the brain leakage" is observed. Analysis of the dynamics of the staff number shows that for 6 years total number of employees is reduced by 2.2 times. The number of scientists has thus decreased by 1.8 times, and that of Drs and Ph.D. scientists by 1.3 times. That means that intensive process "washing away" of scientists takes place, though this process proceeds rather slower, than the decrease of the total number of employees. The qualitative part of the problem is no less important. Appreciable number of perspective young scientists have left science. A part of them change for the work in commercial enterprises established at the territory of the Center. Those who are still working in most cases require retraining in sense of getting skill in new methods of microbiological and biotechnological research.

Despite the tendency of essential reduction of the programmes on scientific staff training in Russia, SRCAM supports these programmes at a sufficiently high level. At present 259 researchers (22 chiefs of laboratories and departments, 35 major and leading researches, 61 senior researchers, 141 junior researchers) are engaged. Of them there are 14 doctors and 131 PhD (that is 56% of the total number of researchers).

The average of the researchers (89 per cent) is of the age of about 50, i.e. they are at the top of their creative activity.

Scientific potential of Obolensk Research Center is concentrated on 4 main directions. The data presented show that the scientific staff is distributed approximately equally between three major scientific directions. The projects provided by ISTC's financial support are concerned with these directions and explore new ones.

As known, the agreement on the establishment of ISTC was signed in Moscow on November 27, 1997.

The main ISTC' objectives are quite familiar to the audience.

Today the following advantages of the cooperation with ISTC can be stipulated:

- ISTC's projects provide new sources of financing the research performed in the Center on the background of heavy economic situation in Russia;
- Cooperation with ISTC provides possibility to extend international scientific relations and to increase the creative potentialities of SRCAM researchers;
- Increase of publications of the Center's researchers in international journals, and their participation in international conferences and workshop meetings is observed;
- 3 International conferences under the sponsorship of ISTC were held, one of these conferences took place in Obolensk. All this made possible to perform a dialogue with leading foreign experts and officers of the financing parties. The last conference in Vyatka is indicative especially in this respect. The decision on the financing one of our project was accepted at the conference;
- It is now possible to perform joint works with the leading scientists of the world in their laboratories. For example, in the fall of 1997 Doctor A. Pomerantzev was on a scientific mission in a Boston medical college. One more example: in October 1997, Doctor Colyer from Boston visited Obolensk in order to get information about the achievements made with in two ISTC's projects.

The other example: the visit of Dr. Meselson from a national academy of sciences of USA has yielded in the cooperation of three projects, including those related to the research in the field of especially dangerous infections.

Within the framework of ISTC's projects the experts of the Center and its guiding masters have a possibility to obtain skills in the field of new organisational form of work, that is system of grants. A new category of scientific and technical workers, project managers, is born.

However, there are some contradictions and misunderstandings in cooperation with ISTC:

From our point of view the period of treating the projects is too long, and the sums submitted do not permit to complete the research and moreover to introduce new technologies developed on the base of the research. It happens so that projects are positively appreciated by the experts from the largest international scientific centers, but the experts from ISTC do not recommend the projects for financing.

The sums submitted by ISTC on concrete research, do not allow to capture all the scientists of the Center. Such sums are insufficient for obtaining reliable results in the research areas new to us. Especially it concerns the projects on the stage of introduction of technologies in practice.

In may report I have already mentioned the possibility of misuse of the advantages gained nowadays by biology.

The urgency and necessity of the most steadfast attention to the given problem were repeatedly stressed at the 3rd Medical Conference on Biological Protection organised by Bundeswer Medicine Academy in October 1996, on which the danger of use of pathogens in biological terrorism was discussed in detail. It was noted, that international services dealing with the protection from terrorism can appear to be powerless in the struggle with terroristic activity related to the use of biological agents because of the following characteristic features of the latters:

- Suddenness of the impact;
- Latent period of the infection development;
- Difficulty of pathogen identification;
- Ability to be spread on large territory;

- Diverse ways of application;
- Aerosol;
- Foodstuff;
- Water;
- Subjects of use.

The problem of the use of pathogens by international terrorism has not been resolved yet, since this theme has appeared rather delicate. However I believe, that for gaining insight of this problem it is quite logical to conduct the analysis of the known cases of application of biological and chemical agents in the local military conflicts, and also the reasons of occurrence of sudden epidemical outbreaks of the infections in humans, animals and plants.

Three hundred years ago during the battle actions in India (Fort Carillon) the British gained practically a bloodless victory because of the fact that Indian troops were destroyed by the use of woollen blankets infected by smallpox causative agent.

We are now entering in a new century, century of biotechnology and, what is important, century of genetic engineering. The genetic manipulations with bacteria and viruses, with trees, fishes many species of animals, food and technical cultures have been already started.

The achievements of modern science in the field of pathogen modification allow to speak about the biological agents as about the weapon of mass destruction. The twentieth century has put mankind before a new aspect of application of the biological agents (BA).

This aspect has received the name "biological terrorism". There is numerous evidence on the facts of preparation, storage and application of BA for physical destruction of concrete persons or groups of people. During 1994 in France there was a home storehouse which had flasks with Clostridium Botulinium; In 1995 a storehouse of the same agent was found in Tokyo; 1995—evidence of application of BA as weapons in criminal world; in Russia—the case of the murder of a known businessman and his secretary, use of microbiological weapons by criminal groups for terrorism or racket; in Germany and Japan the cases of poisoning of foodstuff followed by modification of the population against competitive firms takes place. For this purpose, in any state food-processing industry and trade can be used.

Development and preparation for biological war during the operation " Desert Storm" in Persian Gulf has not but resulted in catastrophic consequences in the Near East.

For the past two years, United Nations inspectors have criss crossed Iraq in pursuit of 25 warheads that are filled with some of the world's deadliest germs and designed to fit atop medium-range missiles. But so far, the inspectors say, each lead has dissolved in a highly dubious tale about the warheads' fate.

Taking into account this situation we tried to classify characteristic features and properties of bio-agents, which determine the probability of their use for terroristic purposes.

Classification of bio-agents (bacteria, viruses, fungi, toxins) is extremely significant for planning and development of measures to prevent and control terrorism.

Among a huge number of microbes, occupying the planet and interacting with organisms of humans and animals, approximately 3,500 are pathogenic or conditional—pathogenic for humans, i.e. are capable of causing diseases. To believe, that all these micro-organisms can be used in terroristic acts would be not only anti-scientific, but also senseless, since it would disorganise the system of observation and control of illegal activity in this sphere.

The bio-agents, which use is possible in terroristic acts, should answer a complex of criteria which are taking into account their biological parameters in a combination to interactions with human organism, environment, and also technological, technical and economy parameters.

For this purpose we applied a criterion approach to evaluate the probability of using micro-organisms as potential destructive agents. The classification we propose is presented on the slide.

According to this classification all bio-agents are divided into 3 groups. In Group 1 characteristic features and properties of bio-agents, which cause mass destruction are enumerated. Group 2 includes bio-agents, which possess properties determining their use for terroristic purposes. In Group 3 there are properties of artificial bio-agents, which can be used for terroristic purposes.

Proceeding from the above stated classification of the bio-agents we have tried to range the concrete bio-agents on a degree of probability of their real application.

Main criteria, determining probability of using micro-organisms for this purpose, were formulated. An expression degree of each parameter was determined on the basis of availability in the literature data given for approximately 30 species of pathogenic bacteria, viruses and toxins.

The accounts obtained on the basis of the table, have allowed "to range" the bio-agents according to a degree of probability of their use as BW.

The first group includes the causative agents having a rating (a sum of marks) 15. To the second group the causative agents having 10-14 marks, lastly in Group 3 there are the microbes having less than 10 marks. Taking into account such distribution of bio-agents, quite certain decision arises to consider, that bio-agents of Group 1 can most likely be used as BW; bio-agents of Group 2 can be surveyed as potentially perspective for terroristic purposes, and bio-agents referring to Group 3 should not be taken into account as the agents, which can be used for mass destruction of people.

Such a criterion approach can be applied to any bio-agents, and not only to those that are indicated in the table. Thus it is necessary only to know (and sometimes to assume), what "ponderability" (mark) should be given to that or other criterion describing the bio-agent. It is necessary also to note, that among them, 10 criteria selected for evaluation of the probability of using bio-agents for terroristic purposes, can be "critical", negatively or positively deciding the question, despite of the general rating.

The complexity of the situation with detection of BA for terroristic purposes is aggravated by the fact, that production of biostuffing to a small ammunition does not require expensive bulky equipment, a small laboratory with a tight isolation ward suffices, the extensive depots are not necessary for a storage, and it is much easier to engage experts familiar with technology of bio-mass production, than developers of a nuclear bomb or ballistic missile. In any country "of the third world" there are educated microbiologists, virologists, biologists.

The modern reality is characterised by the increase of a wave of Islam fundamentalism, inter-ethnic contradictions and local conflicts. The conditions after the disintegration of the systems of Socialist countries and appearance of a large number of multinational countries included in a struggle for the markets and changing a geopolitical climate in regions has become particularly complicated. The small efficiency of traditional methods of military actions (for example, Iran-Iraq war, struggle with Kurds) inevitably attracts attention to the cheaper and, in conditions of the underdeveloped countries, more effective biological weapons. The application of primitive species of biological agents, causative agents of especially dangerous infections or their toxins (in particular, the interests of Aum Sinrikur sect to Ebola virus are known) is the most probable.

Biological agents as weapons are of the greatest danger for countries with the undeveloped system of public health services and anti-epidemiological control. Dangerous pathogens can be used as the most probable agents in a war for natural resources or for easing items of the potential opponent in the global market. One of the dangers of political terrorism is the direct use of BA for drawing injury to alive force and engineering (bio-damage), civilians and national economy. The danger is enlarged in accordance with population increase at a simultaneous attrition of natural resources.

The opportunity of latent application is characteristic for such kind of weapons. The availability of population differences of the constitutional immunity to certain kinds of infection causative agents allows to assume a possibility of rather effective utilisation of this kind of weapons in international conflicts, that confirms historic experience of occurrence of epidemics among aborigens of the New World, Africa and Australia in conditions of transfer of pathogents which were not circulating earlier in the given region from the Europeans (a tuberculosis infection in Negroes and Indians, measles in the inhabitants of Southern America and Oceania). Secret infection of some regions in Korea in 1951 resulted in the occurrence of epidemic of a fever "songo", and use of virus "denge" unknown for Cuba earlier has resulted in outbreaks of (more than 350 thousand persons) haemorrhagic fever.

High level of molecular-genetic research on modification of the agents in conditions of international economic and scientific integration, increase in population migration largely promotes the objectives of international terrorism with the use of chemical and biological substances in terroristic acts. The practice shows, that even highly developed countries cannot deal with this phenomenon.

High level agricultural technologies and technologies of

reproduction of agricultural cultures and animals largely increase the efficiency of biological agents as weapons for terroristic acts. So, the plants infected with pathogens not only reduce the productivity, but also produce toxins, dangerous to human beings and animals. In this respect the serious danger is presented by mycotoxins capable to be accumulated in grain cultures and products of their processing. In modern methods of storage there huge parties of food products and forages can be infected.

For animal and poultry industries in conditions of large scale production the epizooties can be strongly damaged due to the infection by human disease causative agents (salmonellosis, anthrax, aphthous fever etc.).

Thus, possible use of the biological agents with an objective of sabotage or terror against any state will inevitably result in national or international econological catastrophy. Biological terrorism in this connection is a direct call to the international community.

In connection with the aforesaid it seems extremely necessary to unite expendient the efforts of politicians, military and civil experts in the field of genetics, molecular biology, pathology, detection and identification of the agents of dangerous infectious diseases for prevention of threat of international biological terrorism.

Hence, it is necessary to undertake preventive measures to warn about the spread of this phenomenon. Many states being in the avant guard of genetic-engineering, accept a series of regulating measures, which can reduce to a minimum the harm to all living beings.

In the Russian Federation the realisation of works in genetic engineering is regulated by the Federal law on state adjustment in the field of genetic-engineering activity accepted by state Duma 05.06.1996, and series of other normative acts establishing main directions of the legal relations and legal bases of international cooperation in this area. In 1993 the changes, stipulating the criminal responsibility for the development, production, purchase and storage of bio-agents were introduced in the legislation of Russia. (Chapter 25 Criminal Codex of Russian Federation of 1996).

General coordination and development of the system of safety in the field of genetic-engineering activity are carried out in the order determined by Government of the Russian Federation.

However in the majority of countries, especially of developing ones, even the most elementary safety measures are not undertaken. International exchange, trade, use of micro-organisms received by genetic-engineering methods are not regulated by the legal documents. We present for your consideration the list of measures promoting the prevention of unauthorised use of dangerous biological agents and knowledge for terroristic purposes.

First of all, it is a severe system of record-keeping and control of production, storage and transfer of dangerous biological materials and scientific and technical documentation, in combination with criminal prosecution under the law for abuse in this area. An obligatory condition is the maintenance of employment (with corresponding payment of work) of scientists and experts, on training and high qualification of whom many material and intellectual resources were spent. For the majority of scientists the public recognition of the importance of their work is also important. The information by press, radio and television, public ceremonies of the awards are the important factor promoting use of intellectual potential of the biologists in noble objectives. On the other hand, the danger "of biological terrorism" is aggravated by unsufficient sanitary-epidemiological education of the population, especially in countries with a low standard of living.

It is necessary to develop the obligatory normative acts even in a form of protocol to the Convention of 1972 to warn the population of the planet about possible terroristic acts with the use of bio-agents.

For prevention of use of biological agents (BA) for terroristic purposes.

1. Acceptance of the laws pronouncing criminal the production, storage and application of BA as well as transfer of information and (or) training of persons (organised groups etc.) in terroristic objectives.
2. State support of organisations having biological potential, by way of granting the orders on the development of useful bio-technological production.
3. State and public control of financing and realisation of works with dangerous biological agents. Periodic inspection.
4. Restriction of the number of sites (enterprises) where

dangerous BA can be stored, creation of a severe system of the quantitative control of storage and transfer of especially dangerous BA.

5. Personal material and moral support of the scientists and experts working with dangerous BA. Recognition of the exclusive public importance of their work.'
6. International integration of research on BA, mutual visits to Research Centers.
7. Sanitary-epidemiological education of the population in order to train them to detect the infection by BA at early stages, taking of measures to prevent panic and to convince the population in the possibility of complete treatment of the disease by modern therapeutic means.
8. Development of means excluding proliferation and diffusion of BA in the systems of life support (the water-pipe, ventilation, water drain etc.).

20

The State Research Center of Virology and Biotechnology VECTOR for the Prevention of Misuse of Biological Sciences

S. Netesov, Russian Federation

The SRC VB VECTOR derives its history from 1974 when the main Institute of the Center was established—All-Union Research Institute of Molecular Biology. In 1985, Scientific and Production Association VECTOR (NPO VECTOR) was created on its basis which in 1994 was granted a status of a State research center. State Research Center of Virology and Biotechnology VECTOR (SRC VB VECTOR, or just Center) is now one of the biggest in Russian research and production associations which basic activities are oriented towards fundamental and applied studies in the field of molecular biology, virology and biotechnology as well as production of medicinal drugs, veterinary drugs and reagents for research.

Till the beginning of the nineties our Center was one of the few civilian institutes in FSU whose main task was to develop the national programme for development of defense measures from biological weapons. This programme consisted of the basic studying of the most dangerous viral pathogens, the establishing of the relationship between their structure and functioning, pathogenesis mechanisms, studying of the natural diversity and variability of these viruses on the base of the sequencing of its genomes and, finally, receiving the data needed for evaluation of potential usage of these viruses by potential enemy as BW agents. The applied part of the programme included the development of diagnostics means, vaccines and therapy technologies against these diseases.

The real research of the dangerous viruses started in VECTOR since 1984 but the construction of all the main buildings needed for complex investigation of dangerous viruses were completed in 1990. These buildings were constructed according to the national and existing international bio-safety regulations including pressurised working zone with diminished air pressure, double-door autoclaves for autoclaving of all the wastes, a lot of special equipment and control devices, etc. They are extremely expensive facilities. These facilities for virus research are unique in Russia and in the former Soviet Union; similar facilities exists now only in USA, Great Britain, Australia and are being constructed in Canada, France and some other countries.

During the eighties our staff was 4,500 but only a little more than 300 directly worked with pathogens. The conditions of work were hard because it had to be done in the special space suits according to the strong biosafety regulations. About 1,000 employees were involved into the molecular biology studying of viruses; other staff provided the Center by molecular biology reagents and enzymes, cell culture media peptides and oligonucleotides, animal sera. The departments of guards, biosafety, epidemiology and others were organised too. Finally, living facilities for 10,000 center employees, members of their families and auxiliary staff had been constructed and formed a small town named Koltsovo in honour of the famous Russian geneticist Nickolai K. Koltsov. All the employees working with pathogens were granted separate apartments in Koltsovo and were provided with high standard medical service.

During the Socialist period our center was a typical example of research institute engaged with defense investigations and had some important privileges, including the 100 per cent state budget financial support, including construction of living facilities, hospitals, guarding and auxiliary services, large salary and pension privileges. But it was a compensation for a very strong level of state management, strongly classified research work without contacts with foreign scientists.

During the last 15 years many excellent scientific results have been obtained, much of them are the results of world priority. These results have been marked in the FSU by medals and special awards.

The major asset for the Center is its highly qualified personnel, who have been selected and trained for more than 20 years of the Center's history and who are able, on the one hand, to conduct molecular-biological and genetic engineering research, and on the other hand, have experience of practical work on highly dangerous pathogens. This factor, in combination with laboratory facilities providing biosafety levels 3 and 4 (BSL3 and BSL4), creates a unique opportunity for a complex investigation of any infections and their pathogens—from fundamental studies of structure and function of viral genomes and pathogenesis mechanism analysis at molecular-biological level to creation and testing of vaccines, diagnostic test-systems and novel therapeutic drugs.

More than 2,000 employees work at the Center now while 1,252 of them are directly involved in research work including 21 Drs.Sci. and 166 PhDs. The list of publications by VECTOR researchers includes more than 2,000 items during the last ten years.

An important contribution to training of highly qualified personnel is made by the Center's post-graduate course and two Councils on Dissertation Approval (virology and bio-technology). This post-graduate course which is subdivided into three specialities (molecular biology, virology, and bio-technology) was established in 1983 and is actively functioning now.

Since 1985, the Council on Dissertation Approval granting a degree of Doctor of Science has been operational SRC VB VECTOR.

At present, the State Research Center of Virology and Biotechnology VECTOR is a large research and production complex including a number of research institutes such as:

- Research Institute of Molecular Biology
- Research Institute of Aerobiology
- Research Institute of Bio-engineering
- Research Institute of Cell Cultures
- Collection of Micro-organism Cultures

Research Construction and Technology Institute of Biologically Active Substances.

It also includes some production facilities, such as:

- Research Pilot Production Plant
- Production and Sales Company VECTOR-PHARM
- Agricultural Pilot Production Plant.

The Center also includes Department of Managing of Preclinical and Clinical Trials, Quality Control Department, Bio-safety Department, Licensing Department, Laboratory Animal Breeding House, Transportation Department and Energy and Water Supply Division.

All establishments of the Center, which facilities cover an area of 200,000 sq.m., are located in the vicinity of Novosibirsk, the biggest city of Siberia with 1.5 million population.

Until 1991 our Center had mainly research activities; volume of production was small and main financial support from government was given for classified (both basic and applied) research in the interests of the Ministry of Defense. Since 1989, the tendency appeared for diminishing the amount of support from government for classified research which reflected the changes in the world and in Russia and therefore we started to elaborate a special conversion and restructuring programme. This programme envisaged the reorientation of our research activities to the needs of open basic research, public health, veterinary, agriculture and environment protection. This programme also contained the plan for production development of medicines, diagnostic kits and vaccines for public health and veterinary. The main goals of this programme are as folllows:

- maintain our personnel who had an extensive experience in the area of genetic engineering and bio-technology and reorient it to carry out research for civilian needs;
- carry on basic research in the field of bio-technology and virology;
- enhance applied research aimed at the development of new therapeutic, preventive and diagnostic means for public health and veterinary employing the latest achievements in genetic engineering and bio-technology;
- build up production facilities to manufacture a wide range of medicines for public health and veterinary;
- collaborate internationally both in basic and applied research and medicine production.

This programme has been developed by our Center during the last six years and some results of its realisation are presented.

It is obvious that the structure of income had been changed to the great increase of the production part. It must be mentioned that this ratio changed not only because of the production growth but because of the diminishing of financial support of research from different sources.

The research potential of FSU was really very high in spite of a lot of false administrative decisions and very different level of different fields of science and different institutes. The EC countries, USA and Japan knew about it and offered to Russian scientists already in 1993 a few opportunities to use their knowledge and qualifications.

There are various international foundations which participate in the funding of science in Russia. Of course, it was not a simple charity—the price of the similar investigations costs 10-30 times more in western countries in comparison with analogous research made in Russia. Of course there are a lot of other bottoms in this basket. One important thing is that the significant amount of these international funds were designated for former defense institutions and therefore a lot of supported institutes were those which were tightly engaged with the interests of the Ministry of Defense. It can be noted the dynamics of change of the structure of our R & D sources. It is obvious that during this period a great diminishing of defense-related research, the increase of percentage of support from the Ministry of Science and Technology (really it is a stabilisation but on the low level) and growth of the percentage of internal and international grants. As for our center which is working in the field of life sciences, one, formerly large enough source of research money,—the Ministry of Public Health—practically ceased the financing of applied research because of great problems with financing of practical public health. It must be specially underlined that there is a large increase of quota of foreign grants during the last few years. It constitutes now about 15 per cent of all the research financial sources. It must be mentioned, however, that these figures do not include the reimbursement of the money for energy supply from the Ministry of Public Health which is twice as much as money shown here.

Here are presented the main directions of basic and applied research of the Center:

- Fundamental studies of viruses' molecular evolution and their interaction with the organism of humans and animals as well as studies in the field of ecological bio-technology.
- Development of genetic engineering methods of viral genetic material reconstruction, construction of producers of biologically active substances while computer-aided modelling methods are actively used.
- Development of agents for immunoprophylaxis, diagnosis and treatment of viral infections including those that are widely spread such as hepatitis and herpes and those that are "exotic" such as Ebola hemorrhagic fever and Marburg hemorrhagic fever.
- Development of novel production technologies, conducting of trials and production of dosage forms of peptide preparations and those based on nucleic acids obtained by using methods of chemical synthesis as well preparations from natural raw material of vegetable and animal origin.
- Commercial production of medical, diagnostic and preventive drugs for medicine and veterinary as well as various bio-chemical and immuno-chemical reagents, enzymes, nutrition media, sera and other preparations for investigation and practical purposes.

According to these main directions our Center already has made numerous very important studies. The main results of these studies are presented below.

Studies in the field of molecular biology and genetic engineering have always been, and are now, an essential part of the research work conducted at the Center. Speaking of results achieved in this direction, one can mention full sequencing and structure-function analysis of genomes of smallpox virus, Marburg virus, Ebola virus, Eastern equine encephalomyelitis virus, tick-borne encephalitis virus and a number of other viruses. Studies of structure-function organisation of viral genomes (smallpox virus, Marburg virus, Ebola virus, Congo-Crimean hemorrhagic fever virus, human T cell leukosis, etc.), that have been conducted for the last several years, allowed to reveal particulars of genetic information realisation and replication of these viruses. The data obtained provide a good basis for development and creation of drugs and vaccines of a new generation as well as agents for diagnosis and identification of viruses.

Studies of molecular bases of evolution of viruses pathogenic for humans such as measles virus, influensa virus, HIV-1 –2 virus, hepatitis C virus, orthpoxviruses are performed at the Center. The data obtained in the course of studies of evolutionary changes of viral isolates will allow to understand molecular basis of pathogenesis of infections caused by these viruses, in particular, the mechanisms of "escaping" from the host organism immune system, that will be used for development of universal diagnostic test-systems, improved remedies for prophylactic vaccination and therapy of viral diseases.

The scientific personnel of the Center conduct an active search for agents that would be used for treatment of such a dangerous disease as AIDS: at several biological models, a phenomenon of production from human immuno-definciency virus cytopathogenic effect by certain antisense oligonucleotides, membrane-active chemical compounds and preparations based on glycyrrhyzic acid was demonstrated. These materials also reveal their activity with other viruses, for instance, on the basis of glycyrrhyzic acid and its derivatives, for the first time, an inhibitor of Marburg virus reproduction which is closely related to Ebola virus was found which recent outbreak in Zaire was in the focus of attention of virologists and epidemiologists all around the world.

At present, the world medical practice makes more frequent use of cytokines—proteins produced by the organism which modulate the immune response to various penetrating agents such as mitogens, bacteria and viruses. Therapeutic application of cytokines was made possible by development of genetic engineering methods that provided a possibility to construct the preparations by microbiological synthesis. The SRC VB VECTOR participated together with other institutions in creation of human alpha2-interferon (Reaferon) This preparation has been produced at the Center since 1991. To meet the demands for Reaferon that constantly grow, the Center's researchers have constructed a novel producer strain that allows a ten-times' and more increase in the target product output. At present, quite a number of recombinant cytokines such as tumour necrosis factors, human γ-interferon analogues, interleukin 2, granulocyte colony-stimulating factor, erythropoietin hormone, etc. are in the developmental stage.

Immuno-modulators, in particular, endogenic interferon inducers, are efficient anti-viral preparations. At our center preparations Ridostin (for medicine) and Vestin (for veterinary) have been developed on the basis of the double-stranded RNA of yeast. Other preparations such as Polyribonate (for medicine) and Polyvedrim (for veterinary) have been developed on the basis of the highly polymerous RNA of yeast. By using genetic engineering, an anti-viral preparation of new generation, SUBALIN, has been developed. This preparation obtained on the basis of live recombinant bacteria Bacillus subtilis 2335/105 with the inserted human alpha2-interferon gene preserves the anti-bacterial activity of the host strain and additionally reveals anti-viral properties which are provided by interferon secreted.

The authority of the Center's researchers has been recognised also in development and use of "phage display" technique which main idea is that of constructing artificial particles of the filamentous phage carrying foreign proteins on their capsid surface. This technique is considered to be as the most efficient means of natural ligand imitators' construction for production of vaccines and diagnostic assays. For example, this technique can be used for construction of the so-called phage peptide libraries. From such a library, one could easily select phages showing the highest affinity with the peptides under study, amplify them and use for various practical purposes.

High-efficiency and harmless drugs based on synthetic peptide hormones are well-known and find favourable application in clinical practice while in some cases they are a absolutely necessary. On the basis of original synthesis schemes, production technologies of peptide hormones substances such as oxytocin, desmopressin, gonadoliberin, adrenocorticotropic hormone (ACTH (1-24) as well as ready-to-use dosage forms on their basis have been developed at SRC VB VECTOR. The center is starting production of oxytocin dosage form in the near future.

Development of efficient remedies for viral disease prophylaxis is an actual problem. At SRC VB VECTOR, studies are conducted aimed at the creation of recombinant vaccine preparations for prophylactics of tick-bore encephalitis and hepatitis B. Along genetically engineered vaccines, traditional vaccines are developed at the Center, in particular, an inactivated cultural vaccine against hepatitis A has been developed which first batches had been produced in 1997. Studies are carried out aimed at the improvement of live measles vaccine production technology by using human diploid cell culture as substrate for vaccinia virus cultivation.

One of the directions of research work conducted at the Center is carrying out of bio-aerosol studies for medicine, virology and environment protection. Mechanisms of delivery and efficiency of aerosolised administration of drugs for therapy of viral diseases are investigated. For particular many-component drugs, data have been obtained that will allow to determine how deep the aerosol particles get in the respiratory tract and how efficiently they sediment there. Subsequent optimisation of the particles' size will provide a higher efficiency of aerosol therapy.

Methods of biological air pollution express-control have been developed. A device for outdoors taking probes of aerosols of biologically active particles that are present in the atmosphere has been developed. New methods of measuring the dipole moment and polarisability of the particles, methods of their density-dependent classification by using strong magnetic fields have been proposed. A device, i.e. classifier for isolation of monodisperse fractions of particles from arbitrary aerosols, has been developed.

Composite technologies of ointment forms of preparations based on chitosan and collagenase that will be used in burn therapy are developed at SRC VB VECTOR. On the basis of inland raw material, a production technology of Allapinin, a preparation used in therapy of cardio-vascular diseases (arrhythmia, fibrillation), is being developed.

A large collection of cell cultures of vertebrates, insects and plants including more than 100 cell lines and their clones, that are widely used in virology, biotechnology and molecular biology, has been created at the Center. Eight seed and working banks of diploid and heteroploid cells have been certified to be used for production of immuno-biological preparations.

A collection of viral strains and isolates, recombinant viruses as well as industrial strains of micro-organisms (more than 10,000 items deposited) has been established at VECTOR and is constantly replenished. The National Collection of Smallpox virus strains (109 strains) is maintained within the Collection of Micro-organism Cultures at the Center. Another official repository of smallpox virus strains is situated in Atlanta, USA. A WHO Inspection team who visited the Center in July 1995 confirmed that conditions under which the work is to be performed meet the international biosafety requirements. The

Collection of Micro-organisms Cultures has been internationally recognised. In 1995, the Collection became a valid member of the European Cultures Collection Organisation.

A collection of hybridomas secreting monoclonal antibodies to arboviruses, orthopoxvirus and to a number of other antigens has been created at the Center.

The other important aspect of our conversion programme is the development of production and the increase of this direction in our Center. At present, SRC VB VECTOR has production facilities that allow to perform a full cycle of bio-preparation production. Several fermentation departments that can be used for pro-and eucaryotic cells cultivation, equipment for filling of ampoules and vials as well as a number of units for freeze-drying labeling and packaging of the final products, including units for blister packing, are available. Ventilation systems and other technological equipment provide that production is carried out under conditions meeting the current international requirements. The production facilities of the Center are oriented not only towards introduction of own R & D but also towards production of other products necessary for medicine and scientific research. The total nomenclature of the preparations produced at SRC VB VECTOR presently consists of more than 300 items.

Biochemical reagents for laboratory studies (enzymes of nucleic acids metabolism, substrates and inhibitors of enzymes, original carriers for affinity chromatography and ion-exchange chromatography), nutrition media and sera for cultivation of cells and viruses (more than 60 items) are produced.

Production of drugs from imported substances has been started recently. Immuno-biological preparations such as immuno-globulin (liquid) for emergency prophylaxis of tick-borne encephalitis, lyophilised Reaferon (human recombinant alpha2 interferon) for injections are produced. Profezyme, a high-efficiency drug for treatment of purulent and necrotic processes of different aetiology is produced.

The Center has a facility for breeding of laboratory animals which is unique in Asian part of Russia and which provides laboratory animals for medical and biological experiments for its own needs as well as for needs of the whole region. Almost all species of animals used in experiments, from pure lines of white mice and leukosis-free Japanese quails to primates, are kept and raised in the facility.

The Center is one of the main producers of ELISA diagnostic test-systems on the Russian market. There are tests for serological markers of immuno-deficiency (HIV-1,2) viruses, hepatitis A, B,C viruses, measles, tick-borne encephalitis, cytomegalovirus, opistorchiasis, chlamydiosis, and many other test-systems.

We had no special programme for prevention of misuse of biological sciences in our Center. But one of the main purposes of our conversion programme was to avert the possible mass discharges of the most qualified personnel by creation the vacancies in the production field, searching the sources for research grants, reorganising the center with the purpose to find our new place in changing Russia and to give the opportunities for development of our young research managers. We really had no intentions to prevent the misuse of biological sciences but our main goal was to survive in the changing country and to save the personnel and research potential of the Center. With this purpose the following steps which lead in the same time for prevention of misuse of biological sciences had been done:

1. State research center was created in 1994 on the base of previous NPO "VECTOR". This center has non-classified programme of basic and applied research in the fields of virology, biotechnology and some other fields of life sciences and ecology. This programme has been approved by the Ministry of Public Health and Ministry of Science and Technology.
2. Koltsovo is located 8 km outside Novosibirsk (20 km to the center of the city) and therefore there are transportation problems for people to work in the city. We tried to help our employees and even former employees to organise the jobs in Koltsovo in which they could successfully apply their knowledge and skill with the purpose to earn enough money for their families. With this purpose very attractive conditions for renting the offices were offered for companies which would like to rent storage facilities, small industrial premises, etc., and a lot of companies started to do it. Of course, we did not offer P-3 or P-4 facilities.
3. We are trying to use all the opportunities to find the additional sources of support to our scientists and therefore we constantly gather all the information about Russian internal

and international foundations which offer the grants for research in life sciences.

4. We created and try to improve now the database about all our scientists involved in the past into the special pathogen research who left our institute. We really know the addresses and occupations of all our former employees who migrated to another countries.

It would not be true if I tell you that we succeeded. In spite of all these efforts and achievements the real situation is even more dramatic for our Center than 3-4 years ago. More than 40 of our employees migrated to other countries, mainly to the West. Fortunately, we maintain contacts with the most of them and know where they are now. But the steps of a few people were lost. We have large financial difficulties now—the delay of salary for our employees is 5 months now and its average is about $ 150 per month for scientists.

Conclusions

In my opinion, the initiative of carrying out this meeting is extremely timely. It is because of the situation when scientists in Russia understood fully that the government has not enough money to support their work even if it is directed towards peaceful goals. On the other hand, these scientists found the ways of visiting international science meetings abroad, presented their results there showing their knowledge and skills and understood their real value as specialists. At the same time the specialists from some countries of the Third World began to understand from the materials of international science meetings that there are a lot of warfare specialists in FSU countries which are practically unemployed and are willing to be hired for research work.

I know exactly that a few years ago some people from the Third World countries came to Novosibirsk and tried to engage individually the warfare specialists from a few former defense institutes. I know that our employees did not agree to emigrate but I do not know what was the situation in other institutes.

I know exactly that some specialists from Middle East countries have been trying to do the same in Russia and other FSU countries. Therefore it is a real danger of hiring these specialists by these countries and starting there to develop very quickly different sorts of unconventional warfare.

I may propose here a few ways to prevent such things:

1. First of all, in my opinion, the main objective of the international science foundations which actively work now for FSU countries should be changed for supporting defense scientists.
2. The structure of science in Russia does not correspond to the structures of research in other developed countries because there is a large turn to the physical sciences. Therefore it would be better to make life sciences prevailing of other sciences in these foundations.

21

The Role of the International Centre for Genetic Engineering and Biotechnology for the Peaceful Use of Biological Science

—A. Falaschi, Italy

The population of the world of today is faced by a challenge that could threaten even its survival in the near future. The gap between the minority of the privileged population and the great majority that lives in total destitution is increasing and could bring to unforeseen levels of violence, made even more destructive than in the past by the technological means available today. In this context, biologists have a specific responsibility, in two ways: on the one hand, with their science they can contribute to solve some of the greatest problems of the developing part of the world and thus reduce or abolish the reasons for tension; on the other hand, they could offer means of global destruction of human people and of the environment so powerful as they never were in history.

In this presentation, I shall make a brief analysis of the way in which biology can have a positive effect on the future of mankind and describe in particular the Organisation I am responsible for, which intends to play a hopefully significant role in helping the humanity of the near future to avoid those dangers.

Among the problems faced by today's mankind, hunger comes

foremost to the mind as possibly the greatest adversity pervading large fractions of the population, particularly in Africa, South East Asia and certain areas of Latin America. Very often crops suffer from a great variety of stresses, either due to climatic conditions or to qualities of the earth in which they grow as well as the stress due to pests such as insects, fungi and viruses. Also, the scarcity of fertilisers causes very poor yields in the crops, whilst the presence of a biological pest, be it a population of insects or the arrival of a novel virus, may wipe out the crops of entire regions.

Secondly, an element of misery for the developing world is the spread of diseases, some of which are typical of tropical areas and of difficult living conditions. In this context one thinks immediately of malaria, possibly the greatest health problem of mankind: approximately half a billion people suffer from this disease in the tropical belt of the planet, and the number of deaths it causes in one year is estimated to be to the order of three million, half of which are children. Other infectious diseases are particularly rampant in the developing world, such as those due to schistosomes, trypanosomas and other protozoa, and to viruses, like the hepatitis (A, B. C, D and E) and AIDS. However, one must not consider only infectious diseases as problems of the developing world: disorders which are considered typical of industrialised countries, present, in fact, huge problems to the Third World: for instance cervical cancer (the main cause of which is also mostly viral) represents the principal cause of death by cancer in women in Africa and Southern Asia.

In the context of health, and referring to the diseases which are more specific to the tropical world, one observes likewise a great dearth of research addressed to them, in view of the limitation of the health market of those countries, in financial terms: thus, diseases like malaria and hepatitis E (not to mention schistosomiasis and trypanosomiasis) can be considered orphan diseases.

Also, the adversities touched upon briefly above, cannot be taken out of the context of the socio-economical conditions of that part of the world: poverty and unemployment are elements which offer the foundation to those problems and prime, so to speak, a vicious circle that, through the difficulties in nutrition and health, aggravates itself. The causes of poverty and unemployment are many, and among them one can certainly consider the scarcity of natural and energy resources, capital and the presence of entrepreneurial spirit.

Also, pollution of the environment disproportionately plagues many areas of the developing world, worsening therefore the nutrition and health problems that are discussed above.

On top of all that, one has to consider that the biological weapons could unfortunately be relatively easily available also to "poor" economies. They are not difficult to produce, relatively easy to hide, and, in the hands of unscrupulous desperate terrorists could cause incredible damage to large populations.

Against these formidable threats, what remedies can we propose? Biology can offer today very powerful means, even if not the only means, for solving many of the problems that plague such a large portion of mankind. Through science, novel methods can be utilised by the citizens of the countries which suffer most to alleviate and possibly abolish the worst hardships which are due to the present, difficult conditions. Thus, in the field of nutrition, the ability to manipulate, almost at will, the genome of higher plants and animals offers a great variety of possible applications to several aspects important for human nutrition. In agriculture, we can improve the nutritional values of the plants of common use in developing countries: for instance, several plant species are relatively poor in proteins as far as quantity and composition are concerned; one can plan to introduce genes in these plants which allow the production in great quantities of proteins as nutritional as human milk. Similarly, one can think of increasing the production of plant species of greater use (such as rice, wheat and maize) by making them genetically resistant to different stresses, such as climatic conditions (extremes of temperature, drought, high salt concentration) or harmful biological agents (insects, fungi, viruses or weeds).

Animal breeding can also draw great advantages from genetic engineering, both for the protection of animal health and for the possible introduction of foreign genes in higher animals, thus improving their nutritional capacity as well as their resistance to infectious agents or to particular environmental conditions. Furthermore, both plants and animals can be "engineered" in order to allow them to produce molecules of particular value, such as drugs of complex protein structure: in this way, one could extract expensive drugs from plants cultured at low price, or from the milk of engineered animals, to assure a continuous, high level and cheap production of useful molecules.

As far as the health problem is concerned, genetic engineering approaches offer, in the first place, not only a highly increased possibility of studying the infectious agents and their interactions with the human organism, but also, thanks to this knowledge, they allow for the preparation of novel diagnostics, vaccines and drugs for such diseases. Therefore, for these countries, the application of genetic engineering technologies to the diagnosis and therapy of the disease prevalent therein, as well as to the study of their molecular basis, is no less important.

Additionally, drugs of great importance for many widespread diseases can be obtained in a safe, efficient and economic way by the genetic engineering approach, such as insulin, human growth hormone, erythropoietin, etc. These biotechnology based products require limited investments and have a high added value. From this consideration, it follows that biology can give a very effective hand to solve the socio-economic problem of this less developed part of the world by offering new ways to industrial development and therefore to the creation of wealth and employment. In particular, the new biotechnologies based on genetic engineering can offer both new industrial products and new methods to obtain traditional products, in conditions particularly fitting to developing countries. In fact, biotechnology-based products and processes are typically low demanding in terms of capital investment, energy, and strategic raw materials, whereas they depend essentially on qualified personnel and on raw materials of biological origin. In this context we may quote, in the first place, the industrial production of new diagnostics, new diagnostics, new vaccines and new drugs. In a similar way the industrial processing of food or feed and the most effective utilisation of agricultural waste-products can be markedly improved by the new biotechnologies.

The chemical industry may also be partly renewed by the utilisation of bioreactors, that is of chemical reactors based on biological organisms or on biologically derived molecules. This give rise to a totally new chemical industry which is much less demanding in terms of capital investment and energy requirements than the traditional one, while being at the same time environmentally friendly. Also, the mining industry can be positively affected by the new biotechnologies, by improving the properties of micro-organisms currently utilised for the concentration of important metals, typically cooper (mineral teaching).

Furthermore, the protection of the environment can be effectively aided by novel scientific approaches: the use of specific micro-organisms may, for instance, remove heavy metals from the polluted areas in a way similar to those utilised for mineral leaching. More importantly, several micro-organisms or micro-organism-derived molecules (biosurfactants) may prove essential for the reduction or removal of oil spills and for the cleaning of oil residues in tankers. Finally, one must consider the use of biopesticides to substitute the chemical pesticides which widely used in many tropical countries and which may cause several problems for the environment. Another approach is to introduce the genes for biopesticide production directly in the plants, making them genetically resistant to the pest, and thus reducing, or abolishing, the need for chemical pesticides.

Ultimately, the scarcity of energy also plaguing many parts of the Third World can be addressed by the novel biotechnologies: the utilisation of agricultural waste by genetic engineering could offer a way of exploiting important renewable energy sources in an efficient way. Likewise, the improvement in fermentation procedures can offer ways to move efficiently utilise energy sources (such as ethanol, methanol or methane) derived from renewable material or from organic waste. The utilisation of micro-organisms capable of partially metabolising the hydrocarbons can allow for the exploitation of the remaining oil from the well (the so-called tertiary recovery of oil). This can potentially multiply the extent of available energy resources derived from fossil hydrocarbons. Also, the bioleaching approach can also be applied to the cleaning of coal sources rich in heavy metals: this would make an important energy source available by rendering it less aggressive to the environment.

Finally, as far as biological warfare is concerned, the modern biotechnologies offer tools of unprecedented specificity and sensitivity to monitor the respect of the Convention on Biological Disarmament: PCR-based methods, with carefully selected primers, following adequate concentration of the atmosphere and dust of a given environment, could allow the detection of traces at level of the single molecular fragment of the passage of a dangerous organism.

I shall now come to the possible role that the International Centre for Genetic Engineering and Biotechnology, whereof I am responsible, can play in this context.

The International Centre for Genetic Engineering and Biotechnology (ICGEB) was established as a centre of excellence for research and training in modern biology addressed to the needs of the developing world. It commenced operation in 1987 as a special project of UNIDO (the United Nations Industrial Development organisation) and became an autonomous inter-governmental organisation with the entry into force of its Statutes, on 3 February 1994. A total of 57 countries, mainly of the developing world, are signatories to the Statutes, 41 of which are now full Member States.

The ICGEB is organised in two main laboratories: one in Trieste, Italy (where the Directorate of the Centre is also located), and one in New Delhi, India. Furthermore, it features a network of 30 affiliated centres, which are national laboratories in Member Countries, whose research activities are co-ordinated with, and partially funded by, the ICGEB. The Centre operates along the following lines:

Research programmes: specific research programmes of high scientific content have been identified and established in the Trieste and New Delhi laboratories, pertinent to the developing world: e.g., the production of novel malaria and hepatitis vaccines; the study of human viruses (papilloma, rotavirus, HIV); the gentic manipulation of nutritionally important plants; the utilisation of vegetal biomass etc. At present, more than 250 people from 28 different countries are working in the component laboratories.

Lond term training: two year post-doctoral fellowships are awarded to Member Country scientists to participate in the scientific programmes in Trieste, New Delhi and selected laboratories in Italy. During the period 1989-1996 almost 300 fellowships were awarded. A pre-doctoral fellowship programme has also been implemented in collaboration with the International School for Advanced Studies (SISSA) in Trieste and with the Jawaharlal Nehru University in New Delhi.

Short term training: a short-term fellowship programme (max. 3 months) has also been initiated, which enables exchanges between member Country scientists in collaborating institutes. Practical or theoretical courses and workshops are organised in Trieste, New Delhi and in affiliated centres. Over 600 scientists participate annually in the courses and workshops organised by the Centre (approximately

15 per year); 12 meetings/ courses have been scheduled for 1997 and a further 4 conferences and symposia are sponsored by ICGEB.

Collaborative Research Programme: during the period 1988-96, 140 grants, for a total of approximately US$ 7, 200,000, were issued to ICGEB affiliated centres in the developing world.

Scientific Services: ICGEB offers consultation for scientific programmes; distributes polynucleotides and polypeptides upon request and provides access to a bioinformatics network which makes the most important biological databases (including those for the Human Genome Project) as well as the relevant software for retrieval and analysis, available to its Member Countries. The ICGEB can be accessed on internet (http:/ /www.icgeb.trieste.it).

The premises of the Trieste component currently cover an area of 7,000 sq.m., while the New Delhi Component is housed in a new 10,000 sq.m. state-of the-art building.

I shall finally come to the more specific role that ICGEB can play for assuring biological disarmament.

Three main reasons can be advanced for considering a positive interaction between the Convention on Biological Warfare and ICGEB and the instrumental role the latter could play.

1) ICGEB's unique nature and its spectrum of activities
2) ICGEB's increasing role in international fora, of which the BWC is an important aspect.
3) ICGEB's character as a centre of excellence for research and training in modern biology.

1) ICGEB's unique nature and its spectrum of activities

The fact that ICGEB started its operations when the spirit of reform was already blowing throughout the United Nations system has given it a unique advantage in so far as its management has been careful to run its operations in an extremely efficient way while minimising the administrative load and reducing the cumbersome bureaucracy. Moreover, the Centre has maintained the peculiarities of a modern research institution. So, while ICGEB is governed by UN rules and regulations and keeps its formal link with the System, it is able to maintain a high degree of operational flexibility on a day-to-day basis.

In this context, it is useful to recall the objectives spelt out in Article 2 of the ICGEB Statutes and see how, on the one hand, they have been translated into action through the main activities carried out in the last 10 years by the Centre and to consider, on the other hand, the relevance of these objectives and activities for the primary role ICGEB can play in the transfer of biotechnology for peaceful purposes and the role it could play in the implementation of Article X of the Biological Weapons Convention.

"a) To promote international co-operation on developing and applying peaceful uses of genetic engineering and biotechnology, in particular for developing countries;

b) to assist developing countries in strengthening their scientific and technological capabilities in the field of genetic engineering and biotechnology;

c) to stimulate and assist activities at regional and national levels in the field of genetic engineering and biotechnology;

d) to develop and promote application of genetic engineering and biotechnology for solving problems of development, particularly in developing countries;

e) to serve as a forum of exchange of information, experience and know-how among scientists and technologists of Member States;

f) to utilise the scientific and technological capabilities of developing and developed countries in the field of genetic engineering and biotechnology; and

g) to act as a focal point of a network of affiliated (national, subregional and regional) research and development centres."

Moreover, ICGEB has already been quoted by the Convention's Third Review Conference in the context of the international assistance to facilitate biotechnology research and transfer to developing countries. In addition, the substantial role ICGEB could play in support of a verification mechanism of the Convention has been stressed in a study prepared by a Group of Experts on all aspects of verification for the 50th session of the General Assembly of the United Nations.

2) The concept of clearing house and the increasing role of ICGEB in international fora

In the light of the provisions contained in Article X of the BWC and the subsequent transfer of technologies, especially to the developing countries, the need to create a sort of clearing house mechanism entrusted to an international Authority has become more evident in the past few years.

On the other hand, in the present global context, the international vocation of ICGEB as a "two-way" gateway through which the flow of information and know how could be accessible to all member Countries, and eventually to other non-member developing countries, is becoming more and more crucial.

Thanks to its networking arrangement with its affiliated national laboratories, through its short and long term training activities and with its already well established information systems (see below) could have a major role to play within the overall strategy of a clearing house approach. The Centre would be able to effectively contribute to information sharing mechanism through which both developed and developing countries would exchange information on a fair and equitable basis, a mechanism through which ICGEB could also assist its member Countries on matters relating to biosafety and intellectual property rights.

If properly utilised, the importance such an instrument would have for the international community, cannot be under estimated.

3) The character of ICGEB as a centre of excellence for research and training in modern biology addressed to the needs of the developing world

The fact that ICGEB remains first and foremost an international organisation dedicated to research and training in modern biology gives it the particular characteristics of modern scientific institution where the activities already well underway in some fields of specific interest for the implementation of Article X of the Convention would only need to be extended or increased for them to impact significantly on the major issues relevant to Article X of the Biological Weapons Convention, by way of illustration.

Training activities: an enlargement of the annual course organised by ICGEB on biosafety could be instrumental for providing technical assistance aimed at the gradual upgrading of national biological safety practices in the States Party to the Convention while constituting a proper framework for the enhanced involvement of donor countries in the field of biosafety.

Collaborative Research Programme: an increase of this programme, in connection with more precise information on the outbreaks of emerging diseases, production of new vaccines and diagnostic reagents, would allow for an enhanced circulation of information, thereby improving the conditions of life in certain countries and, more generally, raising the level of confidence among the Parties to the Convention.

Scientific Information Services: apart from offering advice for scientific programmes, ICGEB also provides its Member Countries with access to the most important biological data bases through the ICGEBNET, a bioinformatics network and through BINAS (Biosafety Information network Advisory System) developed in conjunction with UNIDO. ICGEBNET and BINAS, apart from being important research tools could also be adapted to the needs of the Parties to the Convention and become major repositories of data on good manufacturing practices, safe laboratory procedures, biological containment. Besides, they would give to the Parties and to the concerned Authority the access to information on releases of genetically modified organisms and on other experiments which could be related to biological weapons or which could entail environmental risks beyond national borders.

To conclude, these are only a few examples to show how, by enlarging the scope of its already well established activities, ICGEB could make its facilities and expertise available through its programmes whilst exploiting its unique nature of a universally recognised independent scientific institution within the UN system.

Building upon its expertise and capability, ICGEB would be in a position to offer a high quality input to the clearing house mechanism and its participation would also minimise the costs involved since it would be possible to keep the general expenditure at a minimum level.

Conclusions

In this presentation, I hope I have given a clear analysis of how biologists in general are burdened with great responsibilities today for the future of mankind, and I have given an example of how an effort is at least being made for facing these responsibilities through an international organisation like the ICGEB. The Centre's scientific and technical capacities are available to contribute to dispelling the dangers of biological warfare and to decreasing the tensions that could threaten the peaceful progress of all the communities of the world.

22

Bioethics and the Prevention of the Misuse of Biological Sciences

A . Guruge, United States of America

I. Introduction

As a term, **"Bioethics"** has been in circulation for hardly four decades. The debate over its meaning, scope and application is being carried on in earnest in academic circles with little agreement. It is as it should be in a discipline so new and important. Tomes have been written to elucidate, illustrate and defend what biological and medical scientists, philosophers, theologians and lawyers regard as principles and rules of bioethics. A case has also been made to expand the term to **"Biomedical ethics"**. Some argue, however, that medical ethics would only be a subsystem of bioethics.

Over the last five years a plethora of books have been published. That in itself is indicative of the compelling need to evolve a discipline which pertains to ethical, moral and religious concerns on life and living. In view of the practical objectives of this Forum, I shall refrain from entering into any of the on-going controversies and settle on a working definition that **bioethics is a subsystem of ethics pertaining to life and living and, therefore to sciences dealing with them**.

What is desired in the name of bioethics are adequate and clear-cut conclusions on what values, norms, principles and rules should govern the burgeoning capacity of biological and medical sciences to affect life from its most initial stage as a sperm, ovum or zygote to the final termination in death.

Never before in history had the community of biological and medical scientists wielded such immense power through knowledge, skills and sophisticated tools to manipulate life in all its ramifications. All indications are that this power is bound to increase exponentially, as biological and medical scientists, encouraged and stimulated by proven success, forge ahead with increasing discoveries in the ever-expanding discipline of biotechnology. Already reality has surpassed the wildest imaginations of science fiction.

The spectre of misuse haunts humanity. The adage that war is too important to be left to generals may as well apply to scientists as regards biotechnology. Should the exercise of this enormous power be left in the hands of scientists alone? The answer, if we listen attentively to the vocal champions of bioethics or biomedical ethics as well as national and international jurists, is an unambiguous "No". The question before us is whether bioethics is the solution and, if so, in what form and manner and through what modalities.

II. Sanctity of Life and Human Dignity: Foundations in Religion

In all measures from the Hippocratic Oath of the fourth century BC to codes of ethics and public laws of today, the basic motivation has been the desire and the intention of society to restrain the autonomy of the scientist. What Hippocrates deemed as ethically wrong were the administration of deadly drugs, procurement of abortion, seduction or, in more modern terms, the sexual harassment of men and women in the course of health care and the violation of physician-patient confidentiality.

The first two concerns stem from his conviction of the sanctity of human life. That a physician's role was to save life and not to harm or endanger it has been the fundamental understanding on which medicine had been practised in every civilisation. In South and Southeast Asian culture, this notion is reemphasised by the very designation of the art of healing as *Ayurveda,* the science of long life. Christian Europe in the Middle Ages applied a blend of religious and virtue-based ethics to medical practice. The Buddha and Jesus Christ uttered identical sentiments when they equated the care of the sick as a personal service rendered directly to them. (*Mahavagga,*

VIII, 25, 3 and Matthew 25, 40). Islam upholds the practice of medicine as a divine service.

Religions in general advocate the sanctity of life even if some do not go as far as to give equal primacy to animal and insect life as human life.

Judaism interprets the statement "I will reside among them" (Exodus 25, 8-9) to mean that a human being is a tabernacle or a temple from which emanates the divine word. The respect for human life for the Jews, therefore, is absolute, sacred and inviolable. To them and the Christians, the belief that man was made by God in his own image and likeness is a further testimony to the sanctity of human life. (Genesis 1, 26ff).

The Buddha taught, {All beings fear violence. Life is dear to all. Compare your love to your own life and do not kill nor cause to kill." (*Dhammapada* 129) In defining the universality of living beings toward whom loving kindness and compassion are to be extended, the Buddha included "*all forms of life without exception, moving or stable, tall, huge, middling or as tiny as an atom or fat, seen or unseen, living nearby or far away and born or seeking to be born*" (Suttanipata II, 1). As regards human life, the Buddhist view is that to be born a human being is a very rare fortune. It is the human who attains the highest spiritual sanctity of a Buddha. In the Mahayana Buddhist tradition of East Asia every human being is believed to be endowed with the "Awakening Mind" and is a potential Buddha.

The Buddha's senior contemporary, Jina Mahavira, the founder of Jainism, practised so stringent a code of non-violence that his followers, besides being strict vegetarians, take ample precautions to prevent the destruction of the tiniest creature.

The Islamic concept is founded on such texts in the Koran as "We have created man in the most perfect form" (Surah XCV, 4), "God breathed into man his soul" (Surah XXXII,9) and "The human being is a noble creature in the eyes of God" (Surah XV, 70).

Hinduism in its sublimest form attributes that every being from man to an elephant or a gnat to be morsel of the universal soul and attributes to all living being a sanctity based on a common origin and shared divinity (*Brhadaranyaka-upanishad* I, 3,22) So did Swami Vivekananda declare, "*The greatest duty of man is to serve that God whom the ignorant call man.*" Advocating vegetarianism, Hinduism upholds non-violence as its basic doctrine.

All prevailing religions have commandments or precepts which prohibit killing. The Western Asian religions have it as "*Thou shalt not kill*" and it is generally interpreted to apply to homicide. The commandments pertaining to theft, adultery, false testimony and coveting seek to maintain the security and dignity of persons from being violated by others. The Eastern religions have more comprehensive precepts as exemplified by the first of the Five Precepts of Buddhists: "*I take upon myself the discipline of not depriving any living being of its life.*" Right Livelihood in the Noble Eightfold Path of the Buddha embodies guidance for virtuous and harmless means of living by avoiding violations of human rights. Specifically forbidden are sale of human beings for various purposes and such commerce as sale of weapons and poison. The discipline enjoined in religions of India and China is founded on restraint in world, deed and thought in dealing with fellow human beings. **If religions are somewhat ambiguous about the totality of life to which sanctity is attached, they are unequivocally unanimous as regards human life and dignity.**

III. Violence: Nature or Nurture?

Despite the weight of religious advocacy of the sanctity of life and human dignity, there had been no time or place in human history without violence. One even wonders whether human beings are genetically programmed to be violent or whether violence is a vestige of the animal ancestry of humankind. A body of multidisciplinary scientists under the auspices of UNESCO considered this question and in a famous document referred to as the Seville Declaration of 1989 stated categorically that human beings were neither genetically violent nor did they inherit violence from an animal ancestry. If the humankind is not born violent, its tendencies toward inhumanity must be learned, acquired or forced behaviour. Are scientists, by reason of their superior intellectual capacity, knowledge and skills immune to such external factors? They certainly have not proved to be immune. In the Trial of the Doctors at Nuremberg in 1947, the world was shocked to hear how a host of doctors and scientists of related disciplines had participated in the most heinous crimes against humanity. Irreversible damage done through human experiments to persons whose consent had never been obtained brought to light the enormity of abuse which was perpetrated by men who were trained to save lives and alleviate suffering. Even as the scare of this gruesome disclosure of another aspect of the Holocaust was waning came the equally stunning revelations of human experiments done elsewhere in peace time in the free world. These too had been without the knowledge and the consent of the victims but with similar deleterious long-term consequences.

In the United States of America, the whistle was blown in 1966 by Henry Beecher who cited as many as twenty-two instances of unethical conduct on the part of medical researchers. Their victims were drawn from the convicted and the mentally retarded inmates of state institutions, the handicapped, the poor and the minorities. The fear of abuse was no longer a fancy; nor was it the product of paranoia. Scientists were on a trajectory of their own. Albert Guigui, the Grand Rabbi of Brussels, explains the causes and the consequences:

> Technical progress, the compartmentalisation of knowledge in sectors—separated, independent and entrusted to specialists who are increasingly isolated—and the excessive use of the analytical mind risk making [scientists] lose the vision of the whole and veer toward generalisations which are often arbitrary, unjust and inhuman, enabling them to abandon the essential moral laws. (Reseau europeen <<Medicine et droits de l'homme>>p. 74)

The 1978 Belmont Report on Ethical Principles and Guidelines for the Protection of Human subjects of Research sought to prevent the misuse of biological and medical sciences. In the world scene in the 1980s, the World Health Organisation (WHO) provided guidelines and standards on drug trials and human experimentation. The United Nations Educational, Scientific and Cultural Organisation (UNESCO) moved in 1991 to set up an International Commission on the Human Genome. This commission has deliberated over four years and drawn a Declaration for adoption by Member States. The Green Movement of Europe expanded the concept of bioethics to apply to the biosphere as a whole and expressed an increasing concern for the environment.

The irrefutable conclusion which all such actions embody is that **biological and medical scientists, as everyone else in the world, have to be sensitised on the sanctity of life and human dignity and the joint socioethical responsibility to preserve them.**

IV. Issues Pertaining to Bioethics and Biological Sciences

Issues pertaining to bioethics and biological sciences can be grouped under five headings as each has its own moral and ethical implications:

1. Means of mass Destruction: Chemical and Biological Weapons.
2. Manipulation of life itself: medically or therapeutically assisted (a) procreation, (b) termination of pregnancy and (c) death or euthanasia.
3. Human Eugenics, Genetic Engineering and Euphenics: prenatal diagnosis; genetic engineering for rectification of congenital malformations, hereditary diseases and genetic anomalies; germline gene therapy; and organ transplantation.
4. Genetic engineering in agriculture and animal husbandry to enhance quality and quantity of food production or extract new pharmaceutical products.
5. Violation of Animal Rights.

Moral and ethical implications arise from the diametrically opposed positions which scientists, on the one hand, and some moralists, on the other, tend take in respect of these issues, namely:

1. The persistent demand of some moralists that scientists desist from manipulating nature and specially human life, purely on grounds of religious faith or conscience.
2. The scientists' singleminded devotion to the pursuit of research and experimentation for an untrammeled advancement of knowledge as a matter of right or duty.

V. A Common Ground for Moralists and Scientists

The "hands off nature" argument of the theologically oriented moralists is not new. An earlier generation considered the experiments of Wright Brothers to develop a flying machine as blasphemous. "If God willed man to fly, wings would have been provided for the purpose," was the argument! It is from a similar standpoint that doctors and scientists are often admonished "Not to play God." It is not my intention to stray into any theological controversy. I raise this matter only to illustrate the enormous gulf which exists—and, according to some, is continually widening—between the autonomous search for knowledge by the scientists and the exclusively theological objections of some moralists.

I have advisedly referred to this group as "some" moralists because all who advocate bioethics and the prevention of the misuse of biological and medical sciences are not guided by a belief in God. The belief in a Creator-God and Creation is not universal. Even among the believers in God and Creation, interpretations vary widely. An eminent scientist of my acquaintance who, by religion and personal conviction, is a very devout believer in God would argue that the vision and skills of a scientist also emanate from God. He explains that the mission of the scientist so favoured by God is to unravel the secrets of the universe. He calls it a divine mission for which God has given humankind a unique brain and free will with infinite possibilities. One may also be reminded of Sigmund Freud's lament in his old age when he had reportedly asked why the God who had chosen him to unravel the working of the human mind did not provide him the audiences to listen to him until his health and faculties were failing!.

What is most important to my mind is that bioethics should aim at a *via media*. A middle ground has to be explored for the benefit of science and for the good of the humanity. There are at least two major concerns shared by everyone including the scientist and the moralist:

1. The recognition of the overriding importance of maintaining and safeguarding biodiversity, subject, of course, to the proviso that viruses, bacteria and protozoa, injurious to health, would continue to be tracked down and eradicated.;
2. The fear of abuses of scientific knowledge and skills by politically or economically motivated irresponsible elements.

These two concerns constitute the ideal meeting ground for the scientist and the moralist. The admonition not to play God certainly applies to anything done directly or indirectly to the detriment of biodiversity. **Whether one believes in God or not, one cannot overemphasise the importance of sustaining the equilibrium of life forms, in both flora and fauna. Every species endangered threatens the biosphere and eventually its most important benefactor the human being.**

Equally disconcerting as endangering biodiversity is the fear of abuse of scientific knowledge and skills for power or profit or both. Often the literature pinpoints dictators of the Third World as potential culprits. But an even far greater threat comes from the global proliferation of terrorists whose propensity for destruction has no limits. One should not lose sight of an equally menacing interest that transnational corporations could have in the results of biomedical research. These corporations have the resources, the access to scientific research and technology and a primary motivation in profit.

VI. Issue I—Means of Mass Destruction:

Chemical and Biological Weapons

The most cherished dream of the humankind is perpetual war-free, conflict-free peace. But never in history has it been so. Nor is it today. Will the world ever be weapon-free?

Asoka the Righteous, the Mauryan Emperor of India of the third century BC eschewed war for ever. He did so after seeing for himself the havoc which a limited war brought in the form of the killed, the deported and the victims of famine and pestilence. In his vast empire, he replaced the war drums summoning soldiers to the battle front with sermons on righteousness. He had left for us some fascinating and impassioned records on stone. He extolled peace and, as I will pinpoint again, upheld the sanctity of all life. His eloquent inscriptions —specially the Rock Edict XIII—elaborate the benefits of "Conquest by Righteousness" which had spread his influence as far as the Greek potentates four thousand miles from his capital. He appealed to his sons and grandsons to emulate him. Yet he was realistic, for he concluded the Edict with a second appeal: *"Yet if you have to engage in a* ***war of weapons****, be considerate and forgiving and inflict minimum punishment."* Over a thirty seven year regime, of which almost thirty years were after his conversion to Buddhism, he did not disband his army or, as the weight of evidence suggests, abolish capital punishment. The case in point is that war and conflict are inevitable, weapons would need to be used and hence produced and **all that one can realistically hope for is a decent minimum-damage war.**

It is but simple common sense to assume that nations will continue to arm themselves. Armament and even the arms race will continue to be defended as a deterrent to war. Arms production, in that sense, will be held as an effective means of securing and ensuring peace.

The show of prowess in the battlefield had been for millennia the goal and outcome of war. It had been the celebrated subject of epics and songs which glorified bravery and chivalry. But war left the battlefield when weapons of hand-to-hand combat were replaced by the bow (specially the cross-bow), the sling and the cannon which could kill from far. Later on, as war was directed against all and sundry in enemy land, the propensity for destruction had a free play. War aimed not at victory but the long-term punishment of the enemy. The medical officer of the British army which invaded the hilly kingdom of central Sri Lanka at the beginning of the nineteenth century was appalled when fruit tress and means of economic production were destroyed willy-nilly by the advancing army with superior fire power. But that was only the beginning of a trend which had grown and continues to grow to frightening dimensions.

We live in an age when nations are poised against one another even as they declare friendship and talk peace. They parade armaments whose capacity to inflict mass destruction is mind-boggling. The world's richest nations are manufacturers and merchants of weapons. The largest single item of expenditure of the Third World as a whole is for buying weapons which become increasingly sophisticated and rapidly obsolete. The pity is that these armaments are the direct result of the unprecedented progress that science and technology had made in recent years. In the process, the scientist in the quiet and peaceful surroundings in higher education—in laboratories and experimental workshops—has become a part of the war machine.

In the industrialised countries alone, an estimated half a million of the world's best scientists are engaged in developing and perfecting weapons, whether they be conventional, nuclear, chemical or biological. It is often with disbelief and utter frustration that the world receives news of their latest 'achievements' such as anti-person devices which would destroy human beings but save buildings and installations or germs and biotechnical processes which would wipe out whole populations with excruciating pain and suffering. One is at a loss to cry or laugh when these scientists or their sponsors try to assuage their conscience by gloating over relatively minimal "spin-off" gains in peaceful applications of military research.

Arms development without exception is a domain in which the application of bioethics has an unequivocal position. It is unpardonable to develop weapons of mass destruction and to use them under whatever provocation. The social responsibility of the scientist is undisputable. The degree or manner of destruction is incidental. It makes little difference as far as this responsibility is concerned whether the weapons are conventional, nuclear, chemical or biological. The escalation of efficiency and effectiveness of weapons systems has opened the proverbial "Pandora's Box." The world's events of the last few weeks are reminders of the dangers facing the humankind. All that is needed is just one man with blatant disregard for human life to unleash indiscriminate destruction. But he could never have been capable of such a dastardly cruel pursuit without the unsuspecting support and complicity of the scientist who devised the weapons, on the one hand, and the transnational corporations which profited by exporting the technology and the materials, on the other.

It has to be admitted that, next to nuclear armaments, chemical and biological weapons inflict irreversible damage to people and the environment far removed from the scene of war. Salvinia introduced to Sri Lanka during World War II and defoliation chemicals used in Indo-China during the Vietnam War illustrate what long term harm could be done to the environment with biological and chemical intervention of the least sophisticated form. What deleterious effects would advanced chemical and biological weapons of today have on the entire human race in the form of permanent genetic damage, mutation and extinction of species and the on-set of uncontrollable diseases and epidemics?

Happily, the international community has taken note of the need for collective action. Legal instruments have been drawn up and nations are being canvassed to accede to them. Agencies and procedures to implement them, though grossly inadequate and ineffective, are in place. **What is still to be accomplished from the bioethical point of view is a commitment from the scientists and their sponsors, the national governments, to divert their attention from destruction to construction, from warfare among nations to welfare of humanity. Fortunately the theological or philosophical differences which hinder the deliberation of the advocates of bioethics do not apply to this issue.**

VII. Issue II—Manipulation of Life

A. Medically Assisted Procreation

Medically assisted procreation is a reality and the scope is unlimited. At the present state of development of related biological and medical sciences, couples or even single persons who would never have dreamt of enjoying the joy of parenthood are beneficiaries of techniques ranging from *in vitro* fertilisation and frozen embryos to surrogate mothers. Moralists and legislators reacted to the perfection of each technique. The result is a body of legislation in such countries as Australia, Britain, France, Germany and U.S.A. (to name a few). These laws lay down the parameters within which it would be lawful to assist procreation medically and biologically.

The latest legal provisions worked out in Canada as a Bill under the title "Human Reproductive and Genetic Technologies Act" provide a clear idea of what is generally regarded unacceptable. This Bill, according to information given to me by the Canadian Embassy in Washington D.C., lapsed in April 1997 and did not become law. Yet its comprehensiveness is noteworthy. It sought to prohibit (i) cloning or splitting a zygote, embryo or fetus; (ii) fertilisation of human ovum with animal sperm or vice versa for purposes of producing a viable zygote; (iii) fusing human and animal zygotes; (iv) implanting a human embryo in an animal or vice versa: (v) germline gene therapy on ova, sperm, zygote or embryo; (vi) retrieving a sperm or ova from a cadaver to create an embryo; (vii) sperm separation of X and Y bearing sperm for selecting sex of the child; (viii) parental diagnosis (including ultrasound) for determining fetal sex; (ix) maintaining an embryo outside the human body for purpose of research; (x) fertilising an ovum outside the human body solely for purposes of research; (xi) commercialisation of the procurement and use of surrogate mothers; (xii) purchase or sale of ova, sperm, zygote, embryo or fetus; and (xiii) use of ova, sperm, zygote or embryo for research, fertilisation or implantation in another woman without the donor's consent.

Exceptions are made on health grounds, forensic purposes and reimbursement of incurred expenses. As regards the fertilisation of animal ova with human sperm, the Bill allows the continuance of the standard procedure for testing male fertility, namely the insertion of human sperm into ova of hamsters.

The list of banned procedures and the exceptions is indicative of the advances made by biological sciences in a sexual reproduction. All its provisions are indicative of assigning priority to the demands of bioethics rather than to the imperatives of scientific research. Could this imbalance be the reason for the Bill's demise?

B. The Growing Controversy on Cloning

Heading the above mentioned list are procedures which occupy the centrepiece in the controversy intensified by the cloning of a sheep through the technique of nuclear transfer in February 1997 at the Roslin Institute in Edinburgh, Scotland. This significant biological breakthrough has also led to a spate of hastily drafted laws, decrees and policy-decisions, especially in countries where personnel and facilities exist for the replication of the Scottish experiment. *"Hands off humans"* is the crux of such legislation. In their wake the entire issue of medically assisted procreation has come for comment and discussion.

Interestingly, a major bioethical offensive has also been launched from Scotland itself. The Church of Scotland itself. The Church of Scotland General Assembly meeting on 22 May 1997 adopted two resolutions urging Her Majesty's Government in Britain (1) "to take necessary steps to prevent the application of cloning as routine procedure in meat and milk production, as an unacceptable commodification of animals;" and (2) **"to press for a comprehensive international treaty to ban (human cloning) worldwide."**

Let us first examine how this differentiation is maintained as regards animal and human cloning. The first resolution commends the production of proteins of therapeutic value in the genetically modified sheep and other farm animals. These animals had already yielded products bringing relief to emphysema and cystic fibrosis sufferers. On that basis, the advantages of biotechnology, in general, and genetic engineering, in particular, are declared "ethically acceptable." Dr. Donald Bruce, the Director of the Church of Scotland's Society, Religion and Technology Project goes further to say that "the ethical question (is) on **how far** we should apply technology to animals" (emphasis mine). The bottomline of his argument is that on a very limited scale cloning animals "would not seem ethically unacceptable." For urging the limitation of scale in animal cloning, the Church of England advances the Christian argument that the world around us is God's creation and variety is one of its characteristic features: *The overall picture in the Bible, in commandments, stories and poetry is of a creation whose sheer diversity is itself a cause of praise to its creator.* (SRT Website on 1997 General Assembly Report —Cloning of Animals and Humans).

The resulting bioethical conclusion is that cloning of animals by itself is not ethically wrong but **"scale and intention play a part."** So, as far as this issue is concerned, we are in the field of not absolute but relative ethics. Biological scientists would have no difficulty in agreeing with the Church of Scotland. It is an irrefutable biological argument that biodiversity should be sustained at all cost. If that is assured, animal cloning would have the moralist's qualified approval.

C. Human Cloning

It is a different proposition when it comes to human cloning. The lapsed Canadian legislation bans the cloning or splitting a zygote, embryo or fetus and no exceptions are considered. The Church of Scotalnd declares that the cloning of human beings is ethically unacceptable as a matter of principle. It explains its position further:

On principle, to replicate any human technologically is **a violation of the basic dignity and uniqueness of each human being made in the image of God, of what God has given to that individual and no one else.** It is not the same as twinning. There is a world of difference ethically between choosing to clone from a known existing individual and the unpredictable occurrence of twins of unknown nature in the womb. The nature of cloning is that of an instrumental use of both the clone and the one cloned as means to an end, for someone else's benefit. **This represents unacceptable human abuse, and a potential for exploitation which should be banned worldwide**. (Ibid. Emphasis mine)

The concession of scale and intention given to animal cloning is denied to human cloning. The only reason discernible for this ambiguity is that the rational bioethical consideration which went into the decision about animals is replaced by a faith-based theological dogma in the case of the humans. It is certainly within the competence of the Church to put forward its dogma on the creation of man in the image and likeness of God in expressing its objection to human cloning. But to demand worldwide banning on the basis of that dogma alone would be to disregard contrary views which might fund support in other religions.

I see no reason why the sound bioethical principle of scale and intention, which had been so rationally developed in favour of limited and well-intended animal cloning cannot be extended to human cloning. Based on the progress made in other branches of genetic engineering, there could be scientists who would wish to investigate whether human cloning has its own set of benefits to humanity: e.g. yet another option open to desperate people who have no other way of having a child of their own; an assured means of supplying custom-made organs for transplanting to save lives; Increasing the sum-total and distribution of supremely endowed persons, whether they be savants and saints or luminaries in arts and sciences. In our present state of knowledge we do not even know whether these constitute reasonable expectations. Nor do we know whether the fears of the prophets of doom are based on fact or fiction. We do not know enough to make an informed decision. We need information and that has to come from experimentation. In this case, such experimentation is not likely to have any ill-effects on the donor. Is there any other way in which we could have answers to the following questions?

1. **Is human cloning feasible or is the humankind biologically or otherwise beyond the scope of manipulation?**
2. **What exactly will be the result of human cloning: A new person—*tabula rasa*? Or, an unaltered and unalterable identical replica of the donor of the clone with his or her own pre-existing personality, memory, habits, inclinations and so forth?**
3. **Are the speculations on the benefits accruing to humanity from human cloning within a reasonable limit of accuracy?**
4. **What are the real dangers against which safeguards should be taken?**

Unless we have scientifically established answers to these question we are in the domain of wild speculation in no way different from wondering whether the earth is flat or global.

I am not questioning the sincerity and the serious concern of individuals and organisations which urge the banning of human cloning. My position is that their campaign is premature and to a very great extent counter-productive for the following reasons:

- *First,* no effective ban can be imposed and supervised for the vast resources necessary for it could never be raised.
- *Second,* if cloning humans is lucrative business, ways would be found by industry to do it any way.
- *Third,* bans may be effective in countries where rule by law

is the standard. The scientists who would be prevented from the experiments will be the very ones who will be objective, disciplined and responsible, on the one hand, and who have the resources to deal with any untoward consequences, on the other. The scene of human cloning could move to places where similar standards might not be enforced and deleterious results could go undetected and unattended.

- *Fourth*, the only result achieved by banning will be a delay by a few years or decades. We have only to recall the campaigns of yesteryears and observe how the things that were dreaded a few years ago are now commonplace.

My recommendation would be for limited experimentation in human cloning overseen by theologians and moralists, biological and medical scientists and national and internal jurists. Let the long term decision be on the principles of bioethics which would emerge from these very experiments.

D. Other restrictions

Though cloning heads the list of procedures proposed for prohibition in the Canadian Bill, several of its other provisions also merit comment.

The **involvement of animals** in the asexual process of human reproduction is one of them. The moralists object to it on the ground that it could lower the dignity of the human race. As a value judgement it is debatable from different cultural standpoints. A more reasonable argument is that the fusion of human and animal reproductive material might set in motion a chain reaction the results of which none could visualise in the present state of our knowledge. Science fiction consists of ample horror stories of monsters to scare most people. Should we still experiment? The answer depends on whether there are any compelling anticipated benefits such as in combating disease or hereditary genetic disorders? In such an eventuality, the bioethical principles of scale and intention should guide the scientist.

Commodification or commercialisation of the intrinsic elements of human procreation (e.g. through sale of ova, sperms, zygotes and embryos; wombs for hire; sperm and ova and frozen embryo banks; and middle men to transact deals) is certainly degrading to human dignity and the measures envisaged in the Bill would receive wider approval. There could even be persons who in the name of bioethics declare that the legal provisions do not go far enough.

On the other hand, the prohibitions relating to **retrieving an ovum or sperm from a fetus or cadaver** and maintaining an ovum outside the human body solely for purposes of research deserve to be examined further. In the first instance, it may be the only scientifically available solace to a grief-stricken person to get back a new life from the lost beloved, be it a spouse or a stillborn baby. If it is feasible, what is the bioethical objection? Does death reduce a human to a lower status? If organs are harvested from the dead to give life and vision, what makes a viable sperm or ovum different? If the objections come in this case from the moralist and legislator, it would be they rather than the scientist who would be seen as "playing God."

In the case of **the maintenance of an embryo outside the womb** for any purpose, prohibitions and controls could only be counter-productive. A determined scientist has a feasible option and that is to conduct the necessary research on a living guinea pig in the form of a naturally developing embryo in a mother's womb. Is it bioethically less objectionable?

E. Continuing Dialogue on Medically and Therapeutically Assisted Procreation

With regard to the issue of medically and therapeutically assisted procreation as a whole, we are at a crucial stage when questions demanding urgent answers are many and complicated. This is the time when it is essential for the scientist to be in readiness to handle unforeseen developments in what is allowed and is in universal practice. Continuing research is indispensable. Indiscriminate prohibitions and controls could be the death knell of research. If this happens, the losers will not be the scientists alone. **Hence my plea that a continuing dialogue among scientists, advocates of bioethics and legislators of national governments and international community.**

F. Medically Assisted Termination of Life

Although the issues pertaining to abortion and mercy killing or euthanasia are important from the point of view of the sanctity of life, they do not relate directly to the question of the prevention of the misuse of biological sciences. It is not my intention to raise issues with which the Forum is not concerned in this session.

VIII. Issue III—Human Eugenics, Genetic Engineering, Euphenics and Organ Transplantation

Spectacular progress has been made in biotechnology in the fields of human eugenics, genetic engineering and euphenics. Benefits to humanity in the form of rectification and cure of congenital malformations, genetic deficiencies and anomalies and hereditary diseases have resulted from research and development since O.T. Avery and his associates discovered the "secret of life" in DNA (deoxyribonuclei acid) in 1944.

Statistically, the magnitude of the task for the biological scientists is enormous. Over two thousand distinct inherited genetic defects have been identified. One in every six newborns is said to have hereditary disorders of varying severity. When seen in the light of relief due to such a large section of the population, one is tempted to welcome the progress made in genetic manipulation.

Eugenics is the branch of biotechnology which aims at enriching the human genetic pool. Eugenics in its negative form reduces the possibility of transmitting defective or inadequate genes by preventing or controlling reproduction on the part of genetically defective persons. Sterilisation is a standard methodology. In the positive form, eugenics promotes preferential breeding with the objective of evolving a genetically enhanced humanity. Its aim is to distribute desirable traits and quality broadly and thereby improve the mental and physical well being of the human race. **Genetic engineering** which is also called gene therapy or "gene surgery" consists of ways and means of affecting the ova and sperms (=gametes) to inhibit transmitted genetic defects and improve the genetic quality of the person to be born out of the treated gametes. Germline gene therapy, which is an extension of genetic engineering, seeks to create genetic alterations which could be transmitted to the next generation. **Euphenics** is a procedure which does not affect the genes themselves but attempts to prevent or correct the detrimental effects of defective genes.

As curative procedures applied to an ailing individual, these three procedures are medical in character. The bioethical concerns, therefore, would relate mainly to competence, reversibility, informed

consent of patient, non-commercialisation or non-commodification of any resulting products and patient-physician confidentiality.

Eugenics and **germline** therapy where the affected parties extend to future generations, the bioethical safeguards have to be far more stringent. Here the biological and medical scientists have the greatest temptation to be do-gooders. They would act with the conviction that the improvement of the overall genetic pool of the human race justified whatever harm that would be done in the process to those whom they identify as the "genetically defective." Here in the first place there is no reversibility. Selective breeding, in the short term, had produced better milk or meat yielding livestock. Is the human race to be treated on par with livestock? Memories are still fresh as regards scientists under a dictator's orders who worked on breeding a blue-eyed blond pure race!

It is eugenics much more than cloning wnich interferes most with nature. Cloning makes a replica of the donor of a gene. Eugenics can change generations to come to fit any whim or fancy as regards height, weight, complexion, colour of hair or other physical and mental qualities. One ethically most troubling step in the process is that which deprives certain individuals of the freedom to procreate simply because of a perceived genetic defect. Its most devastating corollary is genocide through compulsory deprivation of the right to reproduce. Who survives and who perishes for the good of the human genetic pool is too momentous a decision to be made by anyone. The scientists, who may be called upon to make it, might have no autonomy. As in the case of their counterparts making chemical and biological weapons, they may be cajoled, bribed or intimidated by their sponsors, whether political or commercial.

The issues concerned with eugenics are not purely ethical. There are scientific reasons why artificially enhancing the human genetic pool could lead to disaster. The principle of biodiversity assumes a special significance when it comes to the preservation of the incredible diversity in the human kind. Humanity has survived many a vicissitude of nature from Ice Ages and floods to natural and man-made disasters because its diversity provided a wide variety of genetic characteristics to adapt to the changing environment. A standardised humanity would lose this advantage.

Organ transplantation is a science which has come of age.

The advantages are self-evident. Bioethical questions relate to procurement of organs and prioritisation of recipients. Those relevant to procurement of organs centre on

a. use of cloning technology for custom-made organs;
b. exploitation of the poor and third World populations as cheap sources of organs;
c. theft of organs from unsuspecting strangers; and
d. illicit traffic of organs with links to crime

Of these, only the utilisation of cloning technology would fall within the sphere of responsibility of biological scientists. Apart from the consent of the donor and the recipient, no major moral issue should arise, unless in the prioritisation of recipients affordability of high cost becomes the only criterion. **If the need arises to mass-produce organs and it is done as a human service rather than a lucrative industry, bioethical principles, norms and rules could be developed to maintain requisite moral standards.**

Both scientific and ethical reasons would indicate that the procedures discussed here have to be used in a limited scale and directed to specific defects in individuals purely with the intention of effecting a cure or remedy. Particularly apt in this context is the following statement with which V.J. Genovesi concludes his article of Genetics in the Encyclopaedic Dictionary of Religion (eds. Paul Kevin Meagher et al) Corpus, Washington D.C. 1979:

What must be decided is whether human influence upon human evolution will continue to be haphazard or instead be systematically planned. If a sense of responsibility inspires the choosing of deliberate intervention, it must also include the caution to proceed humbly and with the utmost wisdom; hopes for future generations must be balanced by an acknowedgement of, and respect for the personal rights of each presently existing individual, including especially those of the weak, the defective and the defenseless. If the step is taken to modify humanity, according to what specifications, will alterations be made? Who will decide upon such specifications and who will supervise the implementation of the plan? The state of public policy-making in environmental, domestic and foreign affairs is proof enough to that questions about genetic policy-making are not at all idle. In the effort to reshape humanity it is imperative not to undermine those very conditions of possibility for genuine human communication among individuals—personal freedom, the desire for truth, the capacity to recognise goodness, and the courage to love.

IX. Issues IV & V—Genetic Interventions in Plants and Animals and Animal Rights

Much has already been said in passing on the issue of applying genetic and related procedures to plants and animals. Ever since civilisation found a more stable food supply for the primitive human in agriculture and animal husbandry, manipulation of nature has become increasingly complex. Ethical issues connected with land, water, plants and animals vary from culture to culture. Some enjoined the emulation of "the tender care of the bee that draws the nectar without harming the flower." Some asserted the overall right of the humankind to be the sole beneficiary of all that was on earth. In others the exploitation of the environment proceeded with such abandon as to create serious ecological problems. Pollution in diverse forms have assumed the magnitude of a global threat to the survival of life on the Planet.

Increasing awareness of the limit to growth has brought about in our times a keen realisation that nature, if not treated with respect and moderation, could endanger the very existence of the human race. The human responsibility to safeguard the irreplaceable heritage of flora and fauna is being recognised with increasing assiduity. The global agreement on the principle of sustaining biodiversity is a direct result of this consciousness. The approach to sustainable development seeks to rectify errors of the past and treat Planet Earth with the respect that is due to it.

The errors of the past have been exacerbated and complicated by the application of science of science and technology. Their rectification has thus become a concern of the scientists and technologists. It is commendable that they have shown a degree of commitment which is in line with the principles, norms and rules as are identified for the purpose by advocates of bioethics. **One must, however, underline the potential danger which continues to lurk in the form of human cupidity for short-term gain. Needless to say that scientists and advocates of bioethics should be united in their effort to save Mother Earth and co-operate to prevent its further degeneration.**

Directly related is the issue of **animal rights**. To protect the habitat of endangered species is only one aspect. Animals sharing the common space with humans have their rights to life.

A document of special significance on animal rights is the Pillar Edict V of Asoka the Righteous whose non-aggressive military policy was discussed earlier. Issued in his twenty-sixth regnal year, that is, circa 239 BC, it is the world's earliest known decree on the protection of endangered species. In addition to twenty species of land, sky and water who are declared "exempt from slaughter," the decree offers sanctuary to pets, suckling farm animals, newborns up to six months and "all quadrupeds who are neither utilised nor eaten." Rules are promulgated on caponising, castration and branding of animals. So are the slaughter of animals and sale of meat and fish on specific holydays restricted. "Husks with living creatures should not be burnt and forests should not be burnt without a purpose of simply to harm," he decreed. The dictum "Living should not be nourished with the living" was the principle on which he abolished the daily slaughter of hundred thousands of animals in the royal kitchen and disbanded huntsmen and fishermen supplying food to the palace. He appealed to his people through his inscriptions to be kinder to the animals and forbade animal sacrifices of the day, even though he did not proclaim laws urging all to become vegetarians.

Today, a growing awareness of animal rights, specially among the youth, makes a significant impact. Vegans argue that one should not only desist from killing animals for food but also refrain from stealing their milk and eggs. Others adopt a more moderate position and consume eggs and milk products. Only extremely stringent codes of ethics, such as that of Jainism, extend the ban to plant life and forbid the eating of roots, yams, onions, garlic and what could be vegetatively propagated. Ecologists decry the unjustified over-exploitation of land in the process of growing animals for meat. Statistics are quoted to show that more land could be freed to nature if the meat consumption declined dramatically. That would be accompanied by a reduced use of chemical fertiliser, which constitutes the most extensive cause of water and land pollution.

Biological sciences do play a major positive role in the remedial processes set in motion for environmental protection. One has to commend the dedication of the scientists involved in developing high-yielding and disease resisting crops, fruit and vegetables with longer shelf-life and ways and means of enhancing the efficiency of food production and preservation for the growing world population. **The bioethical concern of interfering with biodiversity and its possible consequences does, however, persist in relation to the genetic manipulation connected with these efforts.**

The other issue of equal ethical importance relates to **commercialisation or commodification of transgenetic animals, plants and micro-organisms.** A draft European Council Directive on patenting these as well as sections of the human genome has led to a controversy in which ethical concerns figure prominently. A central argument of the Bioethics Working Group of the European Ecumenical Commission for Church and Society has been its ethical position that living organisms are products where "the distinction between what is God's creation and what is human invention is lost or blurred" or which could be "what God has given as free to all."

An ethical issue of far greater connection with the misuse of biological sciences pertains to **animal testing**. How much of it is absolutely necessary? How much abuse takes place because creatures for testing are expendable? It must be first made clear that no biological science or products and procedures discovered by biological and medical scientists would ever have been possible without experimentation on animals. **Bioethics has to be concerned with avoidable abuse and excess rather than on a total ban. If the use of animals in research and animal testing are banned on principle, the only option open to science is to experiment on humans or to create living organisms genetically.**

In the ongoing debate on cloning, it is said that 277 experiments preceded the successful production of Dolly. Deformed and defective sheep resulted from them and hundreds of embryos had to be thrown away. Apart from declaring that the experiment should not have been undertaken at all, there appears to be no alternative.

The entire question of experimentation with animals and animal testing of industrial products is to my mind an area where the bioethical parameters should realistically be "scale and intention." **It cannot be gainsaid that the bioethical demand for greater respect for animal life and rights, whether on the ground of the belief in God and Creation or on a commitment to the sanctity of all life, is justified and calls for the support of scientists experimenting with life forms for knowledge or products for the society.**

X. Bioethics in action:

Conclusions and Guidelines

The foregoing discussion on bioethical issues, sketchy indeed in view of constraints of time and space, leads us to five conclusions:

1. There are unavoidable ethical questions which must be dealt with in the realm of biological and medical sciences and these must be raised and answers sought;
2. Biological and medical sciences should advance in research and experimentation without loss of momentum as many a serious problem of life and living is yet to be resolved;
3. Between the moralists looking at bioethical problems and scientists intently concentrating on discovery, a *via media* has to be evolved, expanding, as necessary, the principles which UNESCO had identified in the Universal Declaration on the Human Geneome and Human Rights;
4. Legislators at the national and international levels should evolve principles, norms and rules to implement a collectively formulated and approved consensus and install the necessary institutional and procedural infrastructure for this purpose;
5. Nothing would be gained by the adoption of any extreme position and a basic policy which would be conducive to the greater good of the humankind is one in which **scale and intention** form the primary criteria.

Thus the three partners, namely the moralist, the scientist and the legislator, have specific responsibilities to bear. It is gratifying to observe that interdisplinarity has been the hallmark of many groups that have been active in dealing with issues and making representations to governments and the international community. They should continue to cooperate in drawing up mutually acceptable principles, norms and rules for bioethics to be functionally effective. **Unless bioethics reaches the level of action on the basis of widely accepted principles of socio-ethical significance rather than on mere dependence on religious dogma, its promise of guiding science and legislation would be an empty dream.**

In this context, a most forward-looking and practical course of action has been worked out in relation to the application of biotechnology to humans by the International Bioethics Committee which the Director General of UNESCO set up as far back as 1991. After four years of deliberation the Universal Declaration on the Human Genome and Human Rights was produced. **As an inter-governmental declaration, its timeliness is as significant as the pragmatic standpoint from which the need to "reflect on implications of progress before it is too late" is recognised and action urged.**

In considering principles, norms and rules which bioethics should evolve to prevent the misuse of biological and medical sciences, the following, to my mind, are overarching guidelines:

1. **The sanctity of life without exception and human life in particular;**
2. **The inalienable right to human dignity;**
3. **The restraint that is absolutely essential for sustaining bio-diversity and ecological equilibrium as *sine qua non* for long-lasting life on Planet Earth;**
4. **The autonomy of the scientists to unravel the secrets of nature and to utilise their research findings for the good and the benefit of the humankind;**
5. **The need for principles, norms and rules for regulation of improper use and application of scientific knowledge, skills and technology, whether such regulation be through self-policing by scientists, codes of ethics or legal provisions with the direct involvement of the legislatures and the judiciary;**
6. **A minimum programme of action jointly and severally by scientists, advocates of bioethics, governments and international community;**
7. **The active involvement of the international community through *a well - represented pluridisciplinary, dynamic, pace-setting and regulatary body* with UNESCO, the UN organisation responsible for science, as the lead agency and WHO, the corresponding UN organisation for public health and medicine, as a key collaborator.**

Bibliography

Michael A. Grodin (ed.), *Meta Medical Ethics—the Philosophical Foundations of Bioethics*, Kluver Academic Publishers, Dordrecht 1995.

H. Tristram Engelhardt Jr., *The Foundations of Bioethics*, Oxford University Press (OUP) 1996.

Kenneth D. Alpern (ed.), The Ethics of the Reproductive Technology, OUP 1992.

Leroy Walters, *The Ethics of Human Gene Therapy*, OUP 1996.

Martin S. Pernick, *The Black Stork—Eugenics and Death of "Defective" Babies in American Medicine and Motion Pictures since 1915*, OUP 1996.

Reseau europeen<<Medecine et droit de l'homme>>, *La sante face aux droits de l'homme, a l'ethique et aux morales - 120 cas pratiques*. Council of Europe Publishing, Strasbourg 1996.

Ananda W.P. Guruge, *Asoka the Righteous - A Definitive Biography*, Central Cultural Fund, Colombo 1993.

23

Actual Problems of Bioethics

—G. Gerin, Italy

UNESCO's central mission includes the promotion of science and science education throughout the world. Informing people about genetic discoveries is an important challenge and responsibility. These discoveries will shape how future human beings think about themselves in relation to the world around them. In addition, information will lead, through the interrelationship of science with technology, and technology with practice, to major changes in medical diagnosis and treatment.

UNESCO is also an institution concerned with securing fair participation in the benefits that flow from scientific and technical advances to all the peoples of the world. How will that be accomplished with technologies such as gene therapy? The claim to share the benefits from genetic information is a particularly strong one. The genome project will be maximally successful to the extent that it can effectively study the variability of human genetic makeup around the world. Many of the important advances will result from comparisons—asking what genetic factors differentiate one group from another—when two groups are observed to differ strikingly in susceptibility to particular diseases. The wider the participation, the greater the prospect of discovery. When all may contribute all should share. These are claims that UNESCO has a special voice in pressing.

Furthermore, UNESCO is an institution with a special role in promoting human rights. There is a widespread realisation that genetic information and its manipulation poses special dangers to human rights. That understanding is the crystallisation of very unhappy human experience with ideologies that used genetic or supported genetic criteria to celebrate some persons, and denied to others their full human dignity. Claims that one person's genes are better than another's are the late 20th Century's flashpaper.

These considerations require attention to the value issues raised by the Human genome Project as a general matter and gene therapy in particular. This branch of Genetic engineering is developing rapidly, in both government-sponsored and commercial research settings around the world. The justification for its developing is compelling, namely the relief of human suffering through remediation of disease. And yet technology itself, and the theoretical limitations of what can be done, are a function of scientific and engineering principles, and not of social constructs such as what makes a condition a disease and what constitutes permissible therapy for it. Indeed, the very term "human gene therapy" selects out for analysis presumptively beneficial procedures.

The technology of gene therapy will permit humans to direct evolution by choosing the traits and heightening the capacities they wish to engineer into the young.

The term human gene therapy came into use because it identified as presumptively beneficent, through the use of the word therapy, technologies that might have provoked more opposition if called human genetic engineering.

Relying on the term therapy to distinguish desirable ends from undesired ones is problematic for familiar reasons, and particularly so in an international and cross-cultural context. Therapy is a complex socially-constructed idea which has no necessary connection to any universally-agreed upon concept of disease. The point may be illustrated by looking at the differences between societies, and the same society at different times, at whether and to what extent infertility is and was regarded as a medical problem as against a personal condition.

My intervention could, of course, maintain the focus on genetic disease by simply defining human gene therapy as therapy directed at such problems. Yet the main bioethical sources do not define gene therapy that way. (See, e.g. Clothier Commission (1992), Declaration of Inuyama (1991). Like us, those reports define gene therapy in terms of intentional manipulation of human DNA. Indeed, they make glancing mention of other possible uses of gene therapy. The bioethical literature, nonetheless, focuses almost exclusively on the ethics of treatment of genetic disease. This preoccupation is understandable in a historical context. The scientists who developed gene therapy procedures sought therapies for genetic diseases (although they mentioned wide uses from the beginning). (Anderson & Fletcher 1980). The discussion about ethics responded to them and their work. Moreover, gene therapy seemed like the only possible therapy for many genetic diseases.

The ethical structures for analysing gene therapy protocols' costs and benefits were crafted with the model of life-time treatment of children in mind. Is that still the right model?

Second, enabling technologies lead to spin-off developments in many related fields of practice. The question whether social investments in gene therapy may be squared with justice in distribution of resources may appear in a different light when one realises the breadth of applications, and the usefulness of standardised means for controlling gene expression in agricultural and even industrial settings.

Third, the fact that gene therapy is involved should not categorically resolve whether somatic cell enhancements are ethical. Such therapies will frequently be no different that drug therapy.

One final point has particular relevance far UNESCO. Gene therapy is an enabling technology, then research institutes all around the world will want to use it sooner rather than later, each competing for applications others may have overlooked, or for uses particularly important in their own medical environment. How can the complex rules for safety working with viruses and vectors characterising the technology be effectively integrated into international practice? The issue is not one of getting formal adherence to treaty provisions requiring national rules on the safe handling of genetically altered organisms; it also requires education and sponsored cooperation across scientific cultures.

My intervention is not going to deal with the ethical principles related to in vitro fertilisation followed by genetic testing and selective embryo implantation, or simply screening and selective abortion, because these procedures are not to be considered as human gene therapy as we define it.

I wish now to discuss the problems concerning gene insertion and successful gene therapy alternative DNA in human cells.

The technologies for gene insertion can be divided into those that rely upon viral mechanisms and those using viruses.

It is necessary to mention not only the popular viral vectors but also the retroviruses and the adenoma viruses.

The use of retrovirus vectors, however, is at present not so easy to control. The use of adenoma viruses is the most popular alternative because they directly infect any human cells (cystic fibrosis).

Chemical and physical means for inserting novel DNA structures have been much less thoroughly studied. Genes can be bound with liposomes, but the transuction efficiency is limited. Direct physical injection of DNA as a plasmid may be accomplished, but so far the approach works only in muscle cells.

A sense of the extraordinary complexity and heavy demands far validated safety procedures to be met by those who prepare gene transfer systems far human use may be gained by consulting the United States Food and Drug Administration's Document, Points to Consider in human Somatic Cell Therapy and Gene Therapy.

All this complex biological machinery must function not only for a few minutes or a day. For the therapy to work, it must continue to function over time. And if cells that are transduced have a short life until they are replaced by the body's new ones, gene therapy requires repeated readministration of the therapeutic gene product, at least if treatment of a genetic disease is the goal.

The bioethics literature gives surprising little attention to the need to repeat treatments again and again. Writers discuss somatic cell gene therapy from the perspective of a germline approach; as if administration of the therapy changes this patient forever (but unlike germline therapy, poses risks to no one else).

Gene therapy protocols are underway in the United States, China, France, Great Britain, Italy, the Netherlands, and perhaps elsewhere. But the United States have issued the greatest number of them. The future of gene therapeutics may be appraised by considering protocals in the planning stage or approved by the United States Recombinant DNA Advisory Committee.

The easiest gene therapies to conceptualise are those directed to genetic disease caused by the absence or failure of a single gene.

The cellular repair metaphor may also be used where cells are producing "bad" proteins. As far as I know, the problem to day is the difficulty to remove the "bad" genes.

There are many gene therapies' proposals for tricking or enhancing immune functions.

The transgenic animals are used for transplantation to human patients without danger for the immune system.

Many protocols prepared to stimulate immune responses (cancer vaccines, protocols) have been introduced to the special committees in the United States, but as far as I know no news about the effectiveness of the system are available.

Finally, gene therapy approaches may successfully treat disease simply by inserting "suicide genes" into target cells of interest.

The conclusion that gene therapy will be most frequently used for treatments other than genetic disease leaps out from the present evidence and trends. I believe that the scientific barriers that remain confirm that view, as do the market-size issues that till shape commercial work.

At this point I deem it necessary to explain what the International Bioethics Committee (I am a member of) has done to provide information and to suggest ethical principles that are grounded in universal ideas even if it takes count of the heavy diversities of cultural and religious traditions that exist in the world.

Personally I am involved in the promotion of human rights because I am President of the Human Rights Studies Institute. This Institute prepares many recommendations for the United Nations concerning bioethics in that the first right of each man is life.

As regards gene therapy, at an international level the following principles are accepted:

1. The respect for human dignity and worth;
2. The right to equality before the law;
3. The protection of the rights of vulnerable individuals;
4. The right not to be subjected to medical or scientific experimentation without free consent;

5. The right to the highest attainable standard of physical and mental health and associated rights to health care;

6. The right to protection against arbitrary interference with privacy or family;

7. The right to enjoy the benefits of scientific progress and its application;

8. The right to freedom for scientific research.

In as much as all present gene therapies constitute medical and scientific experimentation, (and in a rather extreme form of it), the right "not to be subjected to it without free consent is guaranteed. "Free" consent implies informed consent, with no coercion. The duties imposed on researchers and procedures for implementing them have been spelled out in other internationally significant documents. The Nuremberg case was the foundation. That code was formulated in the unusual context of an international war crimes trial, for purposes of stating the internationally-recognised principles that might permit researchers to engage in conduct that would otherwise be a violation of subjects rights, and indeed where injury was risked or caused, a serious crime. The Declaration of Heksinki, prepared by the World Medical Association, was derived from and built on the Nuremberg Code. In turn, the World Health Organisation and the Council for International Organisations of Medical Science built their influential International Guidelines for Biomedical Research Involving Human Subjects on the Helsinki Declaration. The guidelines purpose is to indicate how fundamental ethical principles should guide the conduct of biomedical research involving human subjects. In its most recent version (WHO,CIOMS 1993), one finds as "general ethical principles", the proposition that: all research involving human subjects should be conducted in accordance with three basic ethical principles, namely respect for people, beneficence and justice.

As far as applications to somatic cell therapy are in my opinion concerned, there are no ethical problems against the use of applications to somatic cells, even though contrary arguments may be raised whose reasons I wish to summarise as follows:

a) there is an impermissibly high change of accidents causing large destruction to people; property or the environment that will ultimately occur through accidental release of viral or other materials;

b) accidents in somatic cell therapy will sooner or later result in the alteration of germline cells;

c) the creation of somatic therapies puts us on the slippery slope to worngful germline therapies, which slope we can not hold.

The first argument has echoed in debates about the construction of nuclear power plants. We believe that the premise is wrong, even on the assumption that, accidents may happen. Furthermore, accepting this argument would imply the impermissibility of the full range of genetic technologies, technologies that will have major impact in the world economic life. This risk argument is not limited (if it is to be thought as a limitation) to treating ill people.

As to the second argument, researchers cannot rule out or deprediate the prospect that some mishape may alter some one's germline cells, particularly as echniques to administer gene insertion in vivo are developed. Is the possible accidental alteration of a germline cell be a basis for the total prohibition of all somatic cell gene therapy?

We do not think so.

Risks may be minimised by maintaining a constant focus on safety. But more important the human genome is by no means stable. It constantly changes through mutation, including many mutations that result from human activities. Most mutations, of course, are selected against. They fail to survive. The same would be true with these kinds of accidents.

The slippery slope argument, by contrast is an argument without a stopping point. It would lead to bans on airplanes because they might be used to drop bombs. (Sass 1988). If people believe maintaining sharp boundaries against germline therapies is important we think social, legal and professional standards will hold the line adequately, particularly given the technical challenge such therapies would have to overcome.

These possible arguments against somatic cell therapy do not respect adequately rights to freedom in scientific research, the duty to protect the vulnerable, and the rights to enjoy the benefits of scientific progress.

From any ethical standpoint we know about, the central

responsibility is to maximise the gains and minimise the risks, and ultimately decide or not the benefits realistically to be gained make legitimate asking others to run the risks involved.

Discussions of consent to somatic cell therapy experiments in the literature seem driven, and driven to excess given the technologies present use, by models of experimenting on children with genetic disease.

Gene therapy has a chance to be an enabling technology with widespread uses. The often expressed concern is that the economically developed world expending its medical resources wastefully in developing them is misguided.

The study of gene processes and the control necessary for therapeutic approaches will have wide application in the control of biological processes in other settings, from agriculture to industry.

Recent gene therapy discussions insist that somatic cell procedure should be reserved for "serious disease". The committee concurs in that view because of the highly experimental nature of procedures, and the lack of sufficient experience for determination of the incidence and seriousness of side-effects that accompany various types of cellular alterations.

A cell performs complex chemical processes, through many pathways. It is far too soon for confidence that imposing on it a new "manufacturing" responsibility with energy requirements that must come from somewhere, will not somehow affect the way it (and through action on it, action on other cells) performs other tasks. Second, how stable is transfection? Do inserted genes always stay put or can they recombine and move to other cells and if so how frequently does this happen and with what effects? It takes thousands of cases and years of experience for confident elimination of all the unhappy possibilities. No responsible ethics committee could at present approve a protocol directed at a disease it did not regard as serious.

What is a serious disease? Despite the diversity of cultures and world views, we believe the international community would reach virtual unanimity on a long list of serious diseases and that list would include current targets of gene therapy protocols.

Germline gene therapy has grabbed the world's attention. Imagining a world in which some people—the state, physicians, parents—have authority to select the genetic characteristics of the next generation, choosing chemical constituents to produce desired traits as if they were baking a cake, provides the conjurer with the occasion to reflect on the true nature of individual rights and human dignity.

On the whole, thinking about ultimate values is a good thing. Yet such a future is often described in a way that ignores the moral values that are connected to the complexity and contingency of individual development. Mozart's genes do not guarantee Mozart's genius.

All major statements about germline gene therapy condemn its present use. This position is clearly correct. That genes can be put into animal germlines, and made to express, does not begin to answer the need for safety issues for human use. Moreover, are there animal models that can predict the impact on the human brain? Enormous technical problems would have to be solved to make technology realistic in light of the risks, particularly the control of gene expression throughout the organism's process of cellular differentiation. Furthermore, one must have basis for confidently predicting the consequences of novel or altered genetic material in the workings of each and every cell type. From what we know, there have been no efforts anywhere in the world to attempt gene line therapy on human beings. The prohibition on germline therapy is a matter of formal legislation in some nations (e.g., Sweden) and is accomplished through regulatory controls in many others, for example Great Britain and the United States. Present prohibition does not deny the possibility of future use. Yet, there are important European documents that condemn germline gene therapy unequivocally. Any form of therapy on the human germinal line shall be forbidden." (Council of Europe Recommendation 1100.) Two important recent reports, the Clothier Commission 1992 and the Declaration of Inuyama (1990), however, do not categorically rule out germline therapy. In the United States, a number of prominent commentators believe that discussion should begin from the development of a gene therapy for the germline. (Anderson and Fletcher; Wivel and Walters 1993). Their call for present discussion reflects no disagreement about its present ethical impermissibility. It recognises that the future policy process for approval will take a long time and that more discussion now will probably allow appropriate policy to be in place in the future when technical capabilities will be at hand.

It may be right to debate the ultimate value questions. But those who suggest the desirability of germline therapy have not, the reporters believe, made a plausible cause that the gains would be worth the extraordinary efforts required. Germline gene therapy sounds most attractive as a technology for permitting a couple or individual to spare their descendents the burden of genetic disease. So, for example, a person at risk for a disease caused by a dominant gene might seek for the procedure able to guarantee that his/her future children would not be affected by it. If it is moral, as it surely would be, to remedy his/her conditions by means of somatic cell gene therapy, why not to cure it once and for all by means of germline techniques? Before considering moral dilemmas related to this question we should ask ourselves whether it is ethical to risk harm if there is a safer alternative. If one had the knowledge sufficient to attempt germline therapy in these settings, it would seem certain that one would know how to accomplish the same end without employing germline techniques.

Treating the adult cells in vivo to remove the gene is a technology not even envisaged. By contrast, sorting sperm (or eggs) using DNA probes to separate those that carry the gene from those that do not fall into the category of not presently feasible. If it could be done, however, one could solve the problem by using "good" sperm and discarding rather than repairing the sperm carrying the gene.

The probable starting point for germline therapy is the human zygote at the fourth cell stage, fertilised through in vitro fertilisation. Certainly, that is the only present way to do it. Animal models suggest that gene insertion that will differentiate into every cell is feasible. But if a particular zygote can be identified as carrying (or lacking) the gene, and therefore be appropriate for treatment why would one try the extraordinary repair procedure rather than selecting for implantation a zygote that does not have the gene? Such screening procedures are in use, on an experimental basis, in a few hospitals around the world. For example, at a hospital in Israel, procedures for selective screening of zygotes, using so-called PCR techniques to test the DNA in one cell of the four cells was used to screen for cystic fibrosis.

There are many who strongly oppose any selection or discarding of zygotes, and for these people, that opposition is in itself a sufficient basis to protest the development of germline gene therapy. Opposition to selecting zygotes goes hand-in-hand with opposition to non-therapeutic experiments on zygotes that will ever be implanted. For example, German law prohibits non-therapeutic experiments an human zygote. While all acknowledge the special respect owed to the fertilised egg throughout it's development, what actions are required or prohibited by that respect is an issue on which the world community is deeply divided. This Report does not attempt to solve the issue. It is obviously crucial for germline gene therapy.

Protection of human dignity and worth puts important limitations on tampering with the reservoir of potentialities inherent in the gene pool. But are categorical prohibitions desirable?

They might be based on:

- The need to experiment an zygotes. We will not comment further on this issue.
- Impermissible risks to the future child.

A final argument against germline gene therapy is that it is somehow beyond human rights to interfere with the fundamental process of life.

Finally, it is wrong to have a group of human beings claiming to foresee what traits the world will need in the distant future.

In the end, our recommendations fit a familiar pattern, although it will prove controversial for some.

1. Somatic cell gene therapy is permissible, regulated as an experimental therapy.
2. Its use for enhancement purposes may be widely prohibited, but it should not be categorically disapproved as unethical in all thinkable circumstances.
3. Germline gene therapy is indefensible at present, but it should not be categorically prohibited.
4. The use of germline gene therapy for enhancement purposes should be categorically prohibited.

Bioethics ranges in many fields concerning the life and health of people. According to Axel Kahn, there are two fields which may influence human life, one dealing with genetics and the other one deriving from the environment (environment to be considered as a whole as indicated by the Nobel prize winner Arber).

Starting from the relationship between the physicians and their patients, bioethics has ranged from physics to the empirical sciences. The responsibility of each person and of governors requires a further discussion about the use of scientific discoveries that may be in favour or against the human beings. It is not possible to only hold researchers responsible for it. It is necessary that the ethical evaluation is faced both by single human beings and by governments and at an international level by the organisations of states.

That is the reason why ethical committees have been created, especially in France but also in other European countries and in the United States (in particular through foundations such as the "Hastings Center", etc.). UNESCO has also created the International Committee on Bioethics.

In Europe, a committee with the ethical problems connected to biotechnologies has been devised by the European Union. The Council of Europe has foreseen a convention on human rights and biomedicine with the possibility of turning to the Court for Human Rights in Strasbourg.

Theoretically the problem of human cloning has still to be solved. It is a new question that is arising but that has already been blocked after the experiences with animals (see Dolly).

Among the international documents dealing with the problem of cloning, I wish to confirm that all international states organisations and Bioethics Committees have decided not to proceed in this way.

1) European Union (former European Commission) in which the group of delegates involved in dealing with ethics of biotechnologies (Bruxelles) replied in a negative way although they did not exclude the necessity of knowing more about the fundamental processes connected to the cloning field. They also invited the different committees of the Union to inform each government of each member state about their decisions. The definitions are as follows:

   a) cloning consists in producing genetically identical organisms both as far as embryo splitting and nuclear transfer are concerned. In the first case the genes, both nuclear and mitocondrial, will, be identical; in the second case, on the contrary, only the genes in the nucleus will be identical, whereas the genetic pack could be different between the two people, in that genes can undergo mutations or disappear during their development.

b) normal reproduction leads to different people, whereas clonage leads to identical people: this may be useful in botanics and agriculture, for instance.

2) The European Parliament, in its resolution on the ethical and legal problems connected to genetic manipulation (Doc. A 2.237/88 in U.G. European Community 16 March 1989), as far as cloning is concerned states: *"It believes that a legal prohibition is the only possible reply to the possibility of creating human beings by means of clones, as well as to all experiments aiming at cloning human beings"*. The European Parliament, in its resolution on human embryo cloning (Doc. B 3-1519/93 of 20 October 1993 in Eur. U.G. of 22.11.1993), issued after the news about the first cloning experiment had been "published" (J. Hall of the George Washington University Medi center, communicated during the Montreal meeting of the American Fertility Society, condemned "any cloning of human beings notwithstanding the reasons for, included research, as it represents a violation of the fundamental human rights and does not respect people. Furthermore it is not acceptable from an ethical point of view.")

3) The Council of Europe, keeping in mind that the "convention on Human Rights and Biomedicine" does not expressly prohibit the cloning of human beings, states:

a) the right of each person to his/ her genetic identity;

b) the need for a prohibition of human cloning, through a world wide explicit interdiction;

c) the request for the member states to forbid the cloning of human beings at the different stages of their development and constitution, notwithstanding the process applied, and to penally punish those who do not comply with this prohibition;

d) the request for researchers, biologists and physicians to avoid human cloning as long as there is no legal definition of the problems connected to cloning;

e) the need for biotechnological research to produce proteins, medicines and vaccines to be administered to the patients in order to fight against some diseases;
f) the need for precise information on the present state of the art of genetic research to be given by scientists and governments;
g) the need for a single legislation on the cloning of animals within the European Union and for information on the new scientific discoveries aiming at safeguarding the health of people, the continuity of the human and animal species and to safeguard biological diversity.

The European Assembly has taken a negative decision according to Art. 13 of the Convention on Human Rights and Bio-medicine, thus forbidding the cloning of human beings, including (non reproducing) somatic cells.

The special committee of experts (CAHBI) responsible for issuing a report on the techniques of artificial procreation stated in Art. 20 that *"the use of techniques for artificial procreation aiming at creating identical human beings through cloning or any other means should be prohibited"*.

The Committee of the Ministers of the Council of Europe, with the mandate n. CM/663 of 14 May 1997, asked the Directive Committee for Bioethics to issue by 30 June 1997—a comment on the problem of human cloning.

The conclusion and the text approved by the Committee was;

"Any interventions aimed at creating a person that is genetically identical to another one, dead or alive, is forbidden. From what stated in this article the expression "genetically identical human beings" means people having the same set of nuclear genes in common".

24

Global Biomedical Knowledge: Misuse in the North, Lack of Access in the South

C. Pasternak, United Kingdom

In the first part I shall refer to certain aspects of recent biomedical knowledge that have been accumulated through scientific advances that have been made predominantly in the developed countries of the North; that is to say, North America (especially the USA), Europe (especially the countries of the European Union that make up Western Europe) and the Far East (especially Japan). In short, in those countries that have the wealth to be able to carry out sophisicated biomedical research. In the second part I shall refer to the rest of the world: those countries that are striving to catch up with the developed countries, and those countries that are still far behind.

One of the most exciting projects in biomedicine, and one that is being carried out through international cooperation between the countries that I mentioned earlier—namely USA, Europe and Japan—is the sequencing of the human genome, or HUGO. You are all familiar with the fact that our entire make up-our looks, our metabolism (how we digest food and turn in into energy), the function of our brain, our susceptibility to disease—all this depends on the function of the molecules we call proteins. And every protein in our body is specified by a particular gene, that is to say by a particular stretch of DNA that is part of the chromosomes present in every one of our cells.

We inherit the chromosomes from our parents: one set from our father and one set from our mother.

Why should we want to know the sequence—the is to say the exact composition—of our genes, when it is the proteins that are responsible for human form and function? There are two reasons. First, it is technically much easier to determine the composition of a gene, that is to say of a stretch of DNA, than that of a protein. Second, not all of the DNA that makes up our genome is expressed as proteins: much of it—more than 90 per cent—does not specify proteins, but specified instead the control of those proteins: when they are to be made, how much is to be made, and in what cells they are to be made. This is important knowledge. Also, we have not yet identified all the proteins that we believe are present in our body—we think there are some 100,000 different proteins, but have identified less than half; knowledge of the composition of our DNA will reveal the structure of the remainder. And finally, much of our DNA may specify functions of which we are not yet aware, but would like to know about.

Once we have this knowledge, how will it improve our lives, particularly in relation to disease? First we need to remind ourselves that all disease is due to one of two causes: to our genes, that is to say to hereditary causes, and to the environment in which we live. Some diseases are predominantly of genetic origin. These, include hemoglobinopathies, in which the protein hemoglobin in our red blood cells is impaired in its ability to carry oxygen from the lungs to the tissues, hemophilia, in which the one of the proteins involved in the clotting mechanism of the blood, that is brought into play when we wound ourselves, is faulty; they include diseases of lipid metabolism such as Tay Sachs disease, in which brain development is impaired because one of the protein enzymes is missing, and diseases like cystic fibrosis in which the function of lungs and pancreas is damaged because of a faulty protein in the plasma membranes of those tissues. These diseases are due to a fault in a single gene so that a single protein, that is responsible for normal functioning, is missing or impaired. The diseases are all rather rare —one of the most common, cystic fibrosis, affects less than 0.1 per cent of the population—and this is because most of the diseases lead to an early death and therefore tend to be eliminated from the population.

Other diseases are predominantly due to environmental causes. These include factors like nutrition—too many calories leading to obesity or too few vitamins leading to diseases like pernicious anaemia (lack of vitamin B12). They include factors like pollution and stress, but most of all they include all the infectious diseases, diseases caused by viruses, like measles, influensa, herpes or HIV, diseases caused by bacteria like tuberculosis or pneumonia, diseases caused by protozoa like malaria or leishmaniasis.

But some of the most common diseases, like heart disease, stroke, cancer, diabetes or asthma, are caused by a subtle inter play between genetic and environmental causes. In the case of heart disease, for example, our diet or lack of exercise plays as large a part as the genes we inherit from our parents; in the case of cancer, viruses play a major role in some types, diet in others; in the case of asthma, pollution as well as a genetic predisposition to hypersensitivity play major roles. In all these cases, the hereditary element is not just a single faulty gene, but several genes all of which contribute to an increased susceptibility to disease. It is these susceptibility genes, a few of which we have only recently come to recognise, that the human genome project will identify for us. That knowledge will help us to avoid some of these most common diseases by taking care of our diet and lifestyle, and most importantly by novel drugs and treatments designed to limit or boost the expression of specific genes.

So what is the problem? Is not such knowledge all to the good? Not necessarily. We must be careful about the use to which we put our knowledge. In the words of the English poet Alfred Tennyson, **Knowledge comes, but wisdom lingers.** In other worlds, we must use, our newly acquired knowledge wisely. I shall not dwell here on the misuse of biomedical knowledge for the purposes of aggression and war. First, because I do not believe that one should stop acquiring knowledge in case it is misused by men of evil. Such men have, and will continue, to massacre thousands upon thousands of innocent people by torture, shooting and starvation; Radovan Karadicz or the war lords of Central Africa do not need sophisticated biomedicine to slaughter innocent women and children. Nor does Saddam Hussein; whether he decides to spray Israel with anthrax or with scud missiles depends largely on political, not on technical considerations. My second reason for not discussing the misuse of biological knowledge in the context of biological warfare is that others will do so during this meeting—indeed immediately following this very talk.

Rather I will refer briefly not to the deliberate misuse of biomedical knowledge, but to unintentional misuse; in particular to the ethical problems created by biological knowledge, even though these, too, will be discussed by others in a separate session this evening.

The ethical impact of genetic knowledge was recognised at the outset of the human genome project and some $ 9 million have so far been spent on the ethical, legal and social implications of the project in what has been referred to as a "full employment programme for bioethicists'. And I would like to stress that the pharmaceutical companies—considered by many to be greedy in their attempts to patent genetic knowledge are contributing generously to full discussion of bioethical issues. Both SmithKline Beecham and Zeneca are sponsoring their employees to participate in programmes run by the University of Pennsylavania's Center for Bioethics, and SmithKline Beecham has donated $ 1 million to Stanford University to fund research on genomics, ethics and society. The way that misuse of knowledge of our genetic make-up may come about is, of course, reminiscent of Hitler and the Third Reich, but in this instance often for perfectly sound reasons. If someone is known to be genetically predisposed to die of cancer or a heart attack at an early age, is it not reasonable to charge them more for life insurance or to limit their credit facilities? If someone is known to be genetically predisposed to a neurodegenerative disorder like Alzheimer's or Huntington's disease, is it not prudent to consider their long term employment carefully? Maybe, may be not. As yet we do not. As yet we do not discriminate even against people known to have such debilitating diseases as cystic fibrosis, that kills at a much earlier age. Should we not strive to maintain that position? Another point concerns certain close-knit communities that have been targeted for genetic research because they are known to be rather in-bred. These include Ashkenazi Jews, Mormons and Finns. Any genetic characteristics that are revealed in these populations are likely to be viewed as aberrant, rather than normal. This, as we all know, leads to social discrimination. And finally there is the problem of the individual himself. Does knowing that one is likely to die at an early age not cause so much stress as to exacerbate the situation? Some people can take bad news well, others not. And it is becoming clear that mental stress is itself a predisposing factor towards heart disease, stroke and cancer. These are all problems that this conference should discuss in its deliberations during the next of the week.

There are two other areas where biomedical knowledge—in this case coupled to treatment—arouses ethical and legal, as well as religious, concerns. The first is embryo research, including in vitro fertilisation and prenatal diagnosis leading to termination of pregnancy. If it is acceptable in one community to terminate a pregnancy because the newborn child will have Down's syndrome, is it acceptable in another community to terminate a pregnancy because the offspring will be a girl and not a boy? The second area is the use of fetal tissue for combating certain disease. Parkinson's is an extremely debilitating disease—the typical tremor in the hands and general muscular problems are well known—and over the past few decades the main underlying cause has become clear: a failure of certain neurons in the brain to secrete the neurotransmitter dopamine. When it became clear that fetal tissue taken at abortion can be implanted successfully into the brains of Parkinson's patients and that the dopamine released halts the degenerative changes, a potential cure emerged on the horizon. But is the use of fetal tissues ethical? Does it not encourage unneccessary abortions in order to obtain dopamine-secreting tissue—or indeed other tissues for diseases like diabetes, for example? Again, these are issues that may profitably be debated elsewhere during this conference.

So far, I have drawn attention to ethical, legal and social issues of biomedical knowledge. But sometimes we misuse biomedical knowledge for scientific reasons. Biomedical knowledge is moving so fast—we have improved medical treatments factor during the last 50 years than during the preceding 500 years—that it is not surprising if accidential misuse—as opposed to deliberate misuse—sometimes occurs. Let me remind you of just two instances that have caused immense suffering and misery throughout the world. The first was caused by the drug thalidomide, which was given to pregnant mothers to alleviate morning sickness; many of the children of mothers given the drug were born without arms or legs and with other horrific deformations of their limbs, until use of the drug was halted.

The other instance concerns haemophilia. Sufferers from the disease—in which there is a failure of the blood-clotting mechanism as mentioned earlier—lack one of the proteins of the blood-clotting cascade, namely factor VIII. This has led to the administration of factor VIII isolated from spent blood obtained from transfusion laboratories. All worked well until the 1980s, when the HIV virus came along, and unsuspected by the producers of factor VIII, samples became contaminated with HIV. Many haemophiliacs—who are generally male as the gene for factor VIII, which is recessive, is on the X chromosome—became infected with HIV and subsequently died of AIDS. Many of these people were Japanese who had acquired contaminated blood from the USA where HIV-infected donors had given blood. There is an interesting side line to this particular misuse, which concerns the transmission of HIV. Very few of the spouses of the Japanese haemophiliacs who had become infected with HIV, became HIV positive—despite regular sexual intercourse. This supports the notion that at least in the developed world—HIV infection is transmitted through blood, like hepatitis virus, not through sexual intercourse. While we are on the subject of HIV infection, I cannot resist telling you of a social misuse of knowledge. A few years ago the good citizens of Madras (now called Chennai), the capital of the state of Tamil Nadu in india, decided in a spirit of good-will and social improvement, that they would rescue the girls from Tamil Nadu who were working the brothels of Bombay (now called Mumbai) and re-employ them in good jobs in their home state. Money was raised for the scheme, the girls were brought home to Tamil Nadu and changed their profession to more socially accepted occupations. Until that time, the state of Tamil Nadu had prided itself on having one of the lowest rates of HIV infection and AIDS in India. What happened when the girls from Mumbai arrived? Since most of them were heavily infected with HIV, the infection spread like wildfire as soon as the girls returned home, even though they were no longer carrying on their previous trade, and Tamil Nadu has gone from being one of the lowest areas of HIV infection and AIDS, to one of the highest. Such is the way that misuse of good—socially correct—intentions results in unforeseen consequences. Too much knowledge, and not enough wisdom.

Now let me give you some examples of the misuses of knowledge for scientific reasons that may turn against us in the future. The first example concerns organ transplants. There is such a shortage of human organs available, that many people have suggested the use of animal organs instead; the organs of a pig, like heart and liver, are the same size as they are in humans. Rejection of tissues —despite the use of immunosuppressants—is not necessarily a drawback. In the case of liver transplants occasioned by liver failure, for example, it does not matter too much if the transplanted liver is rejected after some months or years, for provided some undamaged liver tissue remains at the time of the transplant, this will eventually regrow to provide a fully working liver. The use of animal organs is itself an issue of concern to some people. But that is not the point I wish to make. Rather it is to draw attention to the possibility of transmitting some—as yet perhaps unrecognised—infection to the recipient through the transplanted organ. Such an infection might be a virus—remember HIV probably originated in a monkey—or it might be a prion, like bovine spongioform encephalitis (BSE). The same possibility exists in the case where human proteins have been genetically engineered to be made in the milk of animals like sheep—proteins such as antithrombin that are useful in the treatment of atherosclerosis have been produced in huge quantities in this way. Although the sheep used in this particular project were free of prion disease, having been specially imported from Australia, and although the offspring of these sheep continue to produce healthy milk containing the desired human protein, which undergoes careful purification and sterilisation before being given to patients, the admittedly rather remote possibility of some long-term infections cannot be ruled out.

An entirely different and in this case well-documented, example of the misuse of biomedical knowledge concerns vaccination. A common respiratory disease, especially in children, is caused by a virus called respiratory syncitial virus or RSV. Several studies have shown that vaccination of children against RSV exacerbates the disease, instead of preventing it. The molecular details are not yet fully understood, but is clear that vaccination induces an untoward inflammatory response. A common disease in The Gambia in Africa is trachoma, an eye infection that in this instance caused chlamydia. A vaccine against this particular microbe was prepared and used in a trial involving a large number of people. The result was unexpected and disappointing. Instead of achieving immunity against the disease, those who received the vaccine provided to be more, not less, susceptible to controlling chlamydial trachomitis. why? Because their innate immunity was suppressed not enhanced. The reason for this is probably the following. Protection against the disease is afforded by the type of immunoglobulin known as IgA, that is found in mucosal secretions, not by the major type of immunoglobulin known as IgG, that is present in the blood stream. Vaccination indeed increases the amount of IgG in the blood—as had been predicted—but this is ineffective against the chlamydial infection; worse, IgG is increased at the expense of IgA, so that the recipients of the vaccine are worse off than they would have been if they had never been treated. Vaccination against Dengue virus leads to similar quite unexpected results. These treatments are therefore good examples of the misuse of biomedical knowledge, with nothing but the best intentions in mind. I cite them only as they exemplify the saying ' a little knowledge is a dangerous thing' and causes us to pause at the crossroads between knowledge and wisdom. Many other examples of the misuse of scientific knowledge, for quite uninterntional reasons, could be given, but it is time to move to the second part of my talk, the lack of access to the fruits of biomedical knowledge in the developing and undeveloped countries, many of which typically occupy much of the southern hemisphere.

Three quarters of the world have to make do with but a fraction of the world's resources. We can subdivide these countries into two categories: those countries that represent the most advanced of the developing countries, with relatively vibrant economies and relatively good infrastructure and support for science, and those countries that are still far behind in this regard. In the first category we have countries like China, India and the countries of South East Asia like Thailand, Malaysis, Indonesia and the Philippines, some countries in Africa like South Africa and perhaps Kenya, and the countries of Latin America like Argentina, Brazil, Chile, Cuba (yes, Cuba), Mexico and Venezuela. To this list I have added the restructuring or redeveloping countries of Eastern Europe such as the Czech Republic, Hungary, Poland, Romania, Russia and the Ukraine. All the countries in this category are in a position, given some help, to catch up and indeed to overtake the countries of Europe, North America and the Far East.

In the other category are countries that are still some way behind: countries like Somalia and Ethiopia, Burundi and Rwanda. These countries do not need sophisticated biomedicine: they are food, hygiene and an honest government that helps its citizens instead of harassing them. But there are also countries that could move from the second category into the first category of the most rapidly developing countries. The constraints for this vary from country to country. Lack of resources is a common feature; lack of political will is another; religious fundamentalism is a third. I do not need to remind this audience that the major reason why many of the Islamic countries of the Middle East and Africa are no longer the leaders in the world of science and medicine as they were a thousand years ago is not so much lack of resources as lack of will—scientific endeavour requires a tolerant atmosphere look at what happened to Russian biology during the time of Lysenko—and fundamentalism is inimical to scientific advance; a lack of tolerance among fundamentalists is beginning to have detrimental effects on medicine—at least as far as access to abortion is concerned—even in certain of the otherwise enlightened United States of America. There is another point as far as many countries in the first category are concerned. It is that there is a wide gap between the standard of living, that includes access to medical knowledge and treatment between a small minority of the very rich and the vast majority of the very much more poor: countries like India and Brazil spring to mind. But this, like my earlier comments about fundamentalism, is a matter for the politicians not the scientists.

What, then, can be done to help the most deserving countries in terms of cost effectiveness, namely the rapidly developing countries in the first category, so far as biomedicine is concerned? I believe there are three distinct ways each of equal importance and potential benefit. Although all do involve financial considerations, simply giving money to the countries concerned is not the answer. The first productive way is to make access to Western discussions, seminars and conferences easier. I do not in any way wish to criticise the organisers of this excellent conference, but I note that there are very few scientists from Asia, Africa or Latin America present; that of course may have very much to do with the cost of bringing such people here. The second way is to promote the training, in Western countries, of the future leaders of biomedicine in their home countries. The last thing one wishes to encourage is an increase in the brain drain of the brightest scientists away from their countries of origin. But I note that most of the scientists from Latin America countries who receive training in the USA, do return. This is helped both by the infrastructure in those countries and by the enlightened view of supporting agencies such as the Pew Foundation, who send scientists back to their country of origin with a generous grant with which to set up their own laboratory on return. Countries like India and Russia, that lose their most able scientists to the West should take note. The third way involves the investment, by Western companies, in the research and development—including the filing of international patents—of commercially exploitable inventions and products; there is no reason why such investment should not include the country's own financial institutions; again, in countries like India and Russian Federation many local companies and institutions have the means to do this. It is a fallacy to think that applied science emerges only from fundamental science: often the reverse is true, so that promoting investment in applied science in developing countries can lead to the sustained growth of basic science that many developing countries desire.

Is there a model by which such help can be brought about? There is, and it is the institution of which I happen to be the Director. It is called The Oxford International Biomedical Centre and its mission is to improve health and the quality of life throughout the world by closing the gaps between the level of expertise within developed countries compared with developing ones—as well as by bringing science and its medical applications, and academia and industry, closer together. Although, like the very developing countries we seek to help, we are in need of funds ourselves, we have managed to begin to achieve some of our aims. The Oxford International Biomedical Centre is but one example—a very small one at that—of how a better distribution of biomedical knowledge can lead to growth and stability, and hence to peace. And with that final allusion tot he sponsoring organisation of this conference—the UNESCO International School of Science for Peace—I have come to the end of my allotted time and it remains only for me to thank my colleagues throughout the world for having given me the opportunity to look at science from new and different perspectives.

25

The Fragility of the Humoral and Cellular Immune Responses in Microbial Diseases

—V. Colizzi, Italy

1. The microbial world and its co-evolution with homo sapiens

Microbes (prokariotes) and mammals (eukariotes) have been cohabitants for millions of years and their interaction is still part of our evolution.

Microbes are evolving far more rapidly than Homo sapiens, adapting to changes in their environments by mutating, undergoing high-speed natural selection, or drawing plasmids and transposons from the vast mobile genetic lending library in their environments. On the other hand, mammals have developed by selection a variety of systems able to control microbes and their pathogenicity, allowing our coevolution with the microbial word. In particular, the Immune System (IS) has developed under micro-organism selection pressure and has acquired in mammals an enormous specificity in the recognition of chemical compounds, and in the ability to counteract any perturbation of the homeostasis of our body and then to protect from infections. This concept may be more easily understood by considering the analogy between the nervous system (NS) and the IS. The NS recognises, counteracts and maintains memory of physical stimuli such as light and colour, sound and word, temperature (heat and cold), and so on. Conversely, the IS specifically recognises chemical compounds such as lipid, polysaccarides and proteins. The highly specificity of the recognition of the NS and the IS is evidentiated by the ability of both systems to discriminate between words or proteins with single substitution like a vocable or an aminoacid. The entire repertoire of fine recognition of the IS is around 10 milliards distinct peptide recognition. Finally, both systems maintain memory, and these physical and chemical memories are essential tools for our evolution. The application of the concept of immunological memory is widely known in the developed countries as vaccination, but the naturally acquired immunological memory against the malaria antigen has allowed the evolution of the African population. The complexity of the immune systems, its cellular and molecular actors and their way of acting is still puzzling GOD, the Generator Of (immunological) Diversity.

In this talk I will elaborate the concept that the control of emerging or re-emerging infectious diseases and microbial warfare may be considered two faces of the same coin. Then I will also bring some element of knowledge necessary to understand the complexity of the interaction between microbial pathogens and the IS.

Finally, at the end of my talk I will present some initiatives that my University, the Ministry of Health and UVO are launching in the control of emerging infections that could be extended also to field of the biological warfare and may be integrated in the UNESCO International School of Science for Peace.

2. Analogies between microbial warfare and emerging and re-emerging infections

In the microbial world warfare is a constant. The survival of most organisms necessitates the demise of others. Yeasts secrete antibiotics to ward off attacking bacteria. Viruses invade the bacteria and commandeer their genetic machinery to viral advantage.

Further, every microbial pathogen is a parasite which survives by feeding off a higher organism. The parasites are themselves victims of parasitism. Like a Russian wooden doll—with in-a-doll, the intestinal worm is infected with bacteria, which are infected with tiny phage viruses. The whale has a gut full of algae, which are infected with *Vibrio cholerae*. Each micro-organism is another rivet in the Global Village.

Yet there are times of extraordinary collectivity in the microbial word, when the elbowing yields to combating a shared enemy. Swapping genes to counter an antibiotic threat or secreting a beneficial chemical inside a useful host to allow continued parasitic comfort is illustrative of this microscopic alliance.

Taking these lessons from the nature, now with genetic engineering is a simple matter to insert genes coding for virulence factors into the DNA or RNA of a simple virus, such as measles or influensa with high air transmission capacity. Misuse of genetic engineering may allow the development of bugs that would produce famine by wiping out crops, cause widespread veterinary disease, targeting rival's economy.

The relationship between infectious diseases and biological warfare is an old one, although in most of the cases micro-organisms take advantage of the stupidity of the *Homo sapiens* in doing conventional wars modifying their habit, their nutrition state and hence to be more susceptible to microbial attack. Alexander the Great is an historic example as micro-organisms and war are strictly associated, but the Spanish flu with millions of deaths during the I World War is a tragedy still present in our memory.

Apart from conventional war between distinct subsets of *Homo sapiens*, the microbial war is now even more possible considering that changing in our Global Village. The changes in the natural eco-system and nutrition, the large flows of population, impressive urbanisation and travelling are the principal factors that contribute to the emerging and re-emerging infectious diseases. Now we can say that an Army is ready to challenge *Homo sapiens*, and nobody knows how many deaths will leave on the field at the end of each single battle. Although several microbial diseases are under control, such as smallpox or poliomyelitis, other infectious diseases or syndromes, such as Tuberculosis, Malaria and AIDS are still not controlled at all in a relevant part of the world. Moreover, other micro-organisms are emerging from the reservoir of Nature and their identification and control is the context of our co-evolution with the microbial word. Obviously, the list is not complete and it will become old in few years. There is no time to discuss how these microrganisms start to be infectious, which role *Homo sapiens* play in the diffusion of the disease, and the methods used to control and eradicate some of them.

In this moment, in some part of our planet, in the Nature's laboratory of African deserts or Amazonian forests, new micro-organisms are developing. On the other hand, microbiologists, and molecular biologists are developing appropriate and powerful technologies to detect emerging or re-emerging diseases. Gene amplification technology and Polymerise Chain Reaction (PCR) allow to enormously increase the number of DNA or RNA sequences and are widely used to identify new micro-organisms. It has been suggested that these techniques and molecular approaches could be also used to finger new bio-weapons. This is additional evidence that emerging infection and biological weapons may be considered two faces of the same coin. The possibility to developed low cost PCR technologies is under investigation in my University.

3. The immune system as complex recognition defence network and new approaches for prevention and treatment of emerging and re-emerging infectious diseases

The subject of the title of this talk is fragility of the IS: what this means and how this fragility is related to the microbial world and to the functional and evolutionary complexity in the interaction between microbial pathogens and IS.

In fact, this interaction is considered by molecular pathologist "Pathogenetic mechanisms", and could be divided in virus dependent, host factors and virus induced, which could be also considered the players of this wargame between micro-organisms and *Homo sapiens*. Although this slide refers to HIV infection, similar mechanisms are operative in all infectious diseases.

The first level of activity of the IS is the recognition of micro-organisms, i.e. the antigen. Soluble molecules such antibodies and lymphocytes matured in the Thymus and hence called T cells recognises all possible chemical constituents of our word. While T cells with alpha/beta receptor recognises proteins in the context of our major histocompatibility complex (also called tissue transplantation antigens which mark the differences between individuals) and polysaccharides, other T cells carrying the gamma/delta receptor for antigen recognise phosfolipids. These non peptidic components are present in most of the micro-organisms which causes infectious disease and cancer. The interaction between alpha/beta and gamma/delta T cells is very strong and they play an integrated role in the immune response against micro-organisms. In particular, gamma/delta T cells are the first line of defence against micro-organisms and they act directly be killing the infected or transformed cancer cells, but also indirectly by enhancing the development and activation of protein specific alpha/beta T cells and macrophages.

The well known practical application of the IS in bio-medicine and bio-technology is the vaccination. It is more than one hundred years ago that the vaccination procedures with whole bacteria or viruses started to be introduced in the healthy national site maps in the developed countries. New bio-technological vaccines based on DNA are in progress in several laboratories. A supervaccine for several infectious diseases (transdiseases vaccinology) is part of the European Project in Bio-technology and represents the hope of our generation of scientists. In this cartoon is drawed the concept of using a non pathogenic mycobacteria *M. bovis,* strain BCG, widely used for Tuberculosis vaccination, genetically modified to insert eterlogous DNA sequences coding for other pathogenic micro-organisms such as Leishmania or *M. leprae*. Similar approaches are under investigation using viral vector or diirectly nude DNA molecules. Other approaches targeting the gamma/delta T cells are also in progress in our laboratory to push the natural immunity. This because gamma/delta i) have 10-10,000 higher "precursor frequency" in comparison to conventional alpha/beta T cells responding to proteins, ii) lack genetic restriction, and iii) more relevant for infectious diseases cross-react against viruses, bacteria, protozoa and tumour associated antigens.

Just yesterday the Italian Government has decided to enter into and financially support the Eurofighter Programme, a joint project with other European countries to develop a new series of war aircraft. Why these countries do not develop specific programmes for Vaccine Development? Most of the scientific basis for a new series of vaccines are ready to be exploited and transferred to industries for more advanced investigations on their practical development.

But also Vaccine Development is not sufficient. I will give you a dramatic example. In the field of AIDS, we have considered that the scientific and industrial research has brought to the development of safe and effective treatments of HIV infection capable to inhibit virus replication and to prolong significantly life expectancy. Nevertheless, it has been calculated that 1 month of full treatment with anti-HIV drugs such as reverse transcriptase and protease inhibitors cost several thousand US dollars, while the average amount of money available in developing countries for health care is about 10 dollars per person per year. Considering that only in Africa 14 million people are infected by HIV, this make mandatory the development of research aimed to the discovery and evaluation of alternative drugs that for their limited cost, effectiveness, and acceptance in countries with different cultures can be successfully used. The World Foundation for AIDS Research and Prevention co-founded by the UNESCO Director General Federico Mayor and the discoverer of HIV Professor Luc Montagnier recently signed an agreement with Nelson Mandela, President of South Africa, to develop a network between scientists and clinicians from Africa, and Europe to test natural compounds present in African plants that have been recently found active against HIV in vitro. Unfortunately, we have not sufficient time today to analyse these aspects, but it is important to consider that the Simian Immunodeficiency Virus (SIV) which infect African monkeys and is structurally and functionally similar to HIV, but does not cause epidemy in monkeys perhaps due to the presence of factors in Africa and in their nutritional habitat able to establish a correct relationship between virus and the host. Of course, several laboratories could join in this international programme such as the International Centre for Genetic Engineering and Biotechnology in Trieste and the Russian Centre for Virology and Bio-technology VECTOR in Novosibirsk or the State Research Centre for Applied Microbiology near Moscow.

4. What is the University of Tor Vergata doing?

With this background, the University of Tor Vergata and UNESCO are supporting two new initiatives in the field of emerging diseases and their control. The first is the establishment of a UNESCO Chair in Biotechnology devoted to the development of a specific Vaccine and Natural Drug Project for emerging diseases and cancer. In the next three years, these chairs will call in Rome, foreign professors in distinct disciplines (immunology, molecular biology, virology, economy, bioethics) which will develop a draft of a general plan for vaccine and natural drug development to be submitted to UNESCO and to the Italian Government.

The second initiative that has been recently signed by the Director General Federico Mayor and just few days ago sponsored by the Italian Minister of Health, Hon. Rosy Bindi is the institution of the International Centre for Emerging and Re-emerging Infectious Diseases in connection with the Institute for Scientific Research Hospital Spallanzani in Rome. The possibility to integrate this new Center in the UNESCO School Science for Peace is under consideration with UVO and Landau Network.

26

The WHO Military Network for Surveillance and Control of Emerging and Other Communicable Diseases

R.D' Amelio, Italy

Introduction

Infectious diseases represent a heavy burden for mankind. In 1995 infectious diseases were responsible for more than 17 million (33 per cent) of nearly 52 million deaths worldwide. Among these infectious diseases, 65 per cent were those characterised by person-to-person transmission, such as the sexually and air-bore transmitted diseases; followed by food, water and soil-borne (22 per cent), insect-borne (13 per cent) and animal-borne (0.3 per cent) diseases. They include, in order of importance, acute respiratory infections (4.4 million deaths/year), followed by diarrhoeal diseases (3.1 million deaths/year), tuberculosis (3.1 million deaths/year) malaria (2.1 million deaths/year), hepatitis B (1.1. million deaths/year), HIV/AIDS (>1 million deaths/year), measles (>1 million deaths/year), neonatal tetanus (500,000 deaths/year), whooping cough (355,000 deaths/year) and lastly roundworm and hookworm (165.000 deaths/year).[1]

During the past twenty years it has been shown repeatedly that emerging and reemerging infectious diseases and related public health problems are a challenge which has not been successfully met. For example twenty years ago Legionnaires' disease and Ebola virus were first identified, while at the same time gonococcal resistance to penicillin and chloroquine-resistant malaria were becoming widely dispersed in Southeastern Asia. Though HIV infection was still unknown twenty years ago, retrospective analysis shows that it was already present in Africa in 1976, with a seroprevalence of less than 1 per cent. At the same time, however; smallpox still was eradicated, clearly demonstrating the dynamic state of infectious diseases.

Now, twenty years later, Legionnaires' disease and HIV infection are known to be worldwide and continue to spread, and Ebola haemorrhagic fever has occurred periodically, with human outbreaks in Zaire, Sudan and Gabon; and animal outbreaks in Côte d'Ivoire, Italy and the United States of America. By 1996 gonococcal resistance to penicillin and chloroquine-resistant malaria had spread worldwide, while their resistance to other antimicrobial agents had also developed and spread. Current examples of other emerging and reemerging infectious diseases are the spread of diphtheria in Eastern European countries in the early 1990s, of pulmonary Hantavirus in USA in 1993, of cholera which reemerged in Latin America in 1991 after decades of absence, and the appearance of a new subtype of Vibrio cholerae, 0139, in India and Myanmar in 1993.[2]

The causes of the present situation in infectious diseases are several, including urbanisation (about 45% of the world's total population live in urban settlements, with inadequate provision for water, sanitation and garbage collection), increase in travel and trade (the number of total air-travellers exceeded 1,200 million in 1994) and the breakdown of public health practices in both, developing and industrialised countries, the latter linked to the erroneous idea in the 1960s and early 1970s that infectious diseases no longer represented a danger for public health.[1]

Beside the increase in naturally occurring infectious diseases, the strong development of bio-technologies in the past few years allowed to make readily available large quantities of biological agents, that can be offensively used by States and/or terrorist groups. Several signals in the last few years focused the international attention to the threat that biological weapon may actually be used. First, the Russian Government in 1992 disclosed that the former Soviet Union, despite being a co-depositary of the Biological and Toxin Weapons Convention (BTWC), had continued an offensive programme until 1992. Second, in late 1995, the United Nations Special Commission (UNSCOM) on Iraq had obtained the disclosure by Iraq that, despite being a signatory of the convention, it had developed biological and toxin weapons by the start of the Gulf War in 1991. Third, the Aum Shinrikyo sect in Japan, which used nerve gas in attacks in the Tokyo subway in March 1995, had been actively seeking a biological weapons capability.[3] The strategic response of the World Health Organisation (WHO) to the above reported situation has been focused on global and national surveillance, alert and control of infectious diseases. At a series of meetings conducted by WHO in 1994 and 1995, four goals intended to guide the worldwide response to emerging, re-emerging and other communicable diseases and public health problems were defined as follows:

Goal I: To strengthen the global surveillance of communicable diseases;

Goal II: To strengthen the national and international infrastructure necessary to recognise, report, and respond to emerging communicable diseases;

Goal III: To strengthen national and international capacity for the prevention and control of communicable diseases;

Goal IV: To support and promote research in communicable disease control.

These goals are similar to those adopted by the United States of America[4,5] Canada, Thailand and other countries worldwide. Inadequacies in national surveillance programmes were dramatically demonstrated during the recent outbreaks of plague in India (1994) and of Ebola in Zaire (1995), where a more effective and prompt detection and diagnostic capacity would possibly have limited the dimensions of outbreaks.

To implement global surveillance and alert, WHO builds on existing networks, such as the WHO Collaborating Centres, specialised national laboratories and institutions with expertise in infectious disease diagnosis and epidemiology. Who is strengthening its collaborating centre system by encouraging governments to provide the resources necessary to ensure that collaborating centres are up to date, facilitating exchange of information and reagents among its centres, increasing the number of centres in developing countries and ensuring that all centres are linked electronically and are regularly exchanging information with WHO and among themselves.

A second global monitoring system is based on the WHO network for monitoring and containing antimicrobial resistance. WHO is currently developing a network of laboratory centres in developing countries to perform national antimicrobial resistance testing using standard quality-controlled procedures, participate in the WHO proficiency testing programme and regularly report their results nationally and to WHO. Through this network WHO assists countries in using test results for sound national drug polices and uses the information internationally to demonstrate the problem and advocate for research and development on antibiotics. A back of well-characterised resistant strains of micro-organisms being established for research and development.

A third system is the international Health Regulation (IHR), currently the only international public health legislation which requires mandatory reporting of infectious diseases: cholera, plague and yellow fever. To transform the IHR into a working global alert system, the WHO is revising the IHR to focus on the reporting of clinical syndromes of potential worldwide importance for which standard and internationally accepted responses by countries will be promoted. Syndromic reporting will be followed by etiologic reporting once the diagnosis is known at which time modifications in the response may be made. Accompanying the revision is the development of clear and concise guidelines, which describe appropriate and inappropriate responses once a syndrome is reported.

WHO proposes involving military facilities in these systems and a military involvement is currently based at WHO (R.D'Amelio). Justification for liaison with military laboratories in these monitoring and alert systems is that 1) the military is at special risk for infectious diseases; 2) military laboratories in some developing countries are better equipped than civilian laboratories.

The WHO military liaison officer began collaboration in April 1995, with the goal of creating a network of global surveillance in the military, parallel and complementary to those WHO. The creation of network began with a survey among WHO Member States to identify countries committed to collaboration through a military laboratory capable of diagnosing common infectious and/or notifying infectious diseases in a national reporting system. This was followed by a second survey among military laboratories willing to collaborate in which their diagnostic and reporting practices were assessed. The third stage will be training workshops in developing countries, and the establishment of electronic links for exchange of information.

The survey

Until now, out of 107 countries to which an adhesion to the project has been requested, 76 have replied, involving 25/34 (73%) from Africa, 14/28 (50%) from Asia, 9/17 (53%) from the Americas, 26/26 (100%) from Europe, Australia and New Zealand (**table 1**). A military laboratory able to diagnose endemic infectious diseases and to recognise the presence of unusual ones is present in 19/25 (76%) African countries, in 17/26 (68%) European countries, in 5/9 (55%) American countries, in 12/14 (86%) Asian countries and it is absent in Oceania. A military notification system for infectious diseases is present in 19/25 (76%) African countries, in 24/26 (92%) European countries, in 7/9 (78%) American countries, in 11/14 (79%) Asian countries and in 1/2 Oceanian countries.

TABLE I: PRESENCE OF A MILITARY LABORATORY AND/OR OF A MILITARY NOTIFICATION SYSTEM FOR INFECTIOUS DISEASES.

Continent	*(%) countries replied*	*(%) with Military Lab*	*(%) with Military Notification System*
Africa	25/34 (73)	19/25 (76)	19/25 (76)
Americas	9/17 (53)	5/9 (55)	7/9 (78)
Asia	14/28(50)	12/14 (86)	11/14(79)
Europe	26/26(100)	17/26 (68)	24/26 (92)
Oceania	2/2 (100)	0/2 (0)	1/2 (50)
Total	76/107 (71)	53/76 (70)	62/76 (82)

Six countries among the 76 which replied do not notify their National Health Services about infectious diseases occurring in the military environment.

Present situation

Additional questionnaires were sent to the 76 countries which had already replied, asking for more detailed information about the diagnostic capabilities of the laboratories, the characteristics of the notification system, the mandatory diagnostic schedule for infectious diseases on recruitment if present, and the actual vaccination schedule.

Fifty-two countries replied: 15 from Africa, 5 from the Americans, 7 from Asia, 23 from Europe and 2 from Oceania (table II, fig.I.)

Military Laboratories

Among these 52, 39 have a military laboratory to diagnose infectious diseases, 24 of whom declare to be able to undertake at least 4 of the following activities; bacteriology, Virology, Parasitology, Immunology and Molecular Biology (table II, fig.2). Thirty-five out of 52 declare their willingness to participate in the WHO antibiotic resistance monitoring and containment programme.

TABLE II: RANGE OF ACTIVITIES PERFORMED IN THE MILITARY LABORATORIES AND STATE OF COMPUTERISATION/ POTENTIAL FOR DUPLICATE REPORTING

Continent	N° of countries replied/total		N° of activity areas			Computerised	No system to avoid duplicate notification
			All	1-4	No		
Africa	15/25	(60%)	1	12	2	5	7
Americas	5/9	(55%)	1	3	1	5	1
Asia	7/14	(50%)	3	4	0	2	4
Europe	23/26	(88%)	3	12	8	14	3
Oceania	2/2	(100%)			2	1	1
Total	52/76	(68%)	8	31	13	27	16

FIG.1. COUNTRIES WHICH REPLIED TO THE 4 QUESTIONNAIRES

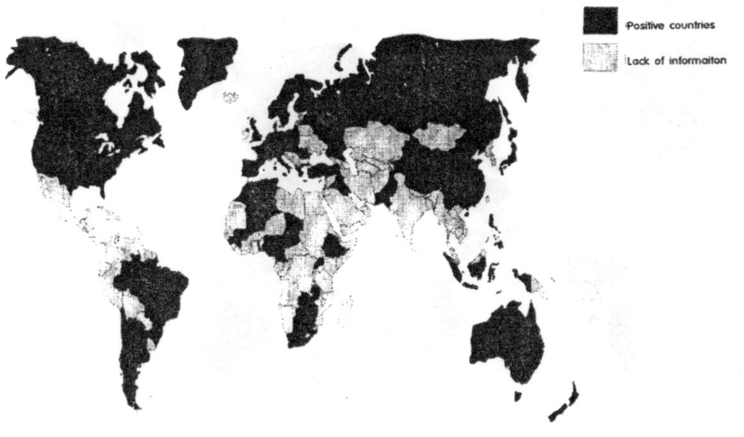

FIG.2. ACTIVITIES OF MILITARY LABORATORIES BACTERIOLOGY, VIROLOGY, PARASITOLOGY, IMMUNOLOGY, MOLECULAR BIOLOGY

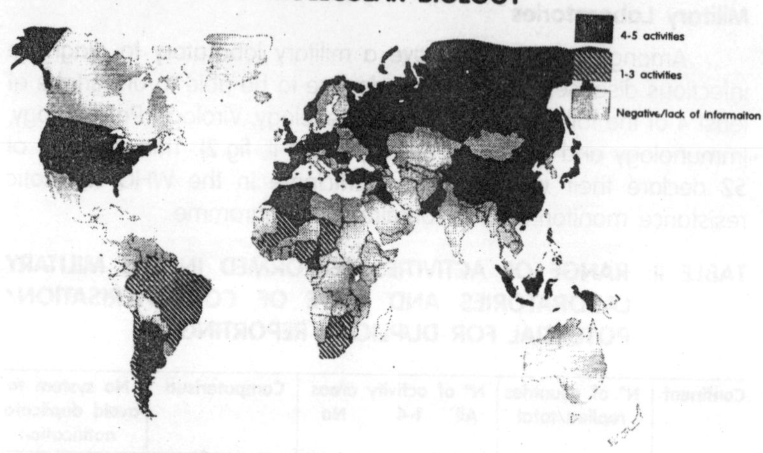

Military Notification System

Twenty-seven (52%) of the countries are equipped with a computerised network for surveillance and 16 countries (31%) report not having a system to avoid duplicate notification of the same disease by the National Civilian Health Service (table II, fig.3).

FIG.3. MILITARY NOTIFICATION SYSTEM FOR INFECTIOUS DISEASES

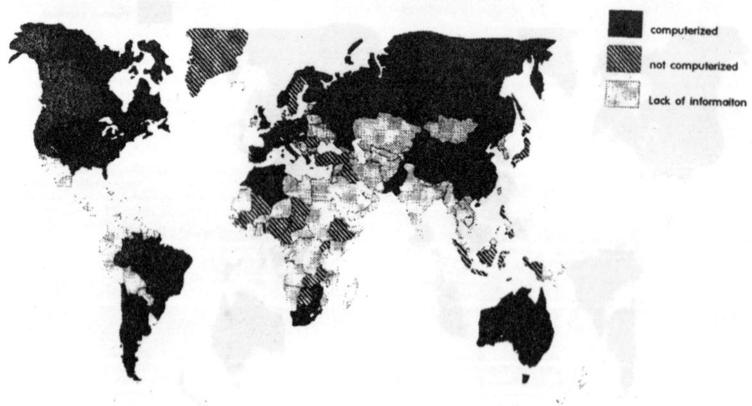

FIG.4. COMPULSORY SCREENING FOR TBC/SYPHILIS ON RECRUITMENT

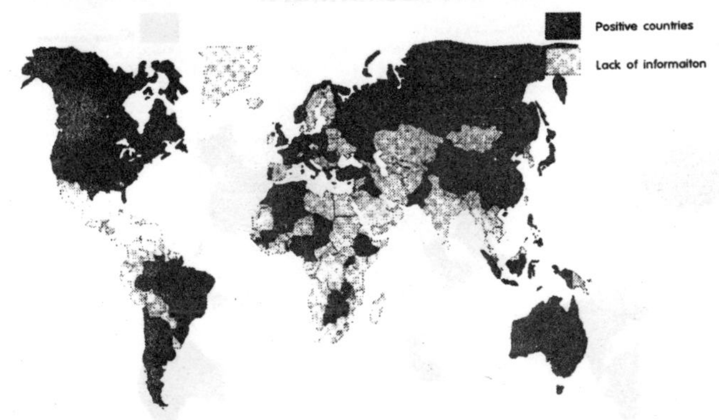

Recruitment Screening

Eighty-three per cent of the countries perform mandatory screening on recruitment for tuberculosis and/or syphilis (fig.4), 27 (52%) for HIV infection, while some screen for other viral diseases, such as HAV, HBV and/or HCV (one country in Africa requires screening for HAV and HBV, but not for HIV) (fig.5). Thirty-eight per cent perform a mandatory screening on recruitment for intestinal, urinary and/or blood-borne parasitic diseases (table III, fig.6).

TABLE III: MANDATORY SCREENING AND VACCINATION ON RECRUITMENT

Continent	Screening on recruitment			Vaccination schedule
	Viral	Tub/Syph	Paras	T/d/Ty/BCG/P/Mn/MMR
Africa	12	14	9	11/9/10/5/2/3/3
Americas	4	5	2	4/3/3/0/2/1/2
Asia	5	7	5	5/2/2/3/2/2/1
Europe	6	16	4	23/14/8/7/8/6/3
Oceania	1	1	0	2/2/0/1/2/0/1
Total	28	43	20	45/30/23/16/16/12/10

Immunisation Requirements

Five of the 52 countries have no mandatory general immunisation schedule; of the remaining 47, 45 require immunisation against tetanus toxoid (fig.7), 30 diphtheria toxoid (fig.8), 23 typhoid fever, 16 BCG, 16 polio, 12 meningococcal meningitis and 10 trivalent measles/mumps/rubella (table III). Two countries still report vaccinia vaccination as compulsory for military recruits.

FIG.5. COMPULSORY SCREENING FOR VIRAL INFECTIONS ON RECRUITMENT

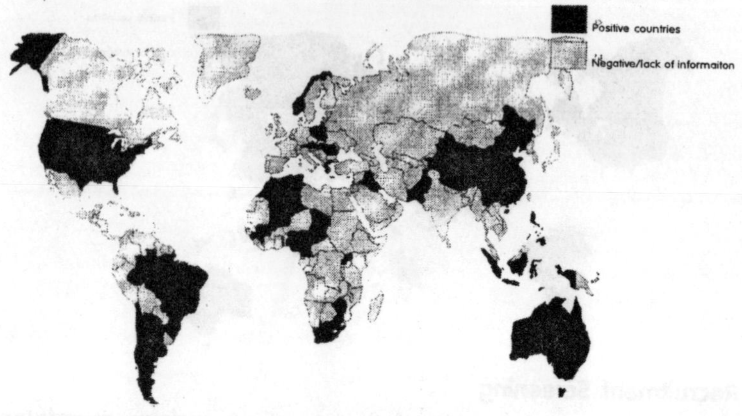

FIG.6. COMPLUSORY SCREENING FOR PARASITIC DISEASES ON RECRUITMENT

Future directions

WHO is making plans to include some developing country military laboratories in training courses on bacteriology, with the aim to bringing them into the network of antibiotic resistance monitoring. This will be the first step to allow the laboratories to actually work together. Once organised, the military network for surveillance of infectious diseases and antimicrobial resistance will be useful for detecting and monitoring both naturally occurring or deliberately caused outbreaks of infectious diseases.

FIG.7. COMPULSORY VACCINATION SCHEDULE IN THE MILITARY—TETANUS

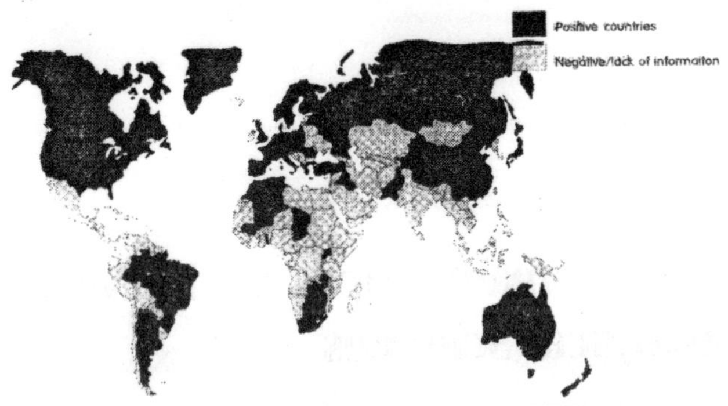

FIG.8. COMPULSORY VACCINATION SCHEDULE IN THE MILITARY—DIPHTERIA

References

1. The World Health Report 1996. Fighting disease. Fostering development. World Health Organisation, Geneva, 1996.
2. Who, Fact Sheet N°122, June 1996. Cities and emerging and re-emerging diseases in the XXIst Century.
3. Dando MR, Pearson GS: Controlling Biological Weapons. The fourth Review Conference of the Biological and Toxin Weapons Convention: Issues, Outcomes, and Unfinished Business. PLS 1997; 16:105-126.
4. Emerging Infections. Microbial Threats to Health in the United States. Edited by: Lederberg J., Shope R.E. and Oaks S.C., jr. Institute of Medicine. National Academy Press, Washington DC, 1992.
5. CDC. Addressing Emerging Infectious Diseases Threats. A prevention strategy for the United States US Department of Health and Human Services. Public Health Service. Centres for Disease Control and Prevention. Atlanta, Georgia, 1994.

27

Possible Consequences of the Misuse of Biological Sciences

The Biological and Toxin Weapons Convention (BWC), which bans the development, production, acquisition, stockpiling and retention of biological (toxin) weapons, entered into force on 26 March 1975. Up to now, (1977) the BWC has been ratified by 140 countries and signed by another 18 States. However this treaty lacks means to verify compliance or to detect non-compliance. The absence of verification provisions is typical of Cold War arms control treaties. Unfortunately the problem of biological weapons proliferation has become more and more complex since the BWC entered into force in 1975, so the lack of verification provisions has demonstrated the weakness of the treaty.

Dual-use technologies, materials and equipment have spread around the globe as the biotechnology and pharmaceutical industries have grown. The community of researchers generating the scientific and technical advances in biotechnology genetic engineering, and other related disciplines has also expanded. In short, new frontiers have opened for those seeking biological weapons, including the possibility of manipulating genes to create novel, treatment and detection-resistant biological agents.

The widespread presence of dual-use technologies, equipment, and materials in countries around the globe makes monitoring of the Biological Weapons proliferation extremely difficult. Telltale signs of a covert biological weapons programme are scarce, and even discrete signs of a covert weapons programme can be hidden if a government is willing to pursue germ warfare in antiquated facilities without such modern safety precautions such as special containment facilities and worker vaccination.

In addition, clean-in-place technologies can quickly eliminate traces of biological agent that inspectors might detect, even if they arrive within hours. While useful, inspections of such facilities might only yield a fleeting impression of what these plants are doing while the inspectors are there, but will result in little confidence of what was going on shortly before or soon after the inspections. The nature of dual-use biological equipment and scientific capabilities is such that not even rigorous shortnotice inspections can ensure high confidence of compliance with the BWC's prohibitions.

In order to strengthen the treaty, in September 1994 the member-states of the BWC have agreed to establish a new negotiating forum —to follow up on the work of the VEREX group—called the Ad Hoc Group (AHG), open to all state-parties, to develop a legally binding protocol to the BWC that specifies the establishment of a biological arms control regime.

Possible elements of the future regime include:

1) confidence building measures (CBM) and transparency measures (TM),
2) measures to promote compliance, including "on-site investigational activities,
3) measures to implement Article X of BWC, i.e. peaceful scientific and technological international co-operation.

Even if the goal to provide a workable verification regime to the BWC is not a simple one due to the enormity of the task facing the AHG, widespread agreement exists that some type of declarations will be needed as an informational foundation for any on-site monitoring activity. At the very least, submission of the data that some countries had been providing since 1987 for confidence-building purposes would become mandatory for all state-parties. Beyond that, the extent and format of any data declarations are still a matter of debate.

The participants of the UNESCO International School of Science for Peace on Possible Consequences of the Misuse of Biological

Sciences at Como, Italy (3-6 December 1997) organised by the Landau Network-Centro Volta (LNCV) with the collaboration of the International Centre for Genetic Engineering and Biotechnology (ICGEB) and of the UNESCO-Hebrew University of Jerusalem-International School for Molecular Biology and Microbiology (HUJ-ISMBM), with the support of the Italian Ministry of Foreign Affairs, of the Lombardia Region and of the Municipality of Como, have identified the following important issues and recommendations:

a) the necessity of routine inspections compatible with the safeguard of highly confidential business and security information in such way as to safeguard intellectual property information;
b) the key role played by the CBMs in building confidence in nation's compliance with the prohibitions of the BWC and as an instrument to prepare countries to join "security regimes"—such as arms control and disarmament—in the area of the biological/toxin weapons;
c) the long-term benefits, and not only in economic concerns, for the compliance protocol provided by fulfilling the goals of BWC's Article X, under which parties undertake to exchange biotechnology know-how, materials and equipment for peaceful purposes.

Concerning CBMs we stress the need of:

a) legally binding declarations that cover facilities and activities particularly suited for an offensive biological warfare capability;
b) supervision of publications and of legislation, i.e. BWC state parties are required to undertake legislation making the possession or transfer of biological/toxin weapons a crime;
c) reports on biological materials and equipment transfer and on requests concerning their transfer;
d) field investigations of human disease outbreaks possibly associated with the covert use of biological/toxin weapons or accidental leak from a clandestine development or production facility;
e) information on pharmaceutical and vaccine production, biosafety capabilities and procedures.

Concerning the spirit of the BWC's Article X we also recommend:

a) to promote an extensive assistance and co-operation between the scientific/academic community and pharmaceutical industries to support the goals and objectives of the BWC. In this regard these communities should have a "seat" at the negotiating table in Geneva in order to facilitate the conclusions of a meaningful and feasible monitoring protocol for the BWC;

b) to encourage in the area of applied microbiology (biotechnology) mutual visits between scientists, participation in international scientific meetings, and funding of joint research projects involving laboratories of all countries. In addition, biological science and biotechnology researchers from countries suspected to support biological weapons programmes should be invited to international conferences and workshops including ethical debates. The UNESCO International School of Science for Peace-LNCV and the ICGEB might lead such an effort offering a permanent link between scientific/academic communities.

28

Universal Declaration on the Human Genome and Human Rights

The General Conference,

Recalling, that Preamble of UNESCO's Constitution refers to {the democratic principles of the dignity, equality and mutual respect of men", rejects all "doctrine of the inequality of men and races", stipulates "that the wide diffusion of culture, and the education of humanity for justice and liberty and peace are indispensable to the dignity of men and constitute a sacred duty which all the nations must fulfil in a spirit of mutual assistance and concern", proclaims that "peace must be founded up on the intellectual and moral solidarity of mankind", and states that the Organisation seeks to advance "through the educational and scientific and cultural relations of the peoples of the world, the objectives of international peace and of the common welfare of mankind for which the United Nations Organisation was established and which its Charter proclaims",

Solemnly recalling its attachment to the universal principles of human rights, affirmed in particular in the Universal Declaration of Human Rights of 10 December 1948 and in the two International United Nations Covenants on Economic, Social and Cultural Rights and on Civil and Political Rights of 16 December 1966, in the United Nations Convention on the Prevention and Punishment of the Crime of Genocide of 9 December 1948, the International United Nations Convention on the Elimination of All Forms of Racial Discrimination of 21 December 1965, the United Nations Declaration on the Rights of Mentally Retarded Persons of 20 December 1971, the United Nations Declaration on the Rights of Disabled Person of 9 December 1975, the United Nations Convention on the Elimination of All Forms of Discrimination Against Women of 18 December 1979, the United Nations Declaration of Basic Principles of Justice for Victims of Crime and Abuse of Power of 29 November 1985, the United Nations Convention on the Rights of the Child of 20 November 1989, the United Nations Standard Rules on the Equalisation of Opportunities for Persons with Disabilities of 20 December 1993, the Convention on the Prohibition of the Development, Production and Stockpiling of Bacteriological (Biological) and Toxin Weapons and on their Destruction of 16 December 1971, the UNESCO Convention against Discrimination in Education of 14 December 1960, the UNESCO Declaration of the Principles of International Cultural Cooperation of 4 November 1966, the UNESCO Recommendation on the Status of Scientific Researchers of 20 November 1974, the UNESCO Declaration on Race and Racial Prejudice of 27 November 1978, the ILO Convention (No.111) concerning discrimination in Respect of Employment and Occupation of 25 June 1958 and the ILO Convention (No. 169) concerning, Indigenous and Tribal Peoples in Independent Countries of 27 June 1989;

Bearing in mind, and without prejudice to, the international instruments which could have a bearing on the applications of genetics in the field of intellectual property, *inter alia,* the Bern Convention for the Protection of Literary and Artistic Works of 9 September 1886 and the UNESCO Universal Copyright Convention of 6 September 1952, as last revised in Paris on 24 July 1971, the Paris Convention for the Protection of Industrial Property of 20 March 1883, as last revised at Stockholm on 14 July 1967, the Budapest Treaty of the WIPO on International Recognition of the Deposit of Micro-organisms for the Purposes of Patent Procedures of 28 April 1977, and the Trade Related Aspects of Intellectual Property Rights Agreement (TRIPS) annexed to the Agreement establishing the World Trade Organisation, which entered into force on 1st January 1995;

Bearing in mind also the United Nations Convention on Biological Diversity of 5 June 1992 and emphasising in that connection that the recognition of the biological diversity of humanity, shall not give rise to any interpretation of a social or political nature which could call into question "the inherent dignity and (...) the equal and inalienable rights of all members of the human family", in accordance with the Preamble to the Universal Declaration of Human Rights;

Recalling 22 C/Resolution 13.1, 23 C/Resolution 13.1, 24 C/ Resolution 13.1, 25 C/Resolutions 5.2 and 7.3, 27 C/Resolution 5.15 and 28 C/Resolutions 0.12, 2.1 and 2.2, urging UNESCO to promote and develop ethical studies, and the actions arising out of them on the consequences of scientific and technological progress in the fields of biology and genetics, within the framework of respect for human rights and freedoms;

Recognising that research on the human genome and the resulting applications open up vast prospects for progress in improving the health of individuals and of humankind as a whole, but emphasising that such research should fully respect human dignity, freedom and human rights, as well as the prohibition of all forms of discrimination based on genetic characteristics;

Proclaims the principles that follow and *adopts* the present Declaration.

A. Human Dignity and the Human Genome

Article 1

The human genome underlies the fundamental unity of all members of the human family, as well as the recognition of their inherent dignity and diversity. In a symbolic sense, it is the heritage of humanity.

Article 2

a) Everyone has a right to respect for their dignity and for their human rights regardless of their genetic characteristics.
b) That dignity makes it imperative not to reduce individuals to their genetic characteristics and to respect their uniqueness and diversity.

Article 3

The human genome, which by its nature evolves, is subject to mutations. It contains potentialities that are expressed differently according—to each individual's natural and social environment including—the individual's state of health, living conditions, nutrition and education.

Article 4

The human genome in its natural state shall not give rise to financial gains.

B. Brights of the Persons Concerned

Article 5

a) Research, treatment of diagnosis affecting an individual's genome shall be undertaken only after rigorous and prior assessment of the potential risks and benefits pertaining thereto and in accordance with any other requirement of national law.

b) In all cases, the prior, free and informed consent of the person concerned shall be obtained. If the latter is not in a position to consent, consent or authorisation shall be obtained in the manner prescribed by law, guided by the person's best interest.

c) The right of each individual to decide whether to be informed or not on the results of genetic examination and the resulting consequences should be respected.

d) In the case of research, protocols shall in addition, be submitted for prior review in accordance with relevant national and international research standards or guidelines.

e) If according to the law a person does not have the capacity to consent research affecting his or her genome may only be carried out for his or her direct health benefit, subject to the authorisation and the protective conditions prescribed by law. Research which does not have an expected direct health benefit may only be undertaken by way of exception, with utmost restraint, exposing the person only to a minimal risk and minimal burden and if the research is intended to contribute to the health benefit of other persons in the same age category or with the same genetic condition, subject to the condition prescribed by law, and provided such research is compatible with the protection of the individual's human rights.

Article 6

No one shall be subjected to discrimination based on genetic characteristics that is intended to infringe or has the effect of infringing human rights fundamental freedoms and human dignity.

Article 7

Genetic data associated with an identifiable person and stored or processed for the purposes of research or any other purpose must be held confidential in the condition foreseen by law.

Article 8

Every individual shall have the right, according to international and national law, to just reparation for damage sustained as a direct and determining result of an intervention affecting his or her genome.

Article 9

In order to protect human rights and fundamental freedoms, limitations to the principles of consent and confidentiality may only be prescribed by law, for compelling reasons within the bounds of public international law and the international law of human rights.

C. Research on the Human Genome

Article 10

No research or its applications concerning the human genome, in particular in the fields of biology, genetics and medicine, should prevail over the respect for human rights, fundamental freedoms and human dignity of individuals or, where applicable, of groups of people.

Article 11

Practices which are contrary to human dignity, such as reproductive cloning of human beings, shall not be permitted. States and competent international organisations are invited to cooperate in identifying such practices and in determining, nationally or internationally, appropriate measures to be taken to ensue that the principles set out in this Declaration are respected.

Article 12

a) Benefits from advances in biology, genetics and medicine, concerning the human genome, shall be made available to all with due regard to the dignity and human rights of each individual.

b) Freedom of research, which is necessary to the progress of knowledge, is part of the freedom of thought. The applications of research, including those in biology, genetics and medicine, concerning the human genome, shall seek to offer relief from suffering and improve the health of individuals and humankind as a whole.

D. Conditions For the Exercise of Scientific Activity

Article 13

The responsibilities inherent to the activities of researchers, including meticulousness, caution, intellectual honesty and integrity in carrying out their research as well as in the presentation and utilisation of their findings, should be the subject of particular attention in the framework of research on the human genome, because of the ethical and social implications. Public and private science policy-makers also have particular responsibilities in this respect.

Article 14

States should take appropriate measures to foster the intellectual and the material conditions favourable to freedom in the conduct of research on the human genome and to consider the ethical, legal, social and economic implications of such research, on the basis of the principles set out in this Declaration.

Article 15

States should take appropriate steps to provide the framework for the free exercise of research on the human genome with due regard for the principles set our in this Declaration, in order to safeguard respect for human rights, fundamental freedoms and human dignity and to protect public health. They should seek to ensure that research results are not used for non peaceful purposes.

Article 16

States should recognise the value of promoting, at various levels as appropriate, the establishment of independent, multi disciplinary and pluralist ethics committees to assess the ethical, legal and social issues raised by research on the human genome and its applications.

E. Solidarity and International Cooperation

Article 17

States should respect and promote the practice of solidarity towards individuals, families and population groups who are particularly vulnerable to or affected by disease or disability of a genetic character. They should foster *inter alia* research on identification, prevention and treatment of genetically-based and genetically-influenced diseases, in particular rare as well as endemic diseases which affect large numbers of the world's population.

Article 18

States should make very effort, with due and appropriate regard for the principles set out in this Declaration, to continue fostering the international dissemination of scientific knowledge concerning the human genome, human diversity and genetic research and, in that regard, to foster scientific and cultural cooperation, particularly between industrialised and developing countries.

Article 19

a) In the framework of international cooperation with developing countries, States should seek to encourage that:

   i) the assessment of the risks and benefits pertaining to research on the human genome is ascertained and abuse is prevented;

   ii) the capacity of developing countries to carry out research on human biology and genetics, taking into consideration their specific problems, is developed and strengthened,

   iii) developing countries can benefit from the achievements of scientific and technological research so that their use in favour of economic and social progress can be to the benefit of all;

   iv) the free exchange of scientific knowledge and information in the areas of biology, genetics and medicine is promoted.

b) Relevant international organisations shall support and promote the measures taken by States for the aforementioned purposes.

F. Promotion of the Principles Set Out in the Declaration

Article 20

States should take appropriate measures to promote the principles set out in the Declaration, through education and relevant means, including—inter alia through the conduct of research and training in interdisciplinary fields and through the promotion of education in bioethics, at all levels, in particular addressed to those responsible for science policies.

Article 21

States should take appropriate measures to encourage other forms of research, training and information dissemination conducive to raising the awareness of society and all of its members of their responsibilities regarding the fundamental issues relating to the defence of human dignity which may be raised by research in biology, in genetics and in medicine, and the applications thereof. They should also undertake to facilitate on this subject an open international discussions ensuring the free expression of various socio-cultural religious and philosophical opinions.

G. Implementation of the Declaration

Article 22

States should make every effort to promote the principles set out in this Declaration and should, by means of all appropriate measures, promote their implementation.

Article 23

States should take appropriate measures to promote, through education, training, and information dissemination, respect for the aforementioned principles and to foster their recognition and effective application. States should also encourage exchanges and networks between independent ethics committees, as they are established, to foster full collaboration.

Article 24

The International Bioethics Committee of UNESCO should contribute to the dissemination of the principles set out in this Declaration and to further the examination of issues raised by their applications and the evolution of the technologies in question. It should organise appropriate consultations with parties concerned, such as vulnerable groups. It should make recommendations, according to UNESCO's statutory procedures, addressed to the General Conference and give advice concerning the follow-up of this Declaration, in particular the identification of practices that could be contrary to human dignity, such as germline interventions.

Article 25

Nothing in this Declaration may be interpreted as implying for any State, group or person any claim to engage in any activity or to perform any act contrary to human rights and fundamental freedoms, including, *inter alia* the principles set out in this Declaration.

Resolution Adopted by the Twenty-Ninth Session of the UNESCO General Conference Implementation of the Universal Declaration on the Human Genome and Human Rights

The General Conference,

Considering the Universal Declaration on the Human Genome and Human Rights, which was adopted on this eleventh day of November 1997,

Noting that the considerations formulated by the Member States at the moment of the adoption of the Universal Declaration are relevant for the follow-up of the Declaration.

1. Urges Member States:

(a) in the light of the provisions of the Universal Declaration on the Human Genome and Human Rights, to take appropriate steps, including the introduction of legislation or regulations, to promote the principles set forth in the Declaration, and to promote their implementation;

(b) to keep the Director-General regularly informed of all measures they have taken for the implementation of the principles set forth in the Declaration;

2. Invites the Director-General:

(a) to convene as soon as possible after the twenty-ninth session of the General Conference an *ad hoc* working group with balanced geographical representation, comprised of representatives of Member States, with a view to advising, him on the constitution and the tasks of the International Bioethics Committee with respect to the Universal Declaration and on the conditions, including the breadth of consultations, under which it will ensure the follow-up to the said Declaration, and to report on this to the Executive Board at its 154th session;

(b) to take the necessary steps to enable the International Bioethics Committee to ensure dissemination of and follow-up to the Declaration, and promotion of the principles set forth therein;

(c) to prepare for the General Conference a global report on the situation world-wide in the fields relevant to the Declaration, on the basis of information supplied by the Member States and other demonstrably trustworthy information gathered by whatever methods he may deem appropriate;

(d) to take due account, in the preparation of his global report of the work of the organisations and agencies of the United Nations system, of other international organisations, and of the competent international non-governmental organisations;

(e) to submit this global report to the General Conference, along with whatever general observations and recommendations may be deemed necessary in order to promote the implementation of the Declaration.